Saving the World's Deciduous Forests

Saving the World's Deciduous Forests

Ecological Perspectives from
East Asia, North America, and Europe

Robert A. Askins

Yale UNIVERSITY PRESS

New Haven & London

Published with assistance from the foundation established in memory of
Philip Hamilton McMillan of the Class of 1894, Yale College.

Yale University Press books may be purchased in quantity for educational,
business, or promotional use. For information, please e-mail
sales.press@yale.edu (U.S. office) or sales@yaleup.co.uk (U.K. office).

Set in Quadraat Roman type by Westchester Book Group.
Printed in the United States of America.

Library of Congress Cataloging-in-Publication Data
Askins, Robert.
Saving the world's deciduous forests : ecological perspectives from East
Asia, North America, and Europe / Robert A. Askins.
       pages   cm.
 Includes bibliographical references and index.
 ISBN 978-0-300-16681-1 (alk. paper)
 1. Forest ecology. 2. Endangered ecosystems. 3. Forest animals—
Conservation. I. Title.
 QH541.5.F6A82 2014
 577.3—dc23   2013028813

A catalogue record for this book is available from the British Library.

This paper meets the requirements of ANSI/NISO Z39.48–1992
(Permanence of Paper).

10 9 8 7 6 5 4 3 2 1

To

Hisashi Sugawa, Reiko Kurosawa, Tae Okada, Kikuo Hisayama,
Hiroyoshi Higuchi,
*and the many other naturalists and ecologists who introduced me
to the ecology and culture of Japan*

Contents

Preface ix

Acknowledgments xiii

CHAPTER 1. Parallel Worlds: Spring Forests in
New England and Kyoto 1

CHAPTER 2. Origins of the Deciduous Forest 8

CHAPTER 3. Deciduous Forests After the Arrival of People 33

CHAPTER 4. Decline of Natural Forests and the Invention of
Sustainable Forestry 55

CHAPTER 5. Giant Trees and Forest Openings 70

CHAPTER 6. Forest Islands and the Decline of Forest Birds 105

CHAPTER 7. Missing Wolves and the Decline of Forests 137

CHAPTER 8. The Global Threat of Rapid Climate Change 166

CHAPTER 9. Another Global Threat: Transport of Species
Between Continents 182

CHAPTER 10. Blending Conservation Strategies from
Three Continents 207

Appendix of Scientific Names 239

Notes 249

References 267

Index 295

Preface

My original plan was to write a book focusing on conservation of wildlife in the deciduous forests of East Asia, eastern North America, and Europe. Animals can be saved only if the plants that sustain them are protected, however, and all of these organisms depend on functioning forest ecosystems. The scope of the book therefore expanded beyond a discussion of populations of particular species of mammals, birds, and other animals to encompass the history and ecology of entire forest ecosystems. Effective wildlife conservation must be grounded in this broader perspective.

Comparison of deciduous forests of East Asia, eastern North America, and Europe is especially intriguing because these ecosystems have a common origin but have evolved in isolation for millions of years. Do these forest ecosystems function in similar ways because of their common origin and similar climates? Or have they evolved in distinctly different ways because of their long separation? If ecological processes are basically similar on the three continents, then ecological insights and conservation strategies learned on one continent can be applied effectively to the other two.

Other books have already catalogued and described the different types of deciduous forests on each continent, so I did not go into detail about variation in deciduous forests within continents. Instead, my goal was to compare the ecology of deciduous forests on different continents. Ultimately I was searching for general ecological patterns and for widely applicable solutions to conservation problems. Although deciduous forests are similar on the three continents, they have distinctly different geological, biological, and cultural histories. A historical perspective at a wide range of time scales is important for understanding the similarities and differences among continents. I therefore discuss the history of these forests in both deep time (at scales of millions of years to tens of thousands of years) and in recent time (at time scales spanning those used by archaeologists, historians, and field ecologists). I emphasize environmental threats that have a global impact on the biological diversity of deciduous forests, such as high densities of deer, forest fragmentation, tree pathogens, and climate change. There are other environmental problems (such as acid deposition, air pollution, unsustainable hunting, and pesticide contamination) that I have not emphasized because their effects tend to be more localized and less pervasive in deciduous forests around the world.

I do not include all of the world's deciduous forests in this discussion. My goal is to compare wildlife ecology of forests that are derived from an ancient mixed deciduous-coniferous forest that once circled the higher latitudes of the Northern Hemisphere. Consequently, I don't consider the deciduous forests of dry tropical climates in places like Thailand, Costa Rica, and Senegal. These tropical deciduous forests are distinctly different from the "summer-green forests" of the north. They are green during the wet season and leafless—to a greater or lesser extent—during the dry season, and they mainly support plants and animals from tropical groups. I also don't consider the temperate-zone forests of the Southern Hemisphere. The deciduous *Nothofagus* forests of Chile and the deciduous forests of Australia evolved separately from forests in the Northern Hemisphere, so the dominant groups of plants and animals are different. In addition, I don't focus on small, disjunct regions of temperate deciduous forest at high elevations in the mountains of Mexico and Guatemala, or the streamside and aspen woodlands of western North America. These are derived from the widespread northern temperate forests of the early Miocene, but in comparison with the great eastern deciduous forest of North America, they are minor and incomplete remnants. It would be impractical to try to capture all of these qualifications in the book's title. Suffice it to say that the book is not intended as a survey of all deciduous forests of the world, or even all deciduous woodlands and forests on the three continents referred to in the title.

Also, given the constraints of my own background and the length of the book, I did not attempt a comprehensive comparison of forest ecology and conservation in all of the nations encompassed by temperate deciduous forests. Instead, I primarily emphasized research in the United States, the United Kingdom, and Japan, using these three countries as the central examples for different approaches to land management and conservation. I also include continental Europe (especially Poland) and China in discussions of forest history and ecological research. Many studies from Canada are also included. Only a small piece of the eastern deciduous forest reaches southeastern Canada, but considerable research has been done on the ecology of this "Carolinian forest" because most of it has been destroyed and the remaining remnants support numerous species that are threatened in Canada.

As with my first book (*Restoring North America's Birds: Lessons from Landscape Ecology*), I have used an approach that should be accessible to amateur naturalists, conservationists, and land managers, the people who take the lead in applying insights from scientific research to conservation problems. I think some of the insights derived from comparing North American, European, and Asian forests will also be useful for professional ecologists, foresters, and wildlife biologists as well, however. Writing a book that is appropriate for both audiences is a challenge. To accomplish this, I have made a concerted effort to avoid the technical terms that would have made writing easier for me but make reading the book more difficult for readers who don't have university degrees in ecology or related areas. I avoid terms such as "disturbance regime," "metapopulation dynamics," "gene flow," "maximum sustainable yield," "short rotation period," and "area-sensitive species." I cover these concepts, but use straightforward English rather than the shorthand terms used by ecologists, foresters, and wildlife biologists. Avoiding technical terms forced me to explain concepts more completely, making them accessible to a wider range of readers. I include comprehensive citations so that readers can locate the original sources of information in the academic literature, using an endnote format to make the book more readable.

This book is not intended as a university textbook, but it would be appropriate as a text for a course in forest ecology or as a supplemental text for courses in wildlife management, ecology, or conservation biology. When I teach courses of this sort, however, one of my goals is to help students become fluent in the technical terms used in scientific journals. If this book is used as a course text, the instructor would need to connect the concepts to technical terms, explaining, for example, which sections illustrate canopy gap dynamics or trophic cascades.

Acknowledgments

This book was inspired by four visits to Japan, three of which lasted for periods of four or five months when I was teaching in the Associated Kyoto Program and working on research. The book could not have been completed without the help of numerous Japanese ecologists and naturalists who were remarkably generous in describing their research and conservation projects, and sharing their insights about conservation. Help and guidance from Hisashi Sugawa and Reiko Kurosawa were particularly important. Sugawa-san introduced me to many researchers and ecologists in the Kyoto and Lake Biwa regions, and invited my wife and me to visit numerous natural areas where we talked with local naturalists and ecologists. My former student Rae Kurosawa introduced us to researchers in Tokyo, and showed us the forests and other natural habitats of central Honshu and Hokkaido. Tae Okada became a close friend, introducing us to the natural history and culture of Kyoto and the Lake Biwa region. Kikuo Hisayama showed us his study area—and the impressive giant flying squirrels he has studied for many years—in the old forest above Honen-in, an elegant Buddhist temple on the eastern edge of Kyoto. Hiroyoshi Higuchi connected me with ecologists at Tokyo University, Kyoto University, and other research centers, and collaborated with me on a research project on forest birds. Dan Smith introduced us to the forests and endangered species of Okinawa. Dozens of dedicated conservationists and researchers aided us when we visited nature centers, research centers, restored agricultural areas, and forest reserves, helping us understand the conservation approaches they were using. I came away with a rich, if admittedly incomplete, understanding of conservation in Japan, and a deep appreciation for the beauty of Japanese forests and the hospitality of Japanese people.

My direct experience with forests in Europe is more limited than for North America or Japan, but my wife and I have explored the woodlands of England and large deciduous forests in the mountains of Galicia and Asturias in northern Spain. We also visited Białowieża Forest in Poland specifically because I was working on this book. This is one of the few extensive areas of relatively natural lowland deciduous forest in Europe, so it is critically important for understanding the ecology of European forests. My visit helped me interpret the extensive literature on the ecology of this forest. Waldemar Krasowski introduced us to the different habitats and species of this beautiful and diverse forest, and shared his knowledge of the numerous species of birds that live there.

While working on research in the deciduous forests of Connecticut, the highly knowledgeable staff of the Connecticut Department of Energy and Environmental Protection (DEEP) provided me with advice, logistical support, and insights about conservation. I particularly benefited from discussions with Jenny Dickson, Emery Gluck, and Shannon Kearney-McGee, as well as with Patrick Comins and Corrine Folsom-O'Keefe of Audubon Connecticut. It was also helpful to learn about conservation issues from staff from numerous national forests when I presented sessions in short courses for the U.S. Forest Service. Glenn Dreyer, the director of the Connecticut College Arboretum, assisted in my research in the Arboretum.

The broad scope of this book took me well outside my own area of research, so I asked experts in different disciplines to review particular chapters to ensure that I was not misinterpreting the research I described. The following people gave me invaluable help with particular topics: Carole Cheah, Connecticut Agricultural Experiment Station (hemlock woolly adelgid); Chris Elphick, University of Connecticut (bird conservation in England); David Ewert, The Nature Conservancy (effects of deer and forest pathogens on deciduous forests); John Kricher, Wheaton College (vertebrate evolution); Reiko Kurosawa (ecology of birds and approaches to conservation in Japan); Frederick Paxton, History Department, Connecticut College (human history of deciduous forests); Seiki Takatsuki, Azabu University (ecological effects of deer); and Scott Wing, Department of Paleobiology, Smithsonian Institution (fossil record of plants).

I also appreciate the sound advice and guidance from my editor, Jean Thomson Black, who worked with me on both this book and my earlier book. John Marzluff and an anonymous reader reviewed the entire manuscript and made numerous recommendations that improved the book, and Phillip King copyedited the manuscript carefully and thoughtfully.

My research on forest birds in eastern North America, the U.S. Virgin Islands, and Japan provided me with the background to write this book.

This research was funded by the following agencies and conservation organizations: Andrew W. Mellon Undergraduate Research Fund, Connecticut College; Associated Kyoto Program; Connecticut College Arboretum; Connecticut DEEP Wildlife Division; Connecticut Forest and Parks Association; Connecticut Light and Power Company; Institute of Tropical Forestry, U.S. Forest Service; National Geographic Society; The Nature Conservancy (Connecticut and Virgin Islands chapters); Northeast Utilities Transmission; Northeast Utilities Foundation; U.S. Department of Agriculture, Forest Service, Northeastern Forest Experiment Station; U.S. National Biological Service; U.S. National Park Service; and the World Nature Association. Bun-ichi Sogo Shuppan granted me permission to modify some of the text from a chapter in the Japanese edition of my book on bird conservation for this book so that I could present this information in English for the first time.

Saving the World's Deciduous Forests

CHAPTER I

Parallel Worlds
Spring Forests in New England and Kyoto

The inspiration for this book was a walk along a forest stream in the mountains north of Kyoto on a clear morning in early spring. After working in the forests of eastern North America for many years, I found Japanese forests a mix of the familiar and the strange. My surroundings were mostly familiar. Leaves were just emerging from buds on the overhanging branches of maples and oaks. Splashes of color—clumps of violets, anemones, and trilliums—dotted the mottled brown leaf litter of the forest floor. Straight gray beech trunks, the shallow angle of spring sunlight, and the songs of returning migratory songbirds evoked the scenes and sounds of a New England forest in spring. The hemlocks, oaks, and Indian pipes were similar to those along a stream near my home in Connecticut.

On closer inspection, however, the details were distinctly different. The bird songs were new and strange. Although the general types of plants were familiar, the particular species were new to me. And there were many more species, making plant identification a greater challenge. Instead of a single species of beech there were two; instead of three types of maple trees there were more than a dozen; and the field guide showed a multitude of different species of similar-looking violets. In this respect, it was as if I were visiting a North American forest 8 million years ago, before the Pleistocene extinctions. The comparable North American forest can only be viewed dimly by examining fossil imprints of plants from that period.

Despite the differences in particular species of plants, however, East Asia and eastern North America have remarkably similar types of woodlands, with subtropical forests in the south, deciduous hardwood forests in the central region, and boreal coniferous forests in the north. The change in vegetation between Key West and Nova Scotia parallels the change between Okinawa and northern Hokkaido. A traveler on either journey begins on white sandy beach fringed by mangroves and ends on a rocky coast surrounded by spruce, fir, and birch. Both journeys traverse diverse deciduous forests for much of the way.

Of course, the geography and biology of Japan and North America differ in many ways. Japan lacks the extensive prairies, savannas, and deserts that dominate much of the North American landscape. Japan consists of a series of islands, so it has some of the biological characteristics of islands, such as lower species diversity than comparable areas on the continental mainland. And the history of land use and style of agriculture are distinctly different in most parts of Japan and North America. But the great expanses of deciduous forest dominated by oaks, maples, and other hardwoods in Japan and eastern North America result in parallel ecological worlds. If you want to see spectacular displays of autumn color, there are few places on earth that will rival the Appalachians of the United States or the mountains of Honshu. Outsiders may perceive both regions as densely populated and long settled, with little remaining natural habitat, but this is a misconception. Forest covers 60–70 percent of the land in both Japan and the northeastern United States. Both regions are dominated by second-growth woodland interrupted by roads and towns, but both have some surprisingly large expanses of continuous forest. And in both regions many forests are maturing and slowly acquiring the large trees, closed canopy, and dead wood of an ancient forest.

Why Are North American and Japanese Forests Similar?

Fifteen million years ago, before the repeated advance and retreat of glaciers across northern Eurasia and North America, deciduous forest encircled the northern continents. These Miocene forests were remarkably diverse, with a great number of plants and animals that are now extinct. Some of these species succumbed to habitat change and hunting pressure as people spread throughout the Northern Hemisphere, but most disappeared long before the evolution of technologically skilled humans. They vanished because they did not survive a changing global climate, especially the severe disruptions of the Pleistocene, when kilometer-deep glaciers and cold, dry winds made much of the North Temperate Zone uninhabitable for all but the most flexible and hardy plants and animals.

Today you can find isolated regions of deciduous forest in the middle latitudes of eastern North America, Europe, and East Asia.[1] The European forests have relatively few surviving tree species because Miocene cooling and Pleistocene glaciation were severe across much of western Europe, and the Mediterranean Sea blocked the spread of forests southward into warmer areas. Forests of eastern North America are more diverse because plants and animals could spread to relatively warm refuges on the Gulf Coast and what is now the continental shelf under the Gulf of Mexico. To experience the true diversity of the original deciduous forest, however, one must travel to eastern China, Korea, or Japan. These regions were not covered with continental glaciers during the Pleistocene, so the disruptions to forests caused by climate change were muted.

Superficially, the forests of Europe, East Asia, and eastern North America appear similar, but of course similar environments can lead to the evolution of similar organisms that are not necessarily related to one another. In the case of temperate deciduous forests, however, the three continents share the same taxonomic groups of plants. The dominant herbs, shrubs, and trees belong to the same plant families and often to the same genera (the plural of "genus"). A genus consists of a group of closely related species with a common ancestor, and the large number of shared genera among the three continents indicates that the forests of these regions were connected fairly recently in geological time. The list of shared genera is even greater if one compares fossils from the Paleogene, the relatively warm period between 65 and 23 million years ago, preceding the major climate changes that began during the Miocene.[2] Numerous genera that are now found only in East Asia are known from the fossil record from this period for North America or Europe, or both. Similarly, genera now found in North America but not Europe are known from fossils in Europe. In fact, a majority of tree genera were once found in two or three of the current deciduous-forest regions. A large proportion of these plant groups disappeared from Europe and a smaller number disappeared from North America. Extinction rates of trees were especially high for western North America, where most broad-leaved forests were replaced by grassland, desert, or coniferous forest.[3] In contrast, very few plant groups disappeared from East Asia.

Low extinction rates for plants in East Asia and eastern North America and high extinction rates in the other parts of Eurasia and western North America resulted in an odd pattern of plant distributions. Particular groups of plants are found only in the deciduous forests of East Asia and the eastern deciduous forest of North America; they are missing (except in the fossil record) from other parts of the temperate zone.[4] The remarkable similarity of the flora of East Asia and eastern North America was thoroughly

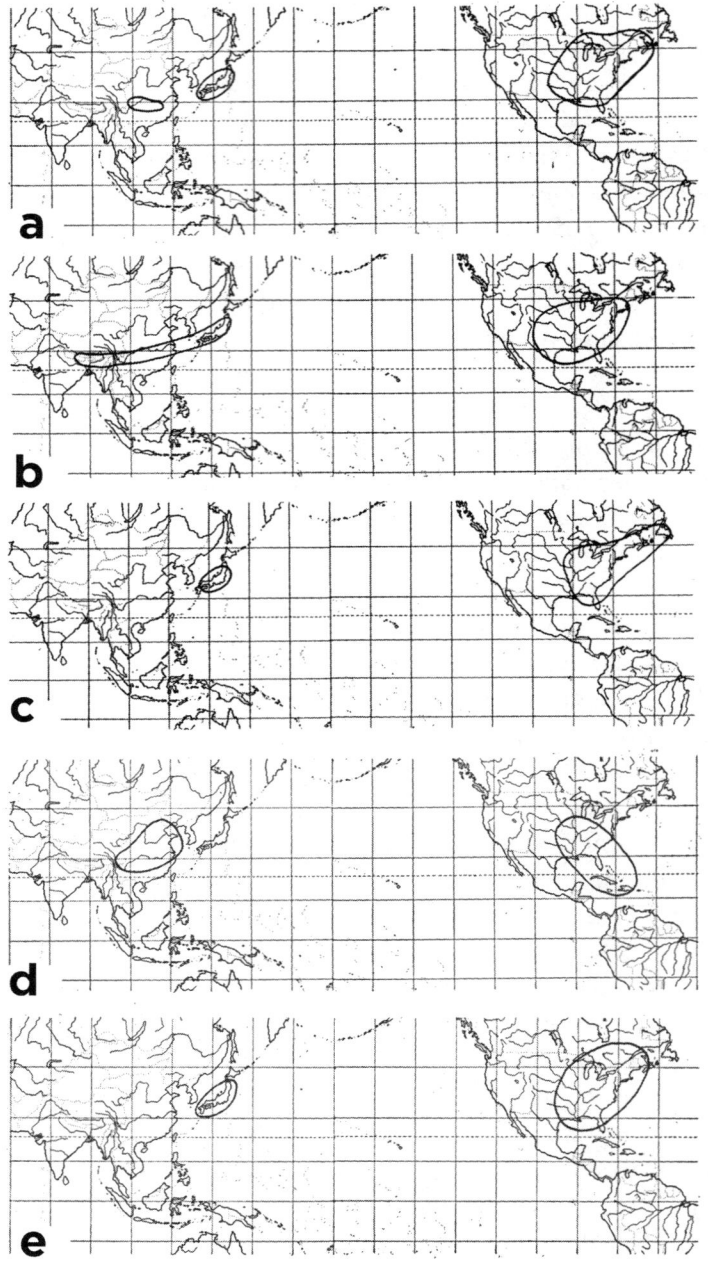

Figure 1. Distribution of a few of the many plant genera that have disjunct geographical ranges in eastern Asia and eastern North America: (a) witch hazels (*Hamamelis*), (b) creepers (*Parthenocissus*), (c) trailing arbutus and relatives (*Epigaea*), (d) catalpas (*Catalpa*), and (e) partridgeberries (*Mitchella*). (Li, 1952, pages 416–422; reprinted with permission of the American Philosophical Society)

documented in the 1950s by Hui-Lin Li, whose monograph includes dozens of maps showing the geographical distribution of genera restricted to these two widely separated regions.[5] Some examples include such familiar North American plants as witch hazel (*Hamamelis*), Virginia creeper (*Parthenocissus*), catalpa (*Catalpa*), trailing arbutus (*Epigaea*), and partridgeberry (*Mitchella*). About 65 genera of plants show this pattern.[6] Comparisons of genetic sequences in different species in many of these genera confirm that the East Asian and North American representatives of each genus evolved from a common ancestor millions of years ago. Hence their similarities do not result merely from convergent evolution in similar habitats.

During the period when the important families of deciduous forest plants were evolving, North America, northeastern Asia, and Europe were intermittently connected by land via Greenland and the Bering Strait, so it isn't surprising that they share or shared many types of temperate-zone plants.[7] These connections ceased about 15 million years ago, after which plant groups that became extinct on one continent were not replaced by immigration of related species from other continents, and some continents lost a high proportion of their original diversity. The current differences between European deciduous forests and the forests of North America and East Asia are mainly due to extinctions resulting from a progressively colder climate and the severe effects of glacial periods during the past few million years. This resulted in the exceptionally low diversity of plant species in European forests.

The differences in diversity between East Asia and eastern North America appear to be more complex.[8] East Asia has higher plant diversity because it has retained more species in ancient groups such as magnolias and ginkgos that became extinct in North America. Other groups are more diverse in East Asia because the rate of evolution of new species apparently was greater in East Asia than in North America. This is partly a product of the mountain or ocean barriers that separate Japan and Korea from mainland China and each other. As described in Chapter 2, this set the stage for isolated populations of a single species diverging into separate species. Also, new deciduous-forest species may have frequently evolved in East Asia due to the close connection between subtropical and tropical forests and temperate deciduous forests in China, permitting many species to move northward and adapt to a more strongly seasonal climate. This may have been facilitated by the mountains of southern China, which support tropical and temperate vegetation in close proximity at different elevations. In contrast, the deciduous forests of eastern North America are connected only tenuously to lowland tropical forests in the West Indies and Mexico. They have been separated by ocean barriers and—for millions of years—by deserts and semi-deserts of the southwestern United States and northern Mexico. The result is that forests in North America are similar to those in East Asia, but are less diverse due to both more frequent extinction and less frequent addition of new species. Although Japan is now separated from China by ocean barriers, it was connected to mainland Asia recently (in geological terms) when sea levels were lower during glacial periods, so it shares a relatively high diversity of plants with the mainland.

Searching for General Patterns in Forest Ecology

The similarities among the deciduous forests of East Asia, Europe, and North America create the conditions for a "natural experiment." Natural experiments are not designed by a researcher, and therefore are not well controlled or replicated. Natural experiments are inferior to designed experiments in every way except one: they test processes on time scales (thousands or millions of years) and geographical scales (entire islands or continents) that are not practical for experimental manipulations by researchers.

During my first visit to Japan I was particularly interested in how the parallel evolution of bird migration had played out on deciduous forests on different sides of the world. The summer forests of both Japan and New England are inhabited by songbirds that migrate hundreds or thousands of miles south each autumn to spend the winter in the tropics. Species on both sides of the world have similar behavior. Males defend small territories that support the breeding pair and their young. They announce their territorial boundaries by singing. Most species feed on insects that they find in the complex, multilayered foliage of the forest. Many species specialize on particular layers of the forest, such as the shrub layer or the tree canopy. Some species glean insects from the undersurface of leaves while others specialize in probing under bark or pursuing flying insects. The overall organization of these bird communities is similar in many ways, but virtually all of the species are different. In many cases, they are not even closely related to one another—they don't have recent common ancestors. Some of the most important families of birds in North America, such as the vireos, New World warblers, and tyrant flycatchers, are missing from Japan. Other bird families, such as the bulbuls and Old World flycatchers, are found in Japan but not in North America. Hence we have the makings of a particularly interesting natural experiment: a similar ecological stage (to paraphrase G. Evelyn Hutchinson), but an almost completely different set of actors.[9] Did the play proceed in the same way? Did similar ecological patterns emerge? Are the conservation problems and solutions similar? If so, then Japanese, American, and Canadian ecologists and conservationists can learn much from one another, and the emerging principles may also apply to deciduous forests in Europe, Korea, and China.

Once I began thinking about parallel evolution in Japanese and North American forests, all sorts of questions occurred to me. How has the relatively recent extinction of wolves in both Japan and most of eastern North America affected natural forest environments? Have recent population explosions of deer in Japan and North America resulted in similar ecological changes? How have species adapted to young forest and forest openings fared as forest has matured in the two regions? Does timber harvesting have similar effects on biological diversity in Japanese and North American forests? If there are similar patterns for forests of Japan and eastern North America, can these be generalized to other temperate deciduous forests in Europe, Korea, and China?

These comparisons particularly intrigue me because some of the basic approaches to conservation—and even the way people perceive natural habitats and natural beauty—differ in Japan and North America. Is one approach to conservation more effective, or do both approaches have strengths and weaknesses? Could we combine these different approaches to better protect temperate deciduous forests, with their many special qualities—colorful fall foliage, spring woodland flowers, four visually distinct seasons—and their high diversity of species (among the highest north of the moist tropics)? And how does this comparison relate to conservation in China and in England and other parts of Europe, which have different traditions of land use and wildlife conservation from either North America or Japan?

CHAPTER 2

Origins of the Deciduous Forest

Arctic Deciduous Forests

The "summer-green" deciduous forests originated in an unexpected time and place. The ancient relatives of many deciduous trees can be traced back to the mid to late Cretaceous Period, 113 to 65 million years ago, when dinosaurs were still the dominant large land animals and most of the earth was tropical or subtropical. Remarkably, extensive deciduous woodlands existed only at very high latitudes, near the North Pole in Alaska, Canada, and Siberia and near the Cretaceous South Pole in Australia and Antarctica. Fossil leaves and vertebrate bones from the North Slope of Alaska, Ellesmere Island, and other far northern sites allow us to piece together a rather murky picture of an ancient deciduous forest that has no modern counterpart.[1] Although flowering plants had already evolved by this period and were common in the understory, the dominant trees were conifers and other gymnosperms.[2] These included needle-leaved conifers such as relatives of the dawn redwood (*Metasequoia*) and a variety of broad-leaved gymnosperms such as the ginkgos and cycads that are familiar to gardeners. Most modern species of conifers are evergreen, so much so that many people think of "conifer" and "evergreen tree" as being virtually synonymous. However, there are conifers such as larches and bald cypresses that lose their needles in the winter. In the arctic forests of the Cretaceous, deciduous conifers and their relatives were the rule rather than the exception, and they dominated the tree canopy. They belonged to plant groups that are extinct or almost extinct, although many appear to be related to modern bald cypresses and sequoias. The most familiar representatives of these ancient groups are the dawn redwoods and ginkgos, both of which survived in relict populations in the forests of China and are now commonly planted in gardens and along streets throughout Europe, North America, and East Asia.

Although geomagnetic evidence shows that Alaska was closer to the North Pole during the late Cretaceous than it is today, the climate was milder, reflecting higher temperatures for the entire planet.[3] This can be deduced from the shape of fossil leaves. There is a remarkably consistent relationship between leaf shape and mean annual temperature for modern plant species.[4] For example, the proportion of species with toothed leaves consistently increases as one moves from warm to cold climates. Based on statistical analyses of a large number of such leaf characteristics, temperatures in the Cretaceous Arctic generally hovered above freezing during the winter. This is confirmed by the lack of geological evidence of permafrost. However, fossilized tree trunks at Ellesmere Island in northern Canada display damaged sections (growth interruptions) consistent with damage caused by severe frost, indicating that occasional frosts occurred.[5] Also, although the winters were generally mild, they were—like modern arctic winters—very dark. Because of the tilt of the earth's axis, very little sunlight reaches high northern latitudes for several months each year, so trees and shrubs of the ancient Arctic could not photosynthesize for several weeks or even months, depending on the latitude.[6] This is apparent from growth rings from fossilized wood of the period; the annual growth rings end abruptly, evidence of a rapid cessation of photosynthesis and plant growth with the onset of winter darkness.

The combination of mild temperatures and dim light during the winter may explain why so many Cretaceous trees shed their leaves at the end of the growing season. This would help reduce water loss and energy use during an extended period of darkness. Because the winters were relatively warm, leaves on an evergreen tree would continue to undergo respiration, resulting in a steady loss of energy during a period when photosynthesis was not possible and energy stores could not be replenished. Thus, the food reserves of the plant would be expended to keep the leaves alive. In contrast, deciduous trees dropped their leaves and entered a period of deep dormancy during which they lost relatively little energy. When the sun reappeared in the spring, new leaves sprang out to take advantage of the short—but almost continuously sunny—growing season.

This explanation for the prevalence of deciduous trees in the ancient Arctic is so logical that it was not rigorously tested for many years. In the early 2000s, however, this hypothesis was undermined by experiments with both evergreen and deciduous species belonging to plant groups that grew in the Arctic or Antarctic forests of the Cretaceous.[7] A team of researchers compared growth rates of saplings for two types of deciduous trees (ginkgo and dawn redwood) and two types of evergreen trees (southern beech and coast redwood) in greenhouses with a simulated Cretaceous Arctic winter (dim light and mild temperatures).[8] Surprisingly, the deciduous trees lost more of their food reserves (as measured by the amount of carbon in the trees) during the simulated winter than did the evergreen trees. This is due to the immediate and massive loss of carbon as leaves are shed, which offsets the carbon saved by winter dormancy. If these modern trees responded in a similar way to their ancient relatives, then it is perplexing that evergreen trees didn't predominate in arctic forests. Deciduous trees may have an advantage during the summer rather than during the winter, however, because their leaves could grow and photosynthesize more rapidly than evergreen leaves during the brief growing season.[9]

Finely detailed imprints of leaves and pieces of fossilized tree trunks provide a hint of the biological diversity and beauty of the deciduous forests of the ancient Arctic, but it is difficult to imagine what it would have been like to stroll through this forest. Based on leaf shapes and fossilized stumps, it was apparently an open woodland of rather low trees. In the summer it would have been sunny and intensely lime green, assuming that the vibrant greens of modern dawn redwoods and ginkgos were shared by their ancient relatives. The ground was covered with ferns and horsetails.[10] To see the precursors of our own deciduous forest, one would need to look closely at the understory, where flowering plants were mixed in with cycads and ginkgos.[11] Here one would find shrubs related to modern birches (Betulaceae), sycamores (Platanaceae), walnuts (Juglandaceae), and elms (Ulmaceae), the ancient relatives of the deciduous hardwood trees that would eventually become dominant species in northern temperate-zone forests. One of the signature plant groups of the modern deciduous forest, the maples (Sapindaceae), was represented by the "Acer" arcticum complex, which is now classified as belonging to an extinct genus of maples.[12] During the autumn the deciduous conifers and ginkgos probably turned rich golden before dropping their leaves, much as tamaracks and European larches do today. Perhaps the deciduous shrubs in the understory added to the range of autumn colors, providing the first traces of the deep scarlets and oranges of modern autumn forests.

Deciduous-Forest Dinosaurs

The most remarkable sight during a stroll through a Cretaceous polar woodland would be large plant-eating dinosaurs moving through the scattered trees. Dinosaurs such as hadrosaurs (the "duck-billed dinosaurs") and ceratopsians (relatives of the familiar three-horned *Triceratops*) browsed on the trees and shrubs.[13] The presence of giant dinosaurs such as the hadrosaur *Edmontosaurus* (which was 18 meters [59 feet] long) fits with the reconstruction of this ecosystem as an open woodland.[14] Dense forests generally supported smaller dinosaurs. For example, fine lake sediments deposited near tropical forests in eastern China in the early Cretaceous preserved fossils of a bewildering variety of small dinosaurs, including many bipedal coelurosaurs, relatives of the infamous *Velociraptor*. There were also primitive birds and other small vertebrates (including mammals).[15] Many of the small dinosaurs were covered with feathers or featherlike filaments, and some could fly or glide. Only a single large herbivorous dinosaur has been discovered at these fossil-rich sites, indicating that smaller vertebrates characterized this tropical forest. Large herbivorous dinosaurs were better adapted to the sparse woodlands of the Arctic. They could walk more efficiently across an open woodland than through a dense forest, and their browsing and movements would help to keep the woodland open. The ultimate cause of widely spaced trees in arctic woodlands, however, may have been the short growing season and periodic fires (which are indicated by charcoal deposits in the sediments of this period).[16] Only trees with full exposure to sun would be able to produce enough carbohydrate by photosynthesis during the short growing season to be able to survive the long, dark winter.[17]

Although large dinosaurs were more common in the open polar woodlands of late Cretaceous Alaska than in the dense, early Cretaceous forests of northeastern China, they were apparently even more abundant farther south in the open plains of what is now Alberta. Here, in the late Cretaceous, low-growing ferns and shrubby flowering plants supported a dense and diverse community of giant dinosaurs, with as many as 15 species of herbivores.[18] Hadrosaurs and ceratopsians were the dominant plant-eaters. Concentrations of massive numbers of skeletons in the same sedimentary layer suggest that they traveled in large herds.[19] Fossils from ancient open plains of the North American West, from New Mexico to Alberta, have formed our primary image of dinosaurs and dinosaur habitat. Fossil beds from other ancient ecosystems, however, show that 70 million years ago the earth was just as geographically diverse as it is today, with sand dune deserts, tropical forests, and swamps, each with a distinct assortment of dinosaurs and other vertebrates. The deciduous woodlands of the Arctic were one of the many distinctive ecosystems.

An open question is how dinosaurs survived the dark, cool winters in arctic woodlands. Hibernation requires a sheltered hiding place, and this seems unlikely for the larger dinosaurs. Like modern arctic caribou, perhaps dinosaurs migrated south.[20] A three-month, 2,000-kilometer trek would have brought them to warmer, sunnier conditions.[21] Another possibility is that dinosaurs simply remained active in the dark, leafless forests. Like modern musk oxen, which do not migrate away from much harsher conditions in the modern Arctic, herbivorous dinosaurs may have foraged on leafless twigs and dead leaves.[22] However, predatory dinosaurs faced a special challenge because they needed to detect and catch their prey. Interestingly, the most common predatory dinosaur fossils in Cretaceous deposits of Alaska belong to a single species, *Troodon formosus*, which was a small, bipedal running dinosaur that was a little less than two meters long.[23] *Troodon* is known for possessing eyes that were exceptionally large for a dinosaur.

End of the Cretaceous Forest Ecosystems

The distinctive deciduous forest ecosystems of the late Cretaceous came to a sudden end—along with much of the biological diversity of the earth—when an asteroid collided with the planet.[24] Of the multitude of dinosaur lineages, only modern birds survived. A large proportion of terrestrial plants of North America also disappeared after the asteroid impact, but many of the deciduous woody plants survived. Compared with many tropical and subtropical plants, deciduous trees and shrubs may have been better adapted to survive the darkness and coolness of a planet shrouded in dust thrown up by the impact. At fossil sites in western North America, deciduous plants had a lower extinction rate than did evergreen plants, and some polar deciduous plants appear to have spread southward after the impact.[25] Deciduous plants eventually became an important component of temperate-zone forests in the Cenozoic, the "age of mammals" that followed the "age of dinosaurs." The asteroid impact set a new trajectory for the evolution of plants and animals on earth, and one result may have been forests dominated by deciduous plants in the northern temperate zone.

This theory of abrupt and widespread extinction caused by an asteroid impact is not accepted by all paleontologists, many of whom argue that species of dinosaurs and other organisms gradually disappeared for a variety of reasons before the end of the Cretaceous.[26] This view has persisted despite overwhelming evidence that an asteroid collided with the earth 65.5 million years ago. The best evidence is a thin layer of clay that is rich in iridium, an element that is scarce in the earth's crust but is much more concentrated in asteroids.[27] This clay layer also includes shocked quartz crystals of the type produced by high-pressure impact. Finally, a crater of the appropriate size and age for the theorized asteroid impact has been found beneath the northern Yucatan Peninsula and adjacent Gulf of Mexico.[28] The Chicxulub Crater is 180 kilometers in diameter and 65 million years old, coinciding with the end of the Cretaceous. However, as the paleontologist Donald Prothero noted, the late Cretaceous was "a bad time to be alive on planet Earth."[29] Other factors probably contributed to extinctions toward the end of the Cretaceous. Widespread volcanoes in western India produced massive lava flows (called the Deccan Traps), adding considerable amounts of ash and noxious gases to the atmosphere. Sea level dropped, leaving many highly productive shallow seas (such as much of the epicontinental sea in central North America) high and dry. These massive changes probably caused extinctions gradually over a long period of time. The key question is whether most of the extinctions in the late Cretaceous occurred abruptly (within months or decades) or gradually (over hundreds of thousands or millions of years). There is increasingly compelling evidence that many species of microorganisms, plants, and animals became extinct abruptly at the time that the iridium-rich clay was deposited by an asteroid impact. These species are common immediately below the clayrock boundary, and completely absent above the boundary. This pattern is not found in all groups of organisms, but has been clearly documented for terrestrial plants and several groups of marine organisms (especially foraminifera and calcareous nanoplankton, which are photosynthetic microorganisms that would quickly die out during a long period of darkness).[30]

Not all extinctions at the iridium-rich boundary would be easy to detect in the fossil record, however. Species with abundant fossils will disappear abruptly from sediments at an extinction boundary, but species that are only sporadically found in the fossil record may not be represented in deposits immediately below the boundary. By chance the most recent fossils for these species are likely to be well below the boundary, giving the appearance of a gradual extinction of these less common species.[31] Fossils of dinosaurs and other vertebrates probably fit this pattern, so even if these species became extinct during an extremely short period following an asteroid impact, they would seem to disappear gradually from the fossil record. The same is often true of fossil leaves, leading to the impression of the gradual loss of species even though the actual pattern may have been simultaneous extinction of many species. On the other hand, pollen is frequently abundant in the fossil record, so the record for a particular species or genus (which is often the finest resolution for identifying pollen) should be continuous right up to the extinction boundary, at which point it disappears. This is exactly the pattern documented in the pollen record for the end of the Cretaceous at many sites in the western United States. Both the diversity and density of flowering plant pollen decline in deposits above the iridium clay layer, and instead these sediments are filled with the spores of ferns.[32] This "fern spike" (referring to a sudden increase in the density of fern spores) is found immediately above the iridium layer at numerous sites, indicating that diverse forests were replaced by an open, fern-covered plain dominated by a single species of fern. Interestingly, a similar pattern was documented on Krakatau, a small volcanic island in Indonesia, following a massive eruption in 1883. Ferns dominated the first vegetation to cover the island after deep ash on the surface cooled.[33] In this case, however, the fern stage lasted only a few years because a variety of flowering plants soon colonized the island from nearby Java and Sumatra. A lengthy period of fern domination following the Cretaceous event indicates a more widespread catastrophe.

Many of the woody plants that survived into the new age following massive extinction were typical of mires (wet, swampy areas) of the northern polar forest.[34] Perhaps this habitat buffered them from the firestorms that may have been generated by the asteroid impact. Also, these deciduous plants were adapted to surviving long winters with low light conditions, so they may have been especially resilient when a cloud of dust cut off the sun over large parts of the earth.[35]

The transition from the Cretaceous to the Paleocene (the earliest epoch of the "age of mammals") has been especially well studied in the Williston Basin, an extensive depression that straddles North Dakota, South Dakota, Montana, and Saskatchewan.[36] The late Cretaceous deposits of the Williston Basin are most famous for their classic dinosaur fossils, including well-preserved *Tyrannosaurus* and *Triceratops* skeletons, but they are also a rich source of fossil leaves and pollen. These deposits are from subtropical habitats, not from polar deciduous forests, but they provide an exceptionally detailed picture of what happened at the end of the Cretaceous because the fossil record here is so rich and so carefully studied. The end of the Cretaceous (the so-called K-T boundary) and the asteroid impact are clearly marked in the Williston Basin by a clayrock layer rich in iridium and shocked

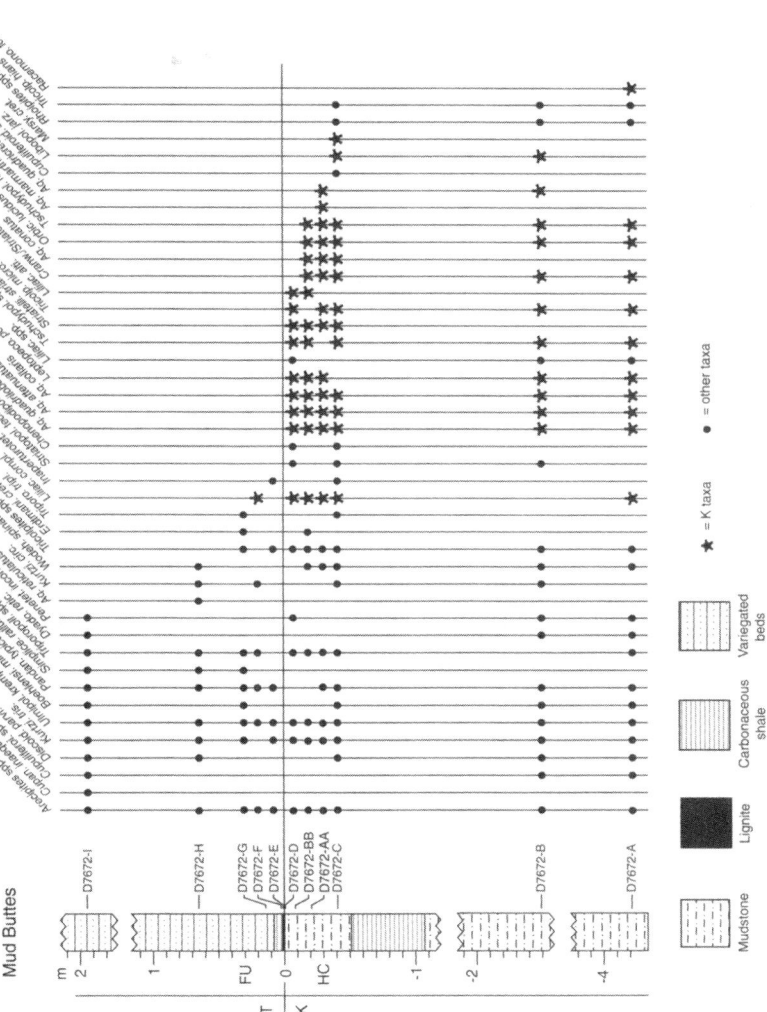

Figure 2. Evidence from Mud Buttes, North Dakota, of a high rate of extinction of plant species at the end of the Cretaceous Period. Pollen records of particular types of plants in strata from before and after the K-T boundary are marked with either circles or stars. Species that have been recorded only in rock strata older than the K-T boundary are denoted with stars; these species apparently did not survive the extinction event at the boundary. (Nichols and Johnson, 2008; reprinted with the permission of Cambridge University Press)

Figure 3. *Triceratops Swamp*, a painting by Jan Vriesen showing a reconstruction of the vegetation in a swamp in the Denver Basin, Colorado, in the late Cretaceous Period. A high proportion of plant groups that survived the extinction event at the end of the Cretaceous were associated with wetland habitats of this sort. A group of *Triceratops* moves through the forest in the background. (Copyright Jan Vriesen and Kirk Johnson; original work owned and commissioned by the City and County of Denver)

quartz crystals.[37] Tens of thousands of leaf fossils and numerous pollen samples have been analyzed in rock formations above and below this boundary. The picture is consistent for widely separated sites across the region: the diversity of plants was substantially lower above the iridium-rich boundary. There is a remarkable change in plant species represented by leaves and other large fossils; 79 percent of these plant species occur only below the K-T boundary, indicating that they became extinct in the late Cretaceous. Many of these species disappear abruptly at the K-T boundary. Moreover, most of the dominant plant species of the Cretaceous sediments are absent from the Paleocene sediments, so there was a major change in the composition of the vegetation. The pollen record shows a similar change, with a loss of 30 percent of the pollen types.

Figure 4. *After Armageddon*, a painting by Jan Vriesen showing a reconstruction of the vegetation in the Denver Basin, Colorado, a few years after the asteroid impact at the end of the Cretaceous Period. The landscape is dominated by ferns and a single species of aquatic plant with heart-shaped leaves. A crocodile is one of the few surviving species of vertebrates. (Copyright Jan Vriesen and Kirk Johnson; original work owned and commissioned by the City and County of Denver)

Although the Williston Basin provides an exceptionally clear fossil record of sudden and massive plant extinctions at the end of the Cretaceous, it is located in North America, relatively close to the asteroid impact crater in southern Mexico. Perhaps the effects of the impact were particularly severe here. Unfortunately it is difficult to find sediments from terrestrial habitats that show the transition from Cretaceous to Paleocene. European and Asian paleontologists have been skeptical that a catastrophic extinction of terrestrial plants occurred at their study sites, but the fossil record for leaves and other plant parts at these sites is generally too fragmentary to distinguish between sudden and gradual extinction of Cretaceous plants, and the boundary between Cretaceous and Paleocene is not always marked by a clear, iridium-enriched clayrock layer, making the transition difficult to interpret.[38]

Recently, however, the analysis of pollen and spore microfossils in coal and sandstone sediments in New Zealand revealed how plants changed across the boundary, which is marked with a coal seam that is enriched with iridium. A large number of plant types, including all of the flowering plants known from the Cretaceous, disappeared at the boundary. Above the boundary there is a distinct fern spike similar to the ones found at sites in the Williston Basin. Ferns apparently composed 90 percent of the vegetation for tens of thousands of years.[39] This detailed record of extinction from the opposite side of the world from Chicxulub Crater indicates that catastrophic extinction affected the entire planet, not just North America.

In addition to loss of biological diversity, the asteroid surely caused massive disruptions in the functioning of ecosystems. Gigantic plant-eating dinosaurs probably had maintained the low vegetation typical of plains, savannas, and open woodlands. These "ecosystem engineers" were suddenly gone. More subtly, the loss of many species of insects probably had a long-term effect on how ecosystems functioned. Most of the plants that survived after the end-of-Cretaceous extinction were wind-pollinated rather than insect-pollinated, suggesting that many pollinating insects were lost.[40] Also, along with dinosaurs, insects were important herbivores on forest plants. When Conrad Labandeira and his colleagues analyzed insect damage on thousands of fossil leaves from the Williston Basin, they discovered a sharp decline in the frequency of damage above the Cretaceous-Paleocene boundary.[41] This was particularly true for damage caused by insects such as leaf miners and gall-producers that specialize on a small variety of host plants in modern ecosystems. This suggests that specialized insects that depend on one or a few plant species were especially vulnerable to extinction. Because about 80 percent of the plant species in the Williston Basin became extinct, it isn't surprising that a large number of specialized insect herbivores also disappeared. The frequency of insect damage on leaves remained low for the next million years.

A New Forest Emerges

For millions of years after the asteroid impact, the mid latitudes in North America were characterized by low plant diversity. Judging by megafossils (fossils of leaves and other large plant parts rather than pollen), almost 80 percent of the species found at mid latitudes during the Cretaceous had become extinct.[42] In contrast, high-latitude forests in central Alberta suffered only about a 25 percent loss of species. Also, at middle latitudes there was a dramatic shift from primarily broad-leaved evergreen vegetation to deciduous vegetation.[43] During the Cretaceous deciduous plants were uncommon at mid latitudes, where they were apparently restricted to frequently disturbed habitats such as streamsides. After many of the evergreen plants became extinct, deciduous trees and shrubs may have spread outward from these restricted habitats and southward from the polar forest where deciduous plants had been dominant during the late Cretaceous. Dawn redwood and other deciduous conifers became common at mid latitudes along with a variety of broad-leaved deciduous plants, including ancient relatives of modern hickories, hackberries, aspens, sweetgums, and maples.[44]

Although numerous well-studied Paleocene forests in western North America fit this profile of a relatively short list of plant species dominated by deciduous trees and shrubs, a stunningly different site was discovered in 1994 at Castle Rock near Denver, Colorado.[45] At the beginning of the Paleocene, only about one and a half million years after the end-of-the-Cretaceous extinction, the Castle Rock site supported a tropical rainforest with much higher diversity than contemporaneous forests at other fossil sites in western North America. The Castle Rock fossils display the distinctive trademarks of rainforest trees. Many had large leaves with smooth edges and "drip tips" (attenuated, pointy tips that permit heavy rain to run off the leaf quickly).[46] This rainforest—and the contribution it may have made to the evolution of North American forests—is still an enigma. It serves as a warning that we have much to learn about the fossil record even in intensively studied regions such as interior western North America. It is tempting to interpret this rainforest as a refugium where a diversity of Cretaceous plants survived into the Paleocene, but most of the plant species at the Castle Rock site are different from the plants found in the same region before the asteroid impact.[47] Instead, the plant fossils from this site indicate that with a favorable climate, plant diversity increased rapidly after the end-of-the-Cretaceous extinctions. At hundreds of other sites in western North America, however, plant diversity remained low for millions of years following the end of the Cretaceous, and deciduous woody plants (not rainforest trees) dominated the vegetation.

Recent studies of insect damage on fossil leaves reveal that the rebuilding of complex food webs was an unpredictable and often lengthy process.[48] Many sites had a low diversity of both plants and plant-eating insects until the late Paleocene, millions of years after the asteroid impact. The

Castle Rock rainforest regained a high diversity of plants remarkably quickly, but the density of leaf-eating insects remained low. At one site in Montana in the early Paleocene, the diversity of plants was as low as other sites in western North America at that time, but the diversity of leaf-eating insects (including specialized insects such as leaf miners) was much higher, indicating that the leaf-eating insects had recovered more quickly than did woody plants. A site in France from the middle Paleocene had a much higher diversity of both plants and herbivorous insects (including miners and other specialized insects) compared with fossil sites from the same period in North America.[49] This suggests a more rapid recovery—or fewer extinctions—from the asteroid impact, perhaps because Europe was farther away from the site of the impact in Mexico.

By the end of the Paleocene, 10 million years after the K-T extinction, evergreen broad-leaved trees were again becoming common in the fossil record across interior western North America, a pattern that would be expected given the relatively warm, mild climate of the period. Interestingly, however, streamside habitats were dominated by deciduous birches, willows, and sycamores, much as they are in many regions of western North America today.[50] During the next geological epoch, the Eocene (which began 56 million years ago), the climate became considerably warmer.[51] The concentration of carbon dioxide increased rapidly during this period, resulting in an increase in global temperatures of 5 to 10 degrees Celsius.[52] Leaf and pollen fossils from the Bighorn Basin of Wyoming show that the composition of the vegetation changed rapidly (within 10,000 years) as the climate became warmer. Previously common species declined and tropical species arrived. Some of these tropical species were known from Paleocene fossil deposits hundreds of kilometers south, indicating that they had expanded their ranges northward. Subtropical forest occurred as far north as Alaska during the Eocene.[53] These forests were dominated by trees with thick, leathery leaves with smooth margins and drip tips, all of which are characteristic of modern tropical rainforest trees. Many of the tree species belonged to families that are now largely restricted to the tropics. At mid latitudes many sites in western North America had a mix of tropical trees that resemble modern figs (Ficus), breadfruits (Artocarpus), and laurels (Cinnamomum), as well as trees that now characterize temperate forests such as elms, walnuts, sweetgums, and ashes.[54]

The prevalence of tropical evergreen broad-leaved trees might have continued if the climate had remained constant, but by the late Eocene the climate began to cool, favoring deciduous trees or (in drier environments) evergreen, needle-leaved conifers. Colder climates in the succeeding Oligocene and Miocene epochs resulted in a greater diversity of deciduous plants throughout the northern temperate zone. By the mid Miocene (about 15 million years ago) temperate deciduous forests reached their peak in terms of both distribution and diversity.[55] In East Asia one can still see deciduous forests with tree and shrub diversity comparable to what we see in the fossil record for northern deciduous forests during the Miocene. For North America and Europe, the late Miocene marked the beginning of a decline in the diversity of deciduous trees and shrubs, and a severe truncation in the extent of deciduous forests.

Climate Change and the Decline of Deciduous Forests

The climate became increasingly cool in the north temperate zone in the Miocene and succeeding Pliocene, leading up to a series of severe glacial periods in the Pleistocene. The reasons for these long-term changes in climate are complex and not adequately understood, but are associated with lower carbon dioxide levels in the atmosphere due to low levels of volcanic activity (leading to the opposite of the greenhouse effect that has caused rising temperatures since the Industrial Revolution) as well as changes in ocean currents due to changes in the size and position of continents. At the same time the uplifting of the Tibetan Plateau and the Rocky Mountains changed the climates in the interior of North American and Eurasia.[56] Air masses cool and lose much of their water as precipitation as they pass over a mountain range, creating a rain shadow that dries up the land downwind. Also, the recently uplifted mountains and plateaus were considerably cooler than the surrounding lowlands. Hence much of western North America was either drier or higher. Deciduous forests were replaced by deserts or grasslands in drier areas or with coniferous forest in high-altitude areas with cold winters or fairly dry conditions or both. Deciduous woodlands remained primarily in wetter sites such as the banks of rivers and streams, and many genera of deciduous trees became extinct.[57] Similar changes occurred across Central Asia, separating the deciduous forests of Europe from those in eastern China, Japan, and Korea.

By the late Miocene, about 8 million years ago, open grasslands covered large areas of the Great Plains of North America.[58] Fossil fruit and seeds indicate that this region was covered with grass and a diversity of prairie herbs with only sparse woodland. This ancient prairie was grazed by camels, rhinoceroses, several species of piglike oreodonts, and five species of horses.

Southern Maryland was distinctly different from the Great Plains at this time.[59] Based on the pollen record, 6 million years ago the deciduous forest in this region would have looked familiar to a modern observer, with pines, maples, dogwoods, beeches, sweetgums, tuliptrees (Liriodendron), sycamores, cherries, three types of oaks, and even poison ivy or a close relative (Toxicodendron). This forest apparently was more diverse than a modern Maryland forest, however, with some additional plant groups that are now extinct in North America. Two of these groups, the zelkovas (Zelkova; Family Ulmaceae) and wingnut trees (Pterocarya, Family Juglandaceae), can still be seen in deciduous forests in Japan and China, but have disappeared in North America. Tall, straight zelkovas, for example, are one of the most impressive deciduous trees in patches of old woodland around Buddhist temples in Kyoto, Japan.

The fossil record for the Miocene in Japan provides us with a detailed view of changes in forests over millions of years. Fossils of leaves, nuts, fruit, and other plant parts permit identification of species.[60] By the Miocene most of the trees of Japan clearly belong to modern genera although most of the species were different. In the Eocene, long before the Miocene, much of southern Japan was covered with broad-leaved evergreen forest, a type of forest that is still found in relatively warm coastal areas in southern Japan. The dominant trees of this forest had thick, leathery leaves that are retained in the winter; these included evergreen oaks and golden chestnuts (Castanopsis). Northern Japan (the island of Hokkaido) had deciduous forest dominated by hickories, elms, sweetgums, sycamores, zelkovas, and maples. As the climate cooled, deciduous forests spread southward, replacing evergreen broad-leaved forests in many parts of Japan. By the late Miocene the forests of Japan were similar to cool, temperate forests found today in Japan and eastern North America. The dominant species were beeches, maples, ashes, hemlocks, elms, zelkovas, and dawn redwoods. Except for the dawn redwoods, this forest probably would have looked familiar to modern observers from Japan and (to a lesser extent) from eastern North America or northern Europe.

The late Miocene forests in Europe were distinctly different from those in either eastern North America or Japan. They were dominated by numerous species of conifers (including a type of Sequoia or redwood) with a smaller component of deciduous trees.[61] More than 30 species of conifers of 16 genera have been recorded from this period in Europe. These mixed coniferous-deciduous forests had a low diversity of deciduous tree species compared with forests in eastern North America or East Asia, and were in many respects more similar to modern forests of the mountains and Pacific Coast of western North America. Deciduous trees may have been most frequent in areas disturbed by fires or floods, as they are in many coniferous forest ecosystems today.

By the beginning of the Pleistocene, about two and a half million years ago, a combination of lower temperatures and wet conditions in the Arctic led to the accumulation of snow that didn't completely melt during the arctic summer, triggering the steady buildup and spread of continental glaciers. These kilometer-deep glaciers moved inexorably southward, destroying forests and bulldozing the landscape. Eventually glaciers covered what are now eastern Canada, the northeastern states and Great Lake states in the United States, and northern Europe. The effects in East Asia were less severe. Most of Japan remained free of glaciers except in the high mountains, and China was free of glaciers at mid latitudes.[62]

Based on landforms and deposits of rock and gravel left by glaciers, geologists originally hypothesized that there were four glacial periods separated by warm interglacial periods during the Pleistocene. It is difficult to decipher the sequence of continental glaciers on land, however, because each glacial advance largely eliminates the evidence of past glaciers. A better record of alternating ice ages and warm periods is obtained from sediments on the ocean floor. The shells of marine organisms in these sediments provide a record of the ratio of two isotopes of oxygen (^{18}O and ^{16}O) dissolved in ocean water at the time that the organisms lived.[63] When the ocean water was colder, more ^{18}O was incorporated into these shells than during warm periods. The oxygen isotope record in marine sediments reveals not just four, but a total of 18–20, alternating glacial-interglacial cycles during the past 1.6 million years.[64] It also reveals that during the past 2 million years, the typical state of the earth was highly glaciated and cold, with considerably lower sea levels because so much water was tied up in deep glacial ice. The implications for Europe, Canada, and eastern North America are stunning. Forests in these regions have repeatedly retreated to the south and then expanded northward with the advance and retreat of glaciers. During glacial periods deciduous forests disappeared not only in the north, where the land was covered by a couple of kilometers of ice, but also from vast regions to the south that were covered with tundra, spruce parkland, or

spruce-fir forest. The dominant species of deciduous forest survived some-
where far to the south, in Louisiana and Bulgaria and southern China, in
relatively small refuges. Even before the Pleistocene, Europe had lost some
groups of deciduous trees as the climate cooled and conifers became dominant. During the repeated glacial advances, many more tree species became
extinct in both Europe and North America.

During each interglacial period—again and again for 2 million years—
the northern woodlands were reconstituted as ice melted and one tree species after another spread northward as a result of seed dispersal. Our best
record for this process comes from the current interglacial period, which
began about 14,000 years ago. As the last continental glaciers melted, they
left behind deep depressions that became lakes and bogs. The lake sediments and peat deposits in these depressions preserve pollen grains that
show how vegetation changed in the surrounding landscape over thousands
of years. The pollen of different genera of trees can be clearly distinguished,
but it is difficult or impossible to accurately identify particular species
within a genus. Hence we can tell whether a landscape was dominated by
oaks or maples or spruces, and determine the general type of forest.

After analyzing hundreds of ponds, lakes, and bogs, ecologists have
worked out the distribution of general types of ecosystems before and after
the most recent continental glacier retreated. During each glacial maximum, when the weather was coldest and the ice extended farthest south,
most of eastern Canada and the northern United States were covered with
ice. South of the ice were bands of tundra, spruce parkland (an open grassland with scattered spruce), and coniferous forests dominated by spruce,
fir, and northern pines (either jack pine or red pine).

Researchers have searched in vain for a broad zone of deciduous forest
or for large islands of deciduous forest in the Appalachian Mountains. Evidence from the pollen record demolished the widely accepted theory that
islands of intact forest have survived since the Miocene on the Allegheny
and Cumberland plateaus of the Appalachian Mountains in Kentucky and
Tennessee.[65] On these high-elevation tablelands one can see the most
diverse temperate forest in North America, the mixed mesophytic forest.[66]
E. Lucy Braun, one of the most influential forest ecologists of the early twentieth century, hypothesized that these plateau forests were "lost worlds"
that have majestically survived for millions of years, and thus represented
the archetype and the source for deciduous forests that grew up on the barrens left by the continental glaciers. Thus the ancient deciduous forest of

the Appalachians should be clearly distinguished from the relatively new forests in regions that were too cold, dry, or ice-covered to support trees during glacial periods.

Hazel Delcourt describes how she set out to test the hypothesis that mixed mesophytic forest had survived the glacial period on the Cumberland Plateau of Kentucky.[67] First, she needed to find lake bottom or bog deposits that dated back to the height of the last glacial period. Most of the Cumberland Plateau consists of sandstone, however, which is too well drained to support long-lasting ponds or lakes that would preserve a record of pollen rain since the last glacial maximum. Fortunately, she and her husband and research collaborator, Paul Delcourt, found a sinkhole with a pond at the bottom that preserved sediments from the past 19,000 years. The lowest layers of clay, which were deposited during the period of maximum glaciation, primarily contained pine pollen rather than the diverse pollen of a mixed mesophytic forest. Instead of being a refuge for deciduous hardwoods, this region was covered with northern pines (probably jack pines judging by pine needle fragments found in the sediments) during the last glacial period.

Hence, deciduous forests did not cover wide latitudinal zones or major plateaus during glacial times. The pollen record indicates that deciduous trees survived in small pockets of habitat in relatively wet locations far to the south of the mountain plateaus. Investigation of sediments from northeastern Texas revealed that maples, alders, hickories, ashes, and hackberries grew in floodplains and lakesides during the last glacial period.[68] Similar glacial refuges for deciduous forest species are known for southwestern Tennessee, south-central Arkansas, and northern Florida, but most of the sites from the southeastern Atlantic coastal plain and the Appalachian Mountains reveal a glacial landscape with spruce, fir, and pine forests.[69] Many of the sites with a diversity of deciduous hardwood pollen are in sediments in river floodplains close to the Gulf Coast.[70] Perhaps some of the deciduous forest refuges were even farther south, on land that is now covered by the Gulf of Mexico. Land extended far to the south of the current shoreline when the sea level was 121 meters lower during the glacial maximum.

Glacial refuges for deciduous trees have also been difficult to locate in Europe. The entire area between the continental ice sheet and the glacier-covered Alps was deforested polar desert and tundra, and treeless steppe extended south of the Alps to the Mediterranean Sea.[71] Based on the way in which tree pollen first appeared in the sediment record at the end of the glacial period, it appears that deciduous tree species survived in refuges in southeastern Europe even though steppe extended south to Turkey and Iran. As in North America, trees may have survived in relatively small pockets of favorable habitat. This resulted in high extinction rates for trees, with one genus after another disappearing with succeeding glacial periods.[72] Based on the fossil record from before the Pleistocene, only 29 percent of the tree genera of Europe survived the ice ages, compared with 96 percent in East Asia and 82 percent in eastern North America.[73] The exceptional diversity of conifers in Europe was lost during this period, and the surviving deciduous trees became dominant during later interglacial periods.[74]

Reassembly of Deciduous Forests

The pollen record for the past 15,000 years reveals how forests spread across the barren landscapes of glacial debris and tundra of northern latitudes in North America and Europe as the glaciers began to recede. Layers of sediment can be dated precisely by measuring carbon radioisotope concentrations, so the sequence of vegetation changes can be followed on the scale of centuries or even decades. The pollen record shows different species of trees moving northward at different rates. As a result, the composition of northern forests has steadily changed.

This view of the history of deciduous forests has implications that go far beyond academic discussions about climate change and evolution of plants. Braun's theory of the continuity of intact islands of mixed mesophytic forest was consistent with a more general theory about how ecosystems work. In this view, most natural communities are ancient associations of species that have evolved accommodations and adjustments to one another to the extent that species are now interdependent. As a result, each natural community has a predictable set of species that interact in predictable ways. The plateau refuges of mixed mesophytic forest would not only protect individual species from glacial conditions, but would also sustain the web of interactions that held the forest ecosystem together. We now know from the pollen record, however, that the Appalachian plateaus were covered with pine and spruce forest during the last glacial period, and there is no evidence for an intact mixed mesophytic forest. Spruce, fir, and northern species of pines (jack and red pine) spread far to the south, even into Florida. The southeastern coastal plain was covered with pine, oak, and hickory forest with a low diversity of deciduous trees.[75] The diverse group of plant species that constitute the mixed mesophytic forest were apparently separated from one another, scattered in widely separated glacial refuges, and then reassembled slowly after the climate became warm again. All of the complex—and supposedly obligatory—relationships among species were disrupted for tens of thousands of years.

Today the most widely accepted theory about the structure of ecosystems is that different species of plants and animals generally respond to climate change and other environmental changes separately from one another, and that temperate forests are not closely interdependent communities, but rather assemblages of species that happen to live in the same place at the same time because of similar environmental requirements and the accidents of history. The older view of forest ecosystems as highly interdependent communities is still deeply embedded in natural history documentaries and popular writing about the environment, but it has been discredited (at least in its pure, unqualified original form) by ecological research.

As I write this paragraph, I'm sitting on a deck surrounded by the red maples, shagbark hickories, white oaks, and flowering dogwoods of a mature forest in coastal Connecticut. This particular forest is not very old; the age of the oldest stand of trees in the valley below me is probably a little more than a hundred years, and most of the nearby trees are younger. Unlike the current woodland, however, the forest ecosystem gives every appearance of being ancient. The tree species in the forest canopy appear to be well adapted to competing with one another for light. The violets and wood anemones that bloom in early spring are well adapted to take advantage of the brief warm, sunny period on the forest floor before the trees leaf out. They produce leaves and flowers during this sunny period of spring, and then set seeds about the time that they are thrown into deep shade as leaves unfurl in the canopy and shut out the sunlight. They flower at the right time to provide nectar for queen bumblebees that are raising their first brood of workers after spending the winter in underground burrows. Throughout the morning I've heard the songs of various migratory birds that nest in the forest around me. These species seem well adjusted to each other, with each species depending on a food supply not much used by the other species. Ovenbirds flip leaves on the forest floor, worm-eating warblers carefully search the leaves of viburnums and other shrubs, American redstarts flush insects out of the foliage in the subcanopy and lower canopy, and red-eyed vireos slowly move through the tops of the tallest trees peering from side to side to inspect leaves. All of these species are looking for insects and spiders to feed themselves and their nestlings, but they are finding different prey species in different places. They don't compete much with one another for food even though they all forage in the same patch of forest. By all appearances, this is a highly organized system that is the product of a long period of evolution.

There may have been a long period of evolutionary adjustments among these species, but it clearly did not occur in Connecticut or in any other static place. There were no forests in Connecticut—only glacial ice—18,000 years ago. Sediments from Rogers Lake, which is only a few kilometers from my deck, provide a precise record of how the vegetation changed since the glacier receded. In the 1960s Margaret Davis, who developed many of the approaches used today to interpret ancient pollen samples in light of recent samples of pollen rain from different habitats, analyzed two cores of the sediments in Rogers Lake.[76] She summarized her results in diagrams displaying changes in the percentage and deposition rates of different types of pollen. Carbon radioisotope concentrations were used to date the sediments at intervals in the 11.5-meter cores. These pollen diagrams show that the vegetation was in constant flux for the past 14,000 years (a short period in geological and evolutionary time). The region was covered with open tundra with few trees from 14,300 to 12,150 years ago. This was replaced with boreal woodland, which persisted for the next 3,000 years. Initially this boreal forest may have been similar to open spruce woodlands that can still be seen in northern Quebec, but with a higher proportion of pine (which was either red or jack pine, neither of which grows naturally in Connecticut today). An abrupt change occurred about 9,100 years ago, when the spruce-red/jack pine forest was replaced with a forest dominated by white pine and deciduous hardwood trees. Hemlocks, poplars, oaks, and maples were frequent in this forest, so it was similar in some respects to forests in the same region today, but many important tree species were still missing. Over the next 9,000 years the composition of the forests around Rogers Lake constantly changed. White pine declined about 8,000 years ago as deciduous trees increased. American beech arrived in the area 6,500 years ago, followed by hickory about 5,500 years ago and American chestnut just 2,000 years ago. Pollen diagrams from Rogers Lake and other nearby sites reveal that even after deciduous woodland became the dominant vegetation, it took thousands of years to assemble the modern forest as species were added one by one. Instead of a "deciduous forest zone" moving northward as a unit, each species of tree had a specific pattern and rate of dispersal that

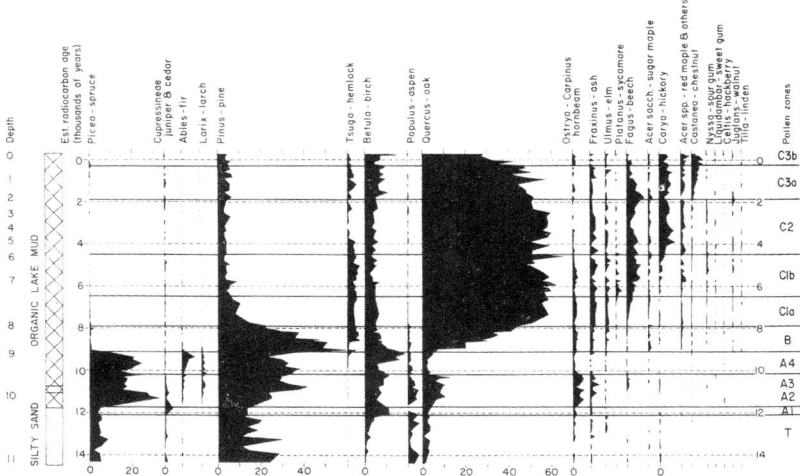

Figure 5. Percentage of pollen for different types of trees during the past 14,000 years in the sediments of Rogers Lake in Connecticut. Percentages are calculated as the proportion of all pollen for terrestrial plants in the sediment. (Davis, 1969; reprinted with permission from the Ecological Society of America)

depended on how its seeds were dispersed (by wind or birds or small mammals) and (ultimately more importantly) its tolerance to cold or dry conditions.[77] Northward rates of dispersal following the last glacial period vary from 287 meters per year for wind-dispersed willows to 70 meters per year for bird-dispersed black tupelo (black gum). Constant flux in the composition of the vegetation was the typical pattern for most regions in eastern North America.[78] Only the pine woodlands and swamp forests of the Gulf Coast plain in the southeastern United States have remained relatively stable throughout the period of Pleistocene climate changes.[79]

The pattern of vegetation change in Europe following the end of the last glacial period was broadly similar to that in North America, but there were some differences.[80] The main mountain ranges of Europe are oriented from east to west, forming a barrier to northward dispersal of trees, which cannot easily pass the treeless alpine zone at high elevations. On the other hand, many large rivers in Europe flow north, which would aid in the dispersal of tree seeds (in contrast to the eastward or southward flow of rivers in eastern North America). Because of differences in the postglacial climate, trees began dispersing northward in Europe about 2,000 years later than in North America, but then they dispersed northward more quickly, with maximum rates of dispersal of 1,000 meters per year, more than twice the maximum rate for the fastest dispersers among North American tree species. Most of the surviving species of deciduous trees and conifers spread from localized refugia in Bulgaria and other parts of the Balkans. Southern Spain also supported deciduous forest during the glacial period, but the Pyrenees form a wall that may have prevented the spread of trees out of Spain. Many groups of trees had become extinct during the Pleistocene glacial periods and the cool, conifer-dominated period preceding the Pleistocene. These included many types of trees that still live in the forests of eastern North America, East Asia, or both, including bald cypress, hemlock, hickory, sweetgum, tuliptree, magnolia (*Magnolia*), tupelo (*Nyssa*), and sassafras (*Sassafras*).[81] The woody plants that survived the glacial periods in Europe were relatively generalized, opportunistic species. They spread rapidly across western Europe as the ice receded, but did not assemble into highly diverse forests with many specialized species of trees and shrubs comparable to the deciduous forests in East Asia and on the plateaus of the Appalachian Mountains in North America.

In East Asia the glacial ice sheets did not move far south, but northern China and northern Japan were covered with tundra and boreal forest.[82] Deciduous forest was restricted to refuges in southern China, but mixed deciduous-coniferous forest was found from southern Japan to southern China, including parts of the East China Sea that were dry at that time. Although the extent of forest with deciduous hardwood trees was much reduced compared with their present distribution, the pollen record indicates that it was more extensive than in either Europe or North America during the past glacial period. As a result, there were relatively few extinctions of plants associated with the deciduous forests in East Asia.[83]

Another likely explanation for the greater diversity of temperate-zone plants in East Asia is more frequent evolution of new species (speciation).[84] In contrast to the almost continuous deciduous forests of Europe and North America, the temperate deciduous forests of East Asia are split into separate regions by ocean and mountain barriers. The deciduous forests of Japan and Korea are isolated from each other and from those in mainland China. This is probably the typical situation during warm interglacial periods, but during glacial periods the sea level is lower and the shallow sea between southern Japan, Korea, and China became dry, connecting these areas. Climate models and the pollen record indicate that the deciduous forests were continuous across these regions during the last glacial period. If the different patches of deciduous forest remained permanently isolated, one would expect distinct species to evolve in each patch and then remain in that patch. The cycles of isolation and reconnection among these landmasses, however, meant that species that evolved in isolation in one patch could colonize other patches, leading to a buildup of closely related species within East Asian forests. The remarkable high diversity of maples and other deciduous forest trees in East Asia may result from this combination of low extinction rates and high speciation rates.

Implications for Conservation

The fossil record provides a deeper understanding that can help us protect and sustain temperate deciduous forests. The groups of deciduous hardwoods that now dominate these forests have persisted for tens of millions of years because they are adapted to relatively harsh, often highly disturbed conditions. They survived the K-T asteroid impact, a catastrophe that eliminated 80 percent of the terrestrial plant species in North America. They dominated the new ecosystems of the Paleocene, persisted during the rapidly warming climate of the early Eocene, thrived with the cooler climates of the Miocene, and rebounded again and again as ice sheets, tundra, and coniferous forests repeatedly covered much of the North Temperate Zone during the Pleistocene. Their resilience is still apparent today. They are frequently the first species of woody plants to colonize areas following fires, floods, landslides, and logging. This is true even when older forests in the same region are dominated by conifers or evergreen hardwoods. In the boreal spruce-fir forests of Canada and Siberia, the pioneer trees are deciduous birches and aspens. Similarly, in the evergreen hardwood forests of southern Japan, disturbed areas are colonized not by the evergreen oaks and golden chestnuts of the old forest, but by deciduous oaks and maples. This resilience and colonization ability probably explains how so many deciduous hardwoods survived the cold and dark of the K-T event and the major climate changes of the past few million years. Also, in Europe and North America (and to a lesser extent in East Asia), modern species of deciduous hardwoods survived in relatively small glacial refuges from which they spread out as the climate warmed. The loss of many species since the Miocene indicates that there is a limit to their resilience, however. Particularly in Europe, the glacial refuges were too small or the barriers to dispersal were too great, and many species of trees became extinct. In the face of rapid environmental change due to human land use and climate change, we need to avoid creating the situation that led to the loss of many of the species of trees, both hardwoods and conifers, in Europe. By understanding how trees survived the last glacial period and then dispersed to restore northern forests as the climate warmed, we can help ensure that they survive massive environmental changes in the future.

CHAPTER 3

Deciduous Forests After the Arrival of People

Before the current interglacial period, temperate-zone forests were molded by changing climate and geological processes such as the rise of new mountain ranges and the emergence and submergence of land bridges due to changing sea levels. Since the retreat of continental glaciers, however, the fate of forests has been increasingly driven by the activities of technologically advanced people. Clearly the impact of people intensified with the development of agriculture, and strengthened even more with industrialization. People may have had a major impact on forests even before agriculture and industrialization, however, because of their use of fire and refined hunting techniques.

Extinction of Large Mammals of Deciduous Forests

In the 1960s, Paul Martin proposed that people had a major impact on ecosystems as soon as they arrived in North America. According to his "prehistoric overkill" hypothesis, experienced hunters with tools and hunting strategies honed in Eurasia invaded a continent filled with mammoths, mastodons, camels, horses, and other large mammals that had no experience with human predators.[1] These "naive" animals were easy prey, and human populations prospered and increased rapidly. The result was a swiftly advancing wave of human hunting groups that spread throughout the Western Hemisphere, to the southern tip of South America, within a few hundred years. This explosive spread of human hunters could account for the loss of most of the large animals of the Pleistocene, when parts of North America had a diversity of large mammals that equaled or exceeded the diversity of the Serengeti Plains in East Africa. Thirty-five genera of North American mammals became extinct, including all native species of elephants, horses, camels, and ground sloths (some of which were as large as elephants).[2] More than 50 percent of species that weighed more than

32 kilograms and all species weighing more than 1,000 kilograms disappeared.[3] Many of the largest mammals of Eurasia also disappeared, but few large mammals became extinct in sub-Saharan Africa at this time, perhaps because hominids had evolved in Africa, allowing enough time for prey species to evolve or learn effective defenses to increasingly effective human hunting practices. Martin and other researchers extended the overkill hypothesis to Australia and the large islands of Madagascar and New Zealand, where extinction of large animals (giant marsupials in Australia, enormous flightless birds in Madagascar and New Zealand) occurred soon after human colonization.[4]

This provocative hypothesis was originally rejected by many (probably most) archaeologists and mammalian paleontologists, who argued that gradual extinction of species due to climate and habitat change—not catastrophic extinction as a result of hunting by a small population of hunters using stone tools—was a more reasonable explanation for the loss of large mammals in North America and Eurasia. Large mammals disappeared during a period of rapid and profound climate change, as the glaciers and tundra retreated northward following the last glacial advance, and then pushed southward again during a thousand-year cold period called the Younger Dryas.[5] The last fossil records for many of these mammals coincide with the Younger Dryas cold period. The arctic steppe and open woodland habitats where mastodons, mammoths, horses, and camels once browsed and grazed must have been greatly disrupted and fragmented by these severe changes in climate.[6] As the large herbivores disappeared, the giant predators and scavengers such as dire wolves, saber-toothed cats, lions, and short-faced bears also declined and disappeared.

The most direct evidence for testing these two hypotheses comes from the fossil record of mammals during the late Pleistocene. Of the 35 genera of mammals that became extinct in North America, only 16 are known to have survived until 13,500–13,000 years ago, the period when there is good evidence of hunters with big-game hunting technology.[7] The Clovis culture of this period is known for long, fluted stone blades that appear to have been used as projectile points on spears. Some of these Clovis points were found near the bones of mammoths, but the last known fossils for many of the large mammal species date from well before the Clovis culture, during the last glacial maximum. This pattern doesn't fit with the hypothesis that giant mammals quickly and simultaneously became extinct soon after human hunters entered the continent (or at least soon after hunters with sophisticated weapons arrived).

Mammal fossils from the late Pleistocene are too infrequent, however, to distinguish a gradual series of extinctions from a short burst of almost simultaneous extinctions. The problem is similar to determining whether many species became extinct during a short period at the end of the Cretaceous; the best evidence for the timing of Cretaceous extinctions comes from the fossil record of pollen, microorganisms, and other abundant fossils, not from the relatively scarce fossils of dinosaurs (see Chapter 2). The extinction at the end of the Pleistocene was selective for large plant-eating animals and their predators and scavengers, and there is no evidence for widespread or numerous extinctions of microorganisms, marine organisms, plants, or smaller animals. Hence, we cannot depend on evidence from abundant fossils such as leaves or marine microorganisms to determine the timing and duration of extinctions.

The lack of a high-resolution fossil record has sustained the overkill–climate change debate for more than 40 years. Scientific debates are exceptionally persistent and polarized when the evidence is incomplete and inconclusive. Recently, however, a new source of information has made it possible to test the overkill and climate change hypotheses more conclusively. *Sporormiella* is a type of fungus that grows on the dung of plant-eating mammals. The spores of this fungus are preserved in lake sediments along with pollen. In the early 2000s researchers analyzed the distribution of these spores in sediments dating back to the end of the last glacial period.[8] The record indicates that this fungus (and the dung it depends on) was common during the glacial period and at the beginning of the current warm interglacial, but that its abundance dropped steeply between 14,000 and 13,000 years ago, which is the period of the big-game hunting Clovis culture. This pattern was documented for sites in upstate New York, Ohio, and Indiana. Although the fungal spores tell us nothing about the fate of particular species of large mammals, they reveal that large mammals as a group declined precipitously. This decline was followed by a major change in vegetation as more deciduous trees grew among the spruce and pine that had previously dominated the area. Also, fires increased in frequency, resulting in more charcoal particles in the lake sediments. It appears that a substantial decline in grazing and browsing animals permitted denser and more diverse vegetation to grow, but the denser vegetation was then prone to more frequent fires. Fires may also have been set by people, either accidentally or purposefully. The result was a major restructuring of woodland habitats after massive plant-eaters such as mastodons and ground sloths were no longer opening and pruning the tree canopy. The pattern fits with the hypothesis that hunting—perhaps combined with burning—by people was a sufficient condition for the extinction of most large mammals. However, the evidence from dung fungus indicates that the decline of large mammals occurred over centuries, with different timing at different sites, which doesn't fit with Martin's hypothesis of a rapid wave of extinctions caused by the leading edge of advancing human hunters. The process of extinction was more gradual, and may have been facilitated and hastened by climate change. However, the drop in dung spore frequency occurred well before the Younger Dryas cold period, so it appears that the mild climate change during the period of large mammal declines was not sufficient, and perhaps not necessary, to cause the extinctions. A similar pattern of *Sporormiella* spore declines followed by an increase in charcoal particles occurs in sediments from 2,000 years ago in Madagascar, soon after that island was first colonized by people.[9] Hip-

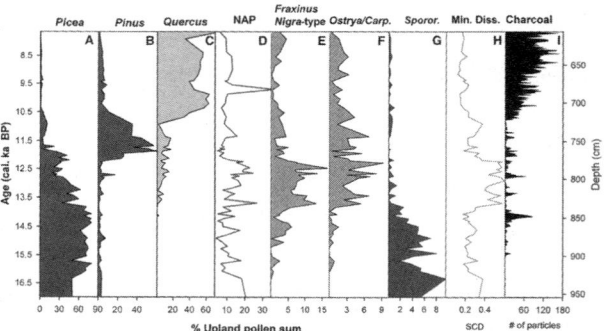

Figure 6. Abundance of pollen and of spores of the fungus *Sporormiella* (an indicator of the prevalence of dung from large mammals) in sediments of Appleman Lake in Indiana, during the past 17,000 years. Pollen abundance is shown for spruces (*Picea*); pines (*Pinus*); oaks (*Quercus*); plants other than trees (NAP); ashes (*Fraxinus*); and hop-hornbeams and hornbeams (*Ostrya/Carpinus*). Abundance of charcoal particles (indicative of fires) is also shown, as well as the level of dissimilarity (Min. Diss.) to the pollen profile produced by modern North American plant communities. The plant community near Appleman Lake was most dissimilar from any modern plant communities during the period after the dung of large mammals declined and before the frequency of fires greatly increased. (From Gill et al., 2009; reprinted with permission from AAAS)

popotami and several species of giant lemurs became extinct during this period.

A major weakness of the climate hypothesis is that the mammals that became extinct at the beginning of the current interglacial period survived similar transitions from glacial to interglacial conditions repeatedly during the past 780,000 years.[10] The last transition was not unusually abrupt or severe, and temperature fluctuations similar to the Younger Dryas apparently occurred at the beginning of other interglacial periods.[11] Also, the pattern and timing of extinction of large mammals differed for different parts of the North Temperate Zone. In Europe extinctions occurred in two phases: 40,000 to 20,000 years ago during the last glacial period, and 14,000 to 10,000 years ago, at the beginning of the warm interglacial period.[12] The first phase involved extinction of species such as straight-tusked elephant and hippopotamus that were associated with warm, deciduous forest. The second phase involved species such as woolly mammoth and woolly rhinoceros that were associated with cold, arctic steppes. In both phases, species became extinct when their habitat had been greatly restricted by climate change, but they had survived many similar climate cycles in the past. Modern humans (*Homo sapiens*) had spread across Europe by the time of the most recent cycle, however, and large mammals may have been especially vulnerable to extinction as a result of hunting during periods when their preferred habitat was restricted due to either glacial or interglacial conditions.

The pattern was different in Japan, where large mammals such as Naumann's elephant and Yabe's giant deer became extinct between 30,000 and 20,000 years ago, which was a period of rapid human population growth as indicated by more than 5,400 Paleolithic archaeological sites.[13] These species disappeared from Japan during a period of relatively mild climate change. Mammoths disappeared from central China, but some large mammals such as Asian elephant and two species of rhinoceros survived in the extensive deciduous forests of central China well into historical times.[14]

Although the pattern and timing of extinction differed on different continents, all of the world's deciduous forests eventually lost their giant mammals (the "megafauna"). The ground sloths, mastodons, and giant beavers of North America; straight-tusked elephants and hippopotami of Europe; and elephants and rhinoceroses of East Asia probably played major ecological roles by creating and sustaining forest openings and more open forest canopies. The data from dung fungus spores reveal that the decline of large mammals preceded major changes in vegetation and fire frequency, indicating that the structure of woodlands became distinctly different after they disappeared. The loss of megafauna probably resulted in a series of cascading changes that produced forests unlike those of earlier interglacial periods.[15] Similarly, the addition or the removal of elephants and white rhinoceros in African ecosystems results in major changes in vegetation and fire frequency, with secondary effects on a large number of other mammals (including medium-sized grazing and browsing mammals).[16] African elephants, in particular, reduce tree cover in open woodland and maintain grassy glades in forests. As African elephant densities increase in East Africa, woodland is converted to shrub coppice, open savanna, or even grassland.[17] The plants, insects, birds, and other organisms of modern deciduous forests evolved over hundreds of thousands of years in forests that were partially engineered by elephants or their relatives and other giant mammals, so the continuous closed-canopy forests that now grow naturally across the mid latitudes of the Northern Hemisphere may not be an ideal habitat for some of these species.

Fire and the Deciduous Forest

The control of fire enabled people in hunting-gathering societies not only to stay warm and cook their food, but also to modify their environment.[18] To some extent this may have been accidental, as fire escaped from human encampments, but much burning was probably purposeful. Fire is used by many hunting societies to drive game and to create more open habitat that attracts grazing animals and provides easier hunting because of greater visibility of prey. Burning can also increase yields of animal forage, boosting prey populations. Purposeful burning may have molded deciduous forests as these forests began to reassemble in the wake of the glaciers. Particularly in drier areas or areas with porous, sandy soil, the result could be conversion of woodland to savannas with widely spaced trees or even to open grasslands. Frequent fires probably maintained the open grasslands of the Shenandoah Valley in Virginia, the Hempstead Plains of Long Island, New York, the oak savannas of Wisconsin and Minnesota, the "bluegrass" region of Kentucky, and much of the "prairie peninsula" (the extensive area of tallgrass prairie that covered much of Illinois, Indiana, and southern Wisconsin).[19] In all of these areas, undeveloped grasslands usually reverted to deciduous forest after fire was suppressed by European immigrants. In areas with more rainfall or less porous soil, periodic burning did not destroy the forest, but may have created an open understory that facilitated hunting and increased prey densities. Some of the first European accounts of the eastern deciduous forest in North America comment on the open understory and attributed this to burning by indigenous people.[20]

The potential role of fire in facilitating prehistoric hunting is especially well documented for eastern and central North America, but may have occurred in other deciduous forest regions. Beginning 9,000 years ago, Mesolithic hunting societies in southern England were concentrated in areas of dry, sandy soil. Stone tools from this period such as axes and microliths (small, carved blades that were mounted on javelins) are concentrated in sandy areas where open oak forest originally grew.[21] Forests of this type have an open understory that would have been conducive to hunting, and are also more prone to burning. Pollen evidence indicates that some of these areas shifted from dense shrub understories to open understories dominated by heaths and grasses, a change that is consistent with the effects of periodic burning by Mesolithic people. Also, between 6,700 and 5,600 years ago, charcoal in lake sediments in Poland indicates extensive burning of forest, and there are many archaeological sites in northern England where layers of charcoal indicate repeated fires and the pollen record is consistent with large brushy clearings.[22]

As Stephen Pyne pointed out, fire not only allowed people to leave the tropical savannas of Africa, but also permitted them to recreate the savanna in other places.[23] This observation applies particularly well to the dense forests of mid latitudes, where fire helped people to survive cold winters and to create more open habitats suitable for their hunting techniques. Ironically, in northern deciduous forests this may have partially compensated for the absence of giant mammals that had previously created and maintained forest openings and open woodlands.

Thus, even before crops and livestock became the main sources of food, people may have had a profound impact on the forest by creating openings and open woodlands. Although there is clear evidence of forest burning, however, the role of hunting and gathering societies as "ecosystem engineers" (organisms that have a profound impact on the structure of ecosystems) is still debated. Burning was associated with people at many archaeological sites, but it is not clear whether this was a widespread or highly local activity in particular regions.[24] We still do not have a clear picture of the mix of relatively undisturbed continuous forest and patchy, open forest across the deciduous forest regions of East Asia, Europe, and North America before the initiation of extensive forest clearing. This is true not only for hunter-gatherer societies but even for people who, like many indigenous cultures of northeastern North America, depended on a mix of small-scale farming, hunting, and fishing.

Farming Spreads and Forests Begin to Shrink

Once people began to cultivate extensive, permanent fields, the record of their impact on forests becomes more obvious. Pollen in lake sediments changed as people cleared extensive areas of forest for large-scale farming and grazing. The pollen of crops, agricultural weeds, and early successional trees became more frequent. Agriculture directly or indirectly affected nearly all temperate deciduous forests. This process began 7,500 years ago in Europe and 8,000 years ago in China, whereas it began about 2,500 years ago in Japan and 1,500 years ago in eastern North America.[25] Ultimately farming, grazing, and harvesting of forest products not only determined where forests survived, but also the structure and composition of the surviving forests. Without considering the impact of agricultural civilizations on forests, we cannot understand the ecology of contemporary forest ecosystems.

Agriculture arose in the Middle East by 7000 BC, spreading into Turkey and Greece by 6000 BC, and into central and western Europe by 5500 BC.[26] From France to the Ukraine, large areas of deciduous forest were cleared to grow barley and wheat and to raise cattle. Archaeological work shows that people of this period built timber "longhouses" in villages that lasted for hundreds of years, indicating that they practiced stable, sustainable agriculture.[27] They created permanent or at least long-term clearings, so the impact on forests was different from the effects of shifting cultivation (slash-and-burn agriculture) practiced in many parts of the tropics today. These early settlements were concentrated on well-drained, fertile soils, but upland areas may have been used for grazing livestock. Hundreds of these Neolithic sites are known from Germany alone. After 4400 BC farmers moved into Great Britain and northern regions of continental Europe, using heavier plows that were pulled by oxen to break the heavy clay soils of these regions.[28] Extensive forest clearing is indicated by the large number of flint axes from this period that have been found in different parts of Europe. Many of these originated in flint mines in France and Poland, from which they were apparently traded over a large area.

Although the common perception is that North America was a pristine wilderness at the time of first European contact, an early form of gardening began at about the same time as European forests were cleared for farming. The best evidence comes from the floodplains of the Mississippi River and its tributaries, where a diversity of native plant species was propagated for food by 5800 BC.[29] These included marsh elder, sunflower, and goosefoot. By 1000 BC these three species of edible plants had become structurally distinct from wild populations as a result of domestication, with larger seeds or thinner seed coats.[30] Although farming was limited and did not entail extensive clearing as it did in Europe, the dense floodplain forests of the central United States were interrupted by garden openings and fairly large settlements. The largest known center was located at Poverty Point in northeastern Louisiana.[31] This site was occupied from 1500 to 700 BC, and is best known for a bird-shaped earthen mound and a set of six concentric semicircular mounds that were two meters high and more than a kilometer long. Nine similar mound-building settlements have been found in Louisiana, Arkansas, and Mississippi. On the basis of house sites, archaeologists estimate that about 5,000 people occupied Poverty Point. This permanent settlement was primarily supported by fishing and harvesting nuts such as hickory nuts, pecans, acorns, and walnuts as well as farming.

Until recently the origin of agriculture and clearing of the deciduous forests of East Asia was not thoroughly understood, but new archaeological research in China gives us a much better picture of the origin and spread of farming.[32] Remains of rice have been discovered on pottery from 8000 BC in the Lower Yangtze (Changjiang) River valley, but it took another 4,000 years to achieve fully domesticated rice.[33] Domesticated rice has longer and wider grains, and the grains remain attached to the stem and thus cannot disperse and propagate without human help.[34] Attached grains slowly became more frequent over a period of thousands of years. This transition may have been delayed because of continual hybridization between domesticated and wild varieties of rice. Hence agriculture preceded the appearance of fully domesticated rice, and land clearing may have expanded as rice cultivation improved.

Excavations at a village site at Tianluoshan in the Yangtze Delta reveal much about the life of these early farmers.[35] This site was occupied between 5000 and 3000 BC. Rice was grown in natural marshes, and there is no evidence of irrigation structures such as water-storage reservoirs (which appear at other sites in the Yangtze Delta by 4000 BC). Cultivation of rice is demonstrated by a high proportion of domesticated rice, a proportion that increased during the 2,000 years that the fields were in use. Pollen in the field sediments revealed a high diversity of wetland plants grew in the fields, suggesting little or no active weeding. High concentrations of charcoal reveal that the fields were periodically burned, however. The rice yields must have been low, but rice farming apparently only supplemented hunting and gathering in the surrounding forest and marshes. Remains of acorns, water chestnuts, fish, and a variety of game animals such as deer and the now extinct short-horned water buffalo show that these people did not depend on farming as their sole or even main source of food. As in North America, there was a slow transition from hunting and gathering to agriculture, and settled villages preceded a dependence on farming.

Unlike early farming in Europe and North America, rice cultivation in China probably had a bigger direct impact on marshes than on forests. Forests were undoubtedly affected by the concentration of rice farmers living along rivers and deltas, however. This is demonstrated by a remarkable record of vegetation change and human activity between 6250 and 5400 BC on Hangzhou Bay south of the Yangtze River delta.[36] Exceptionally detailed and well-dated sediments from the site show the transition from undisturbed forest to settled farmland followed by a decline in agriculture and regrowth of forest after 5400 BC because rising sea levels brought saltwater into the rice fields.

Millet may have been cultivated in China even earlier than was rice. Common millet has been excavated in grain storage pits from various sites in northern China dating between 8000 and 6000 BC.[37] These pits contained thousands of kilograms of stored grain, so millet was probably already a staple food. Two species of millet, common millet and foxtail millet, were grown at several sites along the Yellow (Huanghe) River in northern China between 6500 and 5000 BC. Unlike the rice farms in the Yangtze Valley, these sites are found on foothills between the river and the mountains on plots with deep sediments and a good water supply from the uplands. Thus, millet agriculture may have initially resulted in more extensive forest clearing than did rice farming.

Rice cultivation ultimately had a bigger impact than millet cultivation on deforestation in Asia, however. Rice farming spread quickly out of the Yangtze River valley as efficient irrigation systems with reservoirs and canal systems were developed after 4000 BC.[38] Intensive rice cultivation spread to the Yellow River basin and perhaps even as far as Korea by 3000 BC, eventually reaching Japan by about 400 BC.[39] Rice paddies were highly productive, but the elaborate irrigation systems required centralized planning, resulting in hierarchical societies and urban centers.[40] Rice paddies filled the valleys and marched up the hills on terraces. The dense populations supported by these rice fields required building timbers, firewood, and leaf litter and understory vegetation for fertilizer, so even remote forests were eventually affected by the cultivation of rice. Rice farming is also highly productive and sustainable, so populations became extremely dense, leading to widespread deforestation in many parts of East Asia.

In his book about the people of the Western Hemisphere before Columbus, Charles Mann describes an imagined plane flight in AD 1000 over thriving cities and farming villages in South America, the Maya lowlands of Central America, and the Mississippi River valley.[41] Let's imagine a similar view from above 2,000 years earlier, in 1000 BC, when intensive agriculture had spread to some parts of the temperate zone. A satellite circling the Northern Hemisphere at this time would have photographed a view of extensive deciduous forests interrupted by agricultural openings—some large and some small—that had spread quickly outward from different centers of plant domestication. Farms would already dominate large stretches of central and western Europe and Great Britain, but these farms were restricted to particularly fertile, well-drained soils such as the loess soils of central Europe. In Germany, Britain, and many other parts of Europe, there were still extensive, mostly unbroken forests. In North America, the eastern deciduous forest was interrupted by agricultural openings along the Mississippi River and its tributaries, but much of the forest remained intact. Similarly, in China most agriculture followed the rivers, but forest clearing was much more widespread than in central North America. Japan's forests were barely touched by agriculture; clearing was largely restricted to the southern island of Kyushu, where buckwheat and barley farming had been introduced and irrigated rice cultivation would follow after a few more centuries.[42] In all of these regions large areas of forest may have been opened up by burning even in places not affected by agriculture.

Deciduous Forests Become Open Fields

After 1000 BC the southern deciduous forests of Europe were heavily modified by the spread of Greek and Roman civilizations. In his book *Man and the Mediterranean Forest*, J. V. Thirgood describes in grim detail how the destruction of forests led to severe wood shortages and soil erosion.[43] Greek city-states and the Hellenistic and Roman empires needed large amounts of wood for shipbuilding, heating, cooking, and construction of buildings. Shipbuilding, which was critical for both commerce and war, required large timbers.[44] Because of shortages of ship timber during the Classical Greek period, many cities restricted the export of wood and transported wood from distant locations, first from Macedonia and Thrace, then from Asia Minor and eventually from the Black Sea coast. Clearing of forests for timber and farming often was followed by heavy grazing by sheep and goats, which prevented regeneration of normally resilient deciduous forests. The result, according to Thirgood, was a devastated, eroded landscape that eventually impoverished many parts of Greece, Italy, and Spain. Thirgood quotes Plato's lament that "Attica [the area around Athens] may accurately be described as a mere relic of the original country," and that "originally the mountains of Attica were heavily forested" but "they now have nothing but bee pastures."[45] And in Sicily a large mosaic on the floor of the ruins of an ancient Roman villa depicts the hills surrounding the site covered with rich woodland.[46] The ruins are now located in barren, treeless hills.

The extent of deforestation (especially of tall deciduous hardwood and coniferous forest) in Mediterranean Europe may be overstated by Thirgood and numerous other authors, however.[47] Much of the Mediterranean coast of Europe was probably too dry to support dense, tall trees, which may have been limited to high elevations in the mountains and exceptionally cool, moist areas in the lowlands. Judging by the pollen record from the period before agriculture, savanna (open grassland with scattered evergreen oaks or pines) and maquis (low, scrubby vegetation dominated by shrubs with leathery leaves) covered many sites in southern Spain, southern Greece, and Crete, so the open landscapes of these regions today are not necessarily a product of forest clearing. Forests consisting of deciduous hardwoods such as oaks, beech, hazel, lime, and hornbeams (*Carpinus* and *Ostrya*) occurred in southern France, Italy, and at higher elevations elsewhere, however, and the relatively restricted distribution of these forests probably made them more susceptible to clearing. During Roman times, even the extensive forests of Italy probably disappeared except in the most inaccessible mountains, replaced not only by cropland, but also by artificial woodlands of olive and citrus trees.[48]

In contrast to Mediterranean Europe, continuous deciduous forests covered much of northern Europe after the glaciers receded. Areas with fertile soil had already been cleared by 3000 BC. With the expansion of the Roman Empire to northern Europe, Roman agricultural estates spread to Britain, Gaul (France), Portugal, and some parts of Germany.[49] Northern Europe had large areas of farmland, but there were still extensive forests. The major deforestation of many parts of western and central Europe occurred later, during the medieval period. The estimated forest cover in this part of Europe was reduced from 80 percent to 40 percent between AD 500 and 1300.[50] Many forests in central and eastern Europe were cleared by the Germans between 900 and 1300.[51] The heavy soils created by these old forests could be broken for farming with two new inventions: the heavy wheeled plow and a rigid harness that permitted this plow to be pulled by horses.

By the 1200s the loss of forest to agricultural clearing was so great that the remaining forests were considered valuable as a source of wood. As a result, laws were enacted in France and Germany to protect forests.[52] Before wood shortages and efforts to protect forests became well established, however, forests began to expand in Europe as a result of a sudden drop in the human population. Between 1347 and 1353, bubonic plague killed at least a third of the people in Europe. Villages and farmland were abandoned throughout the continent. The plague was followed by decades of warfare in many parts of Europe, which prevented economic recovery in the countryside. The result was that forests grew back in many areas that had been cleared before 1347.

Forest clearing in the British Isles fit with the general pattern for central and western Europe, but it was more extensive and permanent than in many regions on the continent. Large areas of Britain (and especially England) had been converted to farmland or heath by 2000 BC, and by 500 BC about half of England had been cleared of forest.[53] Thus, when the Romans conquered England the countryside was already largely agricultural. The river valleys were completely cleared for large estates during Roman times, and much of the rest of the country had a mix of farmland and woodland. Also, the demand for firewood for Roman ironworks converted tens of thousands of acres of forest into low, shrubby coppicewood. After the collapse of the Roman Empire, some cleared areas may have reverted to forest, but the pollen record doesn't show a major increase in forest cover. The Norman census of 1086, which is reported in the Domesday Book, gives us a rough picture of the amount of forest because it reports the distribution and size of woods. Based on the Domesday Book, only about 15 percent of England was forested at that time, and most of the woods were small. The amount of forest cover shrank to about 10 percent in 1350, when the decline ended following the arrival of bubonic plague, and some of the remaining woods may have grown larger as trees colonized abandoned farmland.

Forest clearing also increased in eastern North America after 1000 BC. Although farming eventually spread to all parts of the eastern deciduous forest, the most extensive forest clearing was associated with mound-building cultures of the Mississippi River watershed and the floodplains of rivers along the Gulf Coast. Ceremonial mounds were built by the Adena and Hopewell cultures, which flourished in Ohio from 500 BC to AD 400.[54] By the end of this period mound-building settlements had spread throughout much of the eastern forest west of the Appalachian Mountains and south of the Great Lakes.[55] These were relatively large villages centered on massive earthen mounds, many in the shape of animals. These cultures depended on farming for much of their food. Although corn (maize) is known from Hopewell sites from as early as 200 BC, it did not become an important part of the diet. Domesticated varieties of native plants such as sunflower and marsh elder remained important, as did hunting and gathering.[56]

After the decline of the Hopewell culture in AD 550, there was a hiatus in mound-building activity until about 700, when the much more elaborate Mississippian culture arose. Mississippian settlements expanded after beans and a hardier variety of maize were introduced to eastern North America from Mexico between AD 800 and 1000, leading to a major increase in agricultural productivity.[57] For the next thousand years river valleys in the woodlands west of the Appalachian Mountains and on the Atlantic coastal plain of the southeastern United States were occupied by societies that built elaborate earthen mounds capped by wooden buildings. The mounds were centered in large villages or towns in the middle of extensive cornfields. The largest center, Cahokia, was located in a wide floodplain (the American Bottom) along the Mississippi River in Illinois. This city covered 800 hectares (2,000 acres), about 20 percent of which was eventually enclosed in a log palisade with watchtowers.[58] It was centered on a ceremonial plaza and a rectangular mound that was 30 meters (100 feet) high. The floodplain north and south of Cahokia was occupied by dozens of other settlements, including four additional large centers. The region around Cahokia supported 25,000 to 50,000 people by AD 1200.[59] Based on the density of houses, the center of Cahokia supported 10,000 to 15,000 people.[60]

The steady expansion of Cahokia and its satellite villages after AD 1050 was followed by population declines during the 1200s.[61] Tree rings in ancient timbers from archaeological sites indicate that both the rise and decline of Cahokia were tied to climate change. The precise sequence of thick and thin growth rings on a tree trunk can be used not only to date the wood to an exact year in the past, but also to show whether the tree grew during a dry period (resulting in narrow annual growth rings) or a wet period (producing wide growth rings). Tree rings demonstrate that Cahokia expanded during an extremely wet period between AD 1050 and 1150. During the next 150 years, however, the region suffered a series of severe, long-lasting droughts. Upland farming areas were abandoned, the population declined, and by 1350 Cahokia and the surrounding region along the Mississippi River had been abandoned. When French explorers traveled through this section of the Mississippi River in 1673, they found only two small villages of Tuscarora who had recently fled to this area from North Carolina.[62]

Densely populated mound-building centers continued in the southeastern United States, however. These societies were encountered by Spanish military expeditions in the early 1500s. Spanish chronicles describe large, walled towns surrounded by extensive cultivated fields.[63] The Spanish never successfully conquered this region, but the towns were devastated by epidemics that swept across North America, spreading far ahead of coastal European settlements into the interior of the continent. For example, in 1540 Hernando de Soto's expedition discovered that many of the villages around the mound-building center of Cofitachequi in South Carolina were abandoned following an epidemic that had occurred two years previously.[64] Despite the epidemics, however, the last active mound-building center at Natchez was active until 1731, when it was conquered by the French.[65]

The impact of Mississippian societies on the forest is probably best documented for the Little Tennessee River valley in eastern Tennessee, where researchers carefully analyzed plant and pollen remains from the past 11,500 years.[66] Nuts from forest trees (acorns, walnuts, chestnuts, beechnuts, and hazelnuts) were an important part of the diet of the people who lived at this site, but the importance of nuts declined about a thousand years ago when maize became the staple food. Remains of sunflower, marsh elder, and other crops that had been grown at the site since 700 BC also declined at this time. The frequency of charcoal particles and the rate of erosion in-creased markedly after maize became the staple crop, indicating that the surrounding forest was cleared. Extensive clearing is also indicated by the sharp increase in the frequency of pollen from ragweed and other agricul-tural weeds and a decline in pollen from trees that grow in floodplains. The landscape apparently was a mosaic of active and fallow fields. Forest patches dominated by nut-bearing trees remained on hilltops, where they may have provided an important supplemental food source.

Thus, the forests in areas occupied by Mississippian people were heav-ily fragmented and disturbed, and did not fit the romantic image of the for-est primeval for "pre-settlement" North America before Europeans began to clear the forests. These areas were already heavily settled. This type of inten-sive maize agriculture did not extend into New England and southern Can-ada, however. Cultivation of maize, beans, and squash spread to these areas, but the short growing seasons limited productivity. Farming supplemented traditional hunting and gathering from the surrounding forest, but it did not become the main source of food. Perhaps these northern deciduous forests were closer to being undisturbed by people than were southeastern and central deciduous forests. A good test of this hypothesis is provided by exceptionally well preserved sediments in Crawford Lake, Ontario, which reveal how the surrounding forest changed between AD 200 and 1700.[67] After the region was occupied by the Iroquois around 1300, the frequency of American beech, sugar maple, and basswood pollen declined, while the frequency of the pollen of maize, agricultural weeds (such as ragweed), and grass increased. The frequency of charcoal also increased, as did the pollen of oak and white pine. All of these changes fit with a pattern of shifting cul-tivation that involved periodic clearing of trees and burning. The resulting landscape would be a mosaic of active maize fields, recently abandoned fields covered with grass, ragweed, and other low vegetation, and long aban-doned fields colonized by oak and white pine. The Iroquois no longer occupied this area by the time Europeans settled there. The new settlers misinterpreted the groves of tall white pines as "virgin forest" while, in fact, they were probably a byproduct of Iroquois agriculture.[68] White pine fre-quently colonizes abandoned farmland in this region.

While forests were subjected to clearing in river valleys and to more dif-fuse disturbance in the uplands and in more northern areas in eastern North America, much more extensive changes were taking place in East Asia. Al-though the extent and timing of deforestation is difficult to determine, the impact was clearly enormous. As highly productive, sustainable, permanent agriculture spread through China, the population grew rapidly.[69] By AD 1100 urbanization and industrialization had taken place on a much greater scale than in contemporary Europe. There were five cities with populations exceeding 1 million, and iron production reached more than 125,000 tons per year. The demand for fuel and building material depleted the deciduous forests of central China, and agriculture spread southward into the tropical forests of southern China. In his book on the environmental history of China, Mark Elvin summarizes the effects of deforestation from descriptions in ancient Chinese literature.[70] Although his sources do not provide a quantitative measure of the proportion of forest cleared in particular regions or the rate at which agriculture spread, they do provide eyewitness accounts of the effects of deforestation. There are many references to wood shortages and the catastrophic effect of forest clearing on mountainsides, which led to severe flooding and sedimentation in the agricultural fields of the lowlands. Most attempts to protect forests apparently were transient or ineffective, but some forests were protected at sacred sites around Buddhist temples. One of the most enlightening parts of Elvin's analysis is his summary of the contracting range of Asian elephants in China during the past 7,000 years (hence the evocative title of his book, *Retreat of the Elephants*).[71] Several thousand years ago elephants ranged as far north as Beijing. By 1000 BC their range had receded south of the Huai River, and by AD 1000 they were restricted to tropical forests in southern China. They are now found in only a few protected areas on the Burmese border, and their original association with the temperate deciduous forests of the north comes as a surprise to most people. As Elvin points out, Asian elephants depend on forests, so they disappeared quickly as extensive forests were replaced with farmland.

Rice cultivation eventually caused similar destruction of the deciduous forests of Japan. After Korean immigrants introduced rice cultivation and iron tools and weapons to Kyushu around 400 BC, intensive agriculture spread quickly.[72] By AD 700 much of the lowlands of southern Japan—as far north as the Kanto Plain where modern Tokyo is located—were occupied by small, scattered farming villages. The Kinai Basin (the flat coastal plain where Osaka, Kobe, and Kyoto are located) became the political, cultural, and population center of Japan. This basin and the surrounding mountains were the first region in Japan subjected to widespread deforestation. The capital city drew on the surrounding countryside for food, firewood, and timber to build houses and large palaces, Buddhist temples, and Shinto shrines. For example, ancient groves of immense hinoki cypress were harvested to construct such gigantic temples as Todaiji in Nara, which is supported by 84 major pillars made of logs more than a meter in diameter and 30 meters tall.[73] Conrad Totman estimates that construction of this temple (which is still the largest wooden building in the world) required trees that would cover 900 hectares (2,200 acres) of ancient forest.[74]

Initially, the capital was rebuilt periodically at different locations in the Kinai Basin, resulting in deforestation of one region after another, but eventually it moved to the newly constructed city of Heian-kyo (Kyoto), which remained the imperial capital from 794 to 1868, when Tokyo became the capital.[75] Even though old buildings were often disassembled so the wood could be used in new construction, lack of building timber from the Kinai Basin may have ended the tradition of relocating the capital with each imperial succession. Increasingly, timber was transported from distant parts of southern Japan, from the Kii Peninsula, the island of Shikoku, and mountains far to the east of Kyoto. Timber was cut into small lengths for transport, and the smaller size and lower availability of timber forced changes in construction methods even for elite residences and religious buildings. The light, elegant construction now associated with traditional Japanese architecture, with tile or bark roofs, flooring made of sedge mats (tatami), and rice-paper partitions between rooms, began to develop during this period of wood shortage.

Much of the original forest in the Kinai Basin was evergreen hardwood or coniferous forest, not deciduous forest, but the deciduous forests at higher elevations were depleted along with the hinoki cypress and evergreen oaks of the lowlands. When old-growth evergreen hardwood forest was cleared in the lowlands and on low mountain slopes, however, it was often replaced with early successional deciduous forest or a mix of deciduous trees and Japanese red pine. Thus, the famous autumn display of spectacular deep red foliage of maples on the mountain slopes around Kyoto is a product of continual disturbance of forests. Around old temples such as Honen-in where the trees have been protected, evergreen oaks and golden chestnuts and tall conifers such as cryptomeria and hinoki cypress form tall groves. The massive trunks, deep-green winter foliage, and dark, open understories echo the original vegetation of the Kyoto lowlands.[76]

Ironically, the successional forests that resulted from timber harvesting were more conducive to supporting a dense human population than were the ancient evergreen forests they had replaced. Logged areas that were not converted to farmland were colonized by chestnut, beech, and oak. These species provided a good source of firewood, wood for charcoal-making, and small timbers. After cutting, these species resprout from stumps and roots. In some regions stems were harvested every 10 to 15 years, resulting in dense, low coppice woodlands.[77] These provided not only a sustainable source of wood, but also a source of "green fertilizer," leaf litter and under-brush that were harvested in forests to fertilize rice fields.[78] Japan lacked the large herds of livestock that served as a source of fertilizer (manure) for farmers in central and western Europe, so woodlands remained an impor-tant source of fertilizer until the twentieth century, when synthetic fertilizers were introduced. Even in the heavily settled Kinai Basin, however, wood-lands were able to recover, at least in the form of low, bushy coppice. These historically disturbed woodlands have matured since the end of World War II, when making charcoal and harvesting green fertilizer were largely dis-continued.[79] In a few regions, however, the combination of continual har-vesting of wood and underbrush and periodic wildfires on steep slopes resulted in denuded, barren mountains. These can still be seen today in the mountain ranges south of Lake Biwa.

Between AD 1000 and 1600, the population of Japan doubled, reaching about 13 million people. As a result, farming extended to most of the coastal plain and (with the use of terraces) the lower slopes of foothills and moun-tains. The demand for green fertilizer increased as agriculture intensified, and more firewood and timber were needed by the growing population.[80] More intense demand for timber occurred because of continual warfare in the 1400s and 1500s. By this time imperial authority had broken down and Japan was divided among about 250 feudal lords (daimyo) who were con-stantly at war with one another.[81] This led to a proliferation of fortresses that were primarily built with wood, leading to deforestation in many re-gions. Initially these were wooden stockades, but by the 1500s massive cas-tles were constructed from wooden beams. Heavy use of the forest spread well beyond the Kinai Basin to much of Japan south of Tokyo. Some feudal lords instituted laws to protect forests from slash-and-burn agriculture and timber harvesting by villagers, and occasionally forest management went beyond protection of forests for use by the lord, and included active plant-ing of trees. Overall, however, this was a period of rapid and accelerating forest destruction.

Ironically, forest destruction accelerated even more after Oda Nobunaga and Toyotomi Hideyoshi ended the "warring states" period and unified Japan.[82] For the first time a central government could exploit forests in all parts of Japan south of the northern island of Hokkaido.[83] A building boom ensued that involved the construction of massive castles, palaces, mansions, and a large fleet used in an unsuccessful invasion of Korea. The city of Edo (Tokyo), which became the center of the shogunate government, expanded and then was rebuilt repeatedly due to major fires.[84] The government imposed levies of wood on all parts of the country, and the exploitation of forests became increasingly systematized and widespread. The shogun also took direct control of some heavily forested areas, but this was to control harvesting of timber, not to initiate sustainable forestry. Timber was exceptionally important in Japan because even the largest structures were built with wood. Stone was primarily restricted to foundations, perhaps because limestone and sandstone deposits that would provide easily quarried building blocks and a source of mortar were not common on the volcanic Japanese islands.[85] For the first time the forests of large areas of Japan were harvested too quickly to allow for natural regeneration.

A view of the world's deciduous forests in AD 1600 from low earth orbit would reveal considerable change since 1000 BC. Much of the deciduous forest of the Mediterranean coast, Great Britain, and central and western Europe had been converted to farmland. Agricultural clearing had expanded in North America after 1000 BC, particularly in the Mississippi River valley and the southeastern coastal plain, but by AD 1600 forests were reclaiming many of these cleared areas. The central Mississippi Valley was abandoned by intensive farmers by 1350, probably as a result of an extended period of drought, and other agricultural centers were abandoned in the 1500s due to repeated epidemics of smallpox, measles, and other diseases inadvertently introduced by European explorers, traders, and fishermen. In contrast to North America, the deciduous forests in China and Japan were disappearing because of major increases in human populations. This resulted directly and indirectly from the spread of irrigation systems associated with rice farming. Large amounts of food could be produced sustainably on relatively small amounts of land, which fueled population growth. Also, the complexity of irrigation systems led to control of the landscape by centralized governments. Given the demands for wood by growing rural and urban populations, and for residences, temples, fortresses, and ships built by the elite, the prognosis for deciduous forests in East Asia did not look good in 1600.

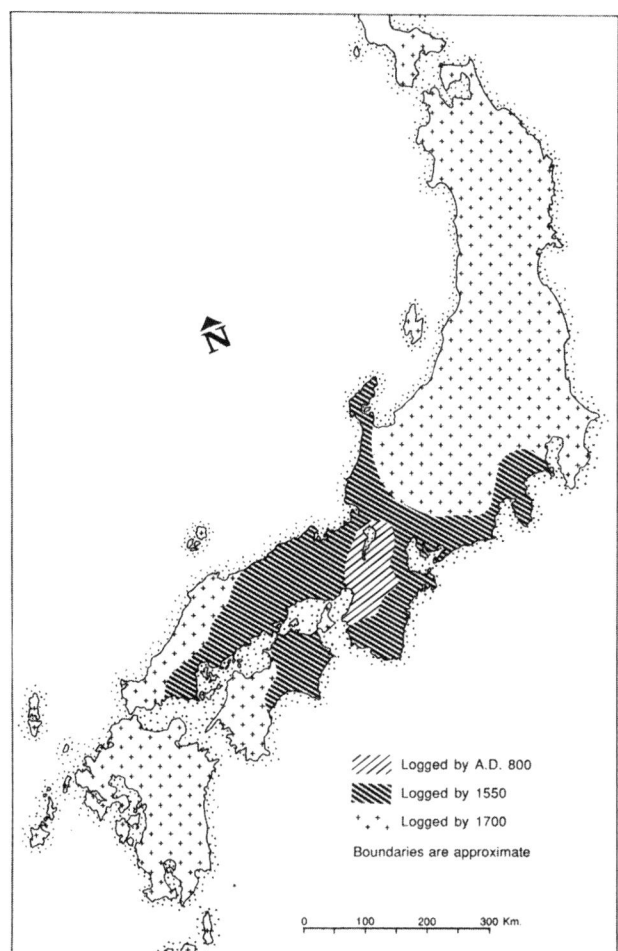

Figure 7. Historical patterns of logging of ancient forests in Japan before 1700. (Totman, 1989; reprinted with permission from University of California Press Books)

Logged by A.D. 800
Logged by 1550
Logged by 1700
Boundaries are approximate

0   100   200   300 Km.

Implications for Conservation

Despite extensive deforestation, deciduous forests survive in many parts of the temperate zone and they are growing back in some regions that were previously cleared for agriculture (Chapter 4). These surviving and regenerating forest ecosystems are incomplete in some profound ways, however. Tropical ecologists use the term "empty forests" for rainforests that are missing the large animals that play key roles in dispersing seeds of forest trees or in controlling herbivore populations. These large animals constitute a small part of the biomass (the weight of living organisms) in a forest, but they have a disproportionate impact on the diversity and stability of the ecosystem. Many temperate forests are empty forests, without large predators or giant herbivores or "ecosystem engineers" such as beavers. They are also missing some of the natural processes, such as occasional wildfires, that once sustained a diversity of different habitats. Restoring these species and natural processes or compensating for their loss will be a major challenge.

CHAPTER 4

Decline of Natural Forests and the Invention of Sustainable Forestry

Origins of Forest Protection

Based on the extent of undisturbed forest and the rate of forest loss in 1600, it would have been difficult to accurately predict the future for deciduous forests during the next 300 years. Who could have anticipated that Japan would become one of the most heavily forested nations while people in eastern North America would worry about future shortages of wood? These outcomes were largely determined by how quickly and efficiently people started to manage their forests to ensure future supplies of wood and timber. Long after Japan and Europe independently developed effective systems of sustainable forestry, North America and China continued to harvest frontier forests with little thought to the future.

Extensive clearing for agriculture had occurred in both Europe and China by the 1600s, but the extent of clearing in China is difficult to gauge. Some of our best evidence for the difference between the two continents comes from Jesuit missionaries who lived in the more developed parts of China in the 1600s and 1700s. Their memoirs and essays often compare the agriculture of China and western Europe. One account contrasts the farmland of France, which is "broken up by woods, meadows, vineyards, parks and lodges for recreation," with farming areas in China where even "enclosures around houses, the avenues of the villages, and the hillslopes" were cultivated.[1] The European missionaries remarked that the land produced several crops each year, and that there were no fallow fields.[2] They attributed the intensive use of the land to the demands for food from a dense population. Land use in China was much more intensive than in Europe because dense populations were sustained by an elaborate system of canals, sluices, reservoirs, and levees that covered a large area and required constant maintenance.[3] Also, instead of permitting fields to recover fertility during a fallow period, fertility was maintained by frequent applications of

fertilizer, including human wastes and animal manure. This intensely cultivated landscape had little space for woods except on mountaintops, steep
slopes, and on the grounds of Buddhist temples. Large areas of intact forest
still existed in southern China, but these were generally subtropical or tropical forests rather than temperate deciduous forests.

Given the high level of government activity in engineering the landscape by creating and maintaining irrigation systems, it is surprising that
sustainable forestry did not become more widespread in China by the 1800s.
Mark Elvin's book presents abundant evidence that Chinese officials and
commentators understood the value of protecting wooded watersheds in
the mountains to prevent erosion and maintain a steady supply of water,
and that there was widespread recognition of the importance of protecting
forests and replanting trees.[4] Perhaps the continual availability of timber
supplies from the periphery of the empire prevented action. At any rate, Elvin ends his book by pointing out that the insights about sustaining forests
had little effect on how forests were actually managed.[5] Forests were protected or restored only in scattered local areas or for short periods of time.
The long-term changes in forests were driven more by the need for food for
a dense population, and for "power and profit."[6]

Later, in the 1950s, as much as 10 percent of China's remaining forest
was destroyed during the Great Leap Forward, the ill-conceived campaign
to boost steel production by building numerous small furnaces in rural areas.[7]
Trees were cut down to provide fuel for these "backyard" furnaces. This was
part of a longer period of environmental destruction resulting from Mao
Tse-tung's conception that the nation could only develop economically by
waging a war against nature.[8] Policies that encouraged deforestation were
reversed after 1998, when catastrophic floods along the Yangtze River were
attributed to deforestation.[9] The Chinese government instituted a nationwide program of forest planting, and by 2011 more than 4 million hectares
had been planted with trees. Much of this woodland consists of eucalyptus,
rubber, and fruit plantations, however, rather than natural forest. In many
cases privatization of forests resulted in the conversion of natural forests
into plantations of non-native trees, leading to a loss of biological diversity
and forest complexity.

Extensive forest destruction also occurred in many parts of Europe after
1600. In the 1500s European economies began to improve and population
growth rates increased as the coastal nations of western Europe began to
trade with Africa and Asia and colonize the Western Hemisphere. Forest

clearing resumed in Europe on a large scale by the 1600s, eventually leading to timber shortages.[10] Forest reserves were common in France, Germany, and Britain, but these were primarily hunting reserves for royalty and nobility. Venice was one of the first states to establish protected forests for timber production. These were set aside to ensure a source of timber for shipbuilding, which was crucially important for an economy that depended on overseas trade. Also, the French government attempted to regulate harvesting of timber from forests. By the late 1600s people in different parts of Europe began to react to timber shortages or the anticipation of shortages by protecting and nurturing forests.[11] In Britain the iron industry managed hundreds of thousands of acres of woodland by coppicing, which works because the stumps of harvested trees resprout and grow into new trees. In England John Evelyn wrote an influential book on protecting and planting trees, and in France an ordinance was passed to regulate harvesting and grazing in the nation's forests. Both Evelyn's book and the French forest ordinance were responses to shortages in shipbuilding timbers, which had become critically important for both trade and defense.

Managing Sustainable Forests in Europe

Extensive tree planting began in Germany in the 1700s, and soon there were numerous forestry manuals as well as instruction on forestry in German universities. Forest management shifted from an emphasis on protection of hunting reserves to a precise focus on increasing the yield of wood while sustaining the woodland.[12] This entailed growing even-aged forests with a single species of fast-growing tree. This scientific approach to forest management spread to other parts of Europe and eventually to North America.

Michael Williams describes how distinctly different philosophies motivated tree planting in England and Germany.[13] After the late 1600s trees were planted on rural estates in England not only as a long-term investment for timber production, but also because of an increasing esthetic appreciation of the beauty of trees and wooded landscapes. The semi-natural forested landscapes created around elite residences reflected a new standard for luxurious living and social status. The German emphasis on wood production and the British emphasis on natural (or semi-natural) beauty both influenced the modern conservation movement. Either approach prevents complete deforestation and the concomitant problems of wood shortages, rapid runoff of water, and soil erosion on steep hillsides. High-yield forestry, however, typically replaces biologically diverse deciduous forests with low-diversity conifer plantations.

Saving the Wood Supply and Watersheds of Japan

Japan also experienced a period of rapid economic and population growth in the 1600s. The amount of cultivated land increased from 1.5 million to 3 million hectares between 1600 and 1720.[14] Conversion of land to agriculture and continual high demand for wood led to severe wood shortages. Central Japan was governed directly by the shogun's military regime, while most of the rest of Japan south of Hokkaido was governed by about 200 barons (daimyo), each with a separate domain.[15] These baronial domains had semi-autonomous governments, but were ultimately subject to regulations from the central government in Edo (Tokyo). Regulations to protect and restore forests arose both at the local level and the national level.

In the early 1600s, forest regulations mainly concerned reserving particular types of large trees or tracts of forests for use by the government.[16] This was accompanied by detailed regulations on the size of houses and the types of building materials that could be used by people in particular social classes. For example, peasants were not permitted to use the two most highly prized woods, hinoki cypress and cryptomeria, for construction. Later in the seventeenth century, protection was extended to forests to prevent the severe flooding and siltation of productive lowland farming areas that would result from denuded hillsides. The shogunate government forbade the further destruction of vegetation along streams and on floodplains, and restricted the use of slash-and-burn agriculture in some mountainous areas.

Sustainable forestry of the sort practiced in Germany arose independently in Japan.[17] As early as the 1650s, farm manuals urged landowners to plant trees and practice rotational harvesting of timber. In the seventeenth and eighteenth centuries, numerous manuals described in great detail how trees should be planted as seedlings or cuttings (slips), and how plantations should be maintained with replacement planting, weeding, and thinning to achieve rapid growth of tall, straight trees.

In the 1700s the Osaka region became a center for nurseries to raise cryptomeria seedlings for tree plantations. These seedlings were shipped to many parts of Japan, and the nurseries exported as many as 300,000 seedlings per month.[18] In response to the high cost of Osaka seedlings, tree

Figure 8. Photograph of a conifer plantation (a pure stand of cryptomeria) in Kyoto, Japan.

nurseries were started in many parts of Japan in the 1800s to support replanting in areas where forests had been harvested. On private and communal land, various incentives were instituted to encourage villagers to plant, nurture, and protect tree plantations, and to forgo the collection of green fertilizer (leaf litter and vegetation) and other forest products.[19] Not only were villagers paid by timber merchants for permission to grow trees on their land, but local governments also shared a substantial proportion of the profits from harvested timber with villagers when they planted and maintained trees on government and communal land. To encourage the spread of tree plantations, forest magistrates frequently provided seedlings, loans, and forestry advice. By the 1800s, many parts of Japan had developed a sophisticated system that protected watersheds and produced wood sustainably on private, communal, and government-controlled land.

     As in Germany, the tree plantations in Japan were monocultures, usually consisting of cryptomeria, hinoki cypress, larch, or other species of conifers. In many parts of Japan, this meant that remarkably diverse deciduous hardwood forests or mixed forests of hardwoods and conifers were replaced by monotonous plantations with a low diversity of plants, animals, and other organisms. The plantations were critically important in protecting the watersheds, streams, and lowland rice fields of Japan, however, and they provided a ready source of excellent building material, removing much of the pressure on remaining areas of deciduous forests. Also, because of their resilience, deciduous forests survived without replanting. As in Europe, they were frequently maintained sustainably by coppicing (in which only a stump is left after periodic cutting) or pollarding (in which a tall, standing segment of the trunk is left after the branches are harvested). In both cases new branches sprout and grow quickly, and these can be harvested after a few years. Oak and beech were frequently managed by coppicing in Japan. Naturally regenerating forests were often protected by restricting access to recently harvested sites and by favoring particular tree species by weeding and thinning.[20] Also, some naturally regenerating forests were managed with selective cutting.

The constant harvesting of leaf litter, understory, and dwarf bamboo from deciduous forests degraded soil fertility, and Japanese red pine (which can grow on dry, infertile soil) became a major component of many deciduous forests. Deciduous forests survived alongside the conifer plantations, however, preserving much of the biological diversity of Japan.

Japan's forests were severely damaged by over-harvesting during the Second World War. Fifteen percent of Japan's forests were logged between 1941 and 1945, and much of the remaining forest was severely degraded because green fertilizer (leaf litter and understory vegetation) was used to replace chemical fertilizers, which were no longer available.[21] During the 1950s the Japanese government sponsored a massive reforestation program to compensate for overcutting of the forests during and immediately after the war.[22] Natural forests were replaced with conifer plantations. During this same period the Japanese shifted to using fossil fuels and artificial fertilizers, however, and the intense use of the forest for fuel and green fertilizer ended. By the 1970s importation of less expensive timber from North America and Southeast Asia led to a major decline in plantation forestry. Expansion of cities and roads destroyed or damaged many of the remaining deciduous forests, but the surviving forests began to recover from intensive use and coppice woodland grew into mature forest.

A good way to experience these recovering forests is to take a cable car from Kyoto to the top of Mount Hiei, and then take a trail southward along

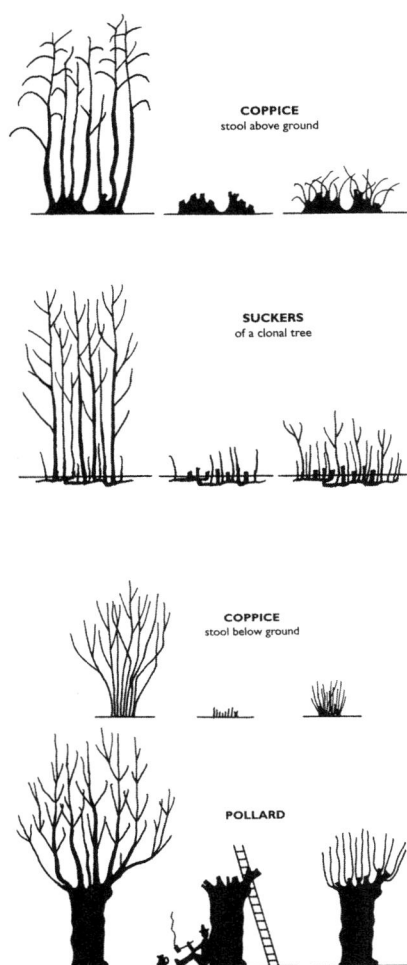

Figure 9. Illustration of traditional European methods for managing trees to provide a constant supply of young sprouts for firewood, woodcrafts, and animal forage. Trees may be harvested near the ground (coppicing) or high enough above the ground to provide protection from browsing animals (pollarding). Similar methods were used in traditional woodland management in Japan. (Rackham, 2006; reprinted by permission of HarperCollins Publishers Ltd., © 2006 Oliver Rackham)

the ridgeline that borders the eastern side of the city. After leaving the tall, dark conifer plantations at the top of the mountain, the trail traverses second-growth deciduous forest. These forests were intensively exploited for 1,200 years, so there are many stands of scrubby deciduous trees and Japanese red pines. Particularly in the valleys, however, tall hardwoods have grown up. Here oriental cuckoos, narcissus flycatchers, and other forest birds sing during the summer, and Japanese green pigeons feed on berries at the top of tall deciduous holly trees during the winter. Japanese giant flying squirrels roost in cavities in tall trees at the base of the hills, and wild boar, troops of Japanese macaques (the native monkey of Japan), and even the occasional Asian black bear wander the heavily forested slopes above the city.

Decline and Rebound of North American Deciduous Forests

During the centuries that the Japanese were restoring forests in even the most remote regions of their domain, European immigrants to North America were destroying deciduous forests at a remarkable rate, with no attempt to replant forests or to ensure future supplies of wood. The first English settlers frequently occupied cleared land that had recently been abandoned by indigenous people who had suffered massive mortality due to introduced epidemic diseases. English colonists at Plymouth, Massachusetts, for example, settled on cleared land that had once supported Indian farms.[23] They were aware that these lands were available because so many of the local people had died during epidemics, and they even interpreted this as a sign of divine providence. English colonists witnessed the effects of a smallpox epidemic that killed 95 percent of the people in some Indian villages in coastal Massachusetts.[24] Smallpox was one of the most deadly diseases for indigenous people in the Western Hemisphere, but numerous other contagious Old World diseases such as chickenpox, mumps, whooping cough, measles, typhus, dysentery, cholera, malaria, and tuberculosis contributed to the disruption of Indian societies. The total population of North America may have declined by 74 percent, leaving many previously populated regions almost empty.[25]

European settlement was concentrated east of the Appalachian Mountains between the early 1600s and 1750. By the time European settlers moved farther west into the forests of the Mississippi River valley, many of the areas cleared by the Mississippian culture had grown up into mature forest, and the settlers were confronted with a wilderness with a low density of indigenous people. When they encountered the ruins of massive earthen works, they had difficulty associating these with the local indigenous people, who lived in small villages and did not create monumental architecture. Consequently, the mounds were attributed to earlier immigrants from the Old World, who, in different accounts, were Vikings, ancient Israelites, or even immigrants from India.[26] Subsequently, archaeological excavation revealed not Old World artifacts, but the pottery, copper ornaments, and stone tools of a technologically sophisticated indigenous culture.[27]

In northeastern North America, the process of clearing the forests proceeded gradually during the 1600s and 1700s as farming families worked hard to create small subsistence farms. The remarkably well integrated and sustainable farming systems developed over millennia in Europe were transplanted successfully to the new continent. The forest was converted to productive farmland that supported people for many generations. Only population growth and the steady arrival of new immigrants from Europe required the clearing of additional forest.

In the 1800s better roads and canals and the rapid expansion of railroads encouraged farmers to grow more crops for the market, so the size of farms and the rate of forest clearing increased. The deciduous forests of Ohio, Michigan, and southern Wisconsin were converted to farmland in a few decades. In the late nineteenth century, the combination of railroads and portable sawmills led to the rapid destruction of the remaining forest, first in the Great Lakes states and then in the Southeast. Much of this logging was directed at conifers, old stands of white pine in the north and tall longleaf pine in the south, but mature deciduous forests were also harvested. Most of the old-growth, floodplain hardwood forests of the Southeast, for example, were cleared during this period. By the early twentieth century only isolated remnants of undisturbed "virgin" forest remained in eastern North America, and the logging companies were beginning to focus their activities on the forests of the West Coast and Rocky Mountains.

This story of forest clearing is well known, but the story of dramatic forest recovery is often ignored or underappreciated. As the wave of forest clearing was still passing through the Midwest, forest regeneration was already beginning to follow. Less productive farmland in New England and the Appalachian Mountains was abandoned as more fertile and less rocky farmland became available in Ohio, Indiana, and Illinois. By the 1820s, non-farm jobs were available in water-powered mills along the rivers of New England, drawing young people away from the never-ending chores on subsistence farms. Farming of perishable produce such as milk and vegetables continued in many of these regions, but eventually faster trains with refrigerated cars meant that even these products could be transported across great distances, leading to another wave of farm failures.

The evidence for this change is particularly obvious in New England, where dry stone walls created by farmers mark off the cornfields, orchards, hay meadows, and pastures of the nineteenth century. Particularly when they are constructed on firm soil or rocky ledge, these walls can stand for centuries. Looking out my study window, I see a beautifully constructed wall, more than a meter high, that was built in the early 1800s. It was constructed with large stones, indicating that the land was once a pasture. Farther up the hill there are walls with a mix of large and small stones, the sign of cropland that required plowing; the small stones were removed and used for the wall so that they would not interfere with the plow.[28] Deep in the forest behind the house one can walk among the ruins of a farm: an intact root cellar with stone walls and a rock slab roof; the foundations of a large barn and a farmhouse; and stone walls that enclosed livestock pens. A large black walnut (perhaps planted) and a ground cover of common periwinkles (a European species often used in gardens) grow in front of the house foundation. The farmstead is now surrounded by a tall forest of black oak, shagbark hickory, white ash, red and sugar maple, and black birch, providing habitat for forest animals such as barred owls, red-shouldered hawks, pileated woodpeckers, southern flying squirrels, and fishers.

This type of forest recovery is not unusual. During the mid 1800s about 50 percent of the land in southern New England was in cultivated fields, hay meadows, or pasture, but by 2000 this declined to 7 percent.[29] In many regions of New England the change was even more dramatic. In his thorough analysis of land-use change in Petersham, Massachusetts, David Foster showed that 85 percent of the forest in this township had been cleared by 1850, and that the remaining woodland was restricted to steep slopes, wetlands, and narrow valleys.[30] After the 1850s, farms were abandoned, and by the 1960s more than 90 percent of the township was covered with forest.

Farmland in northeastern North America was not necessarily abandoned because it had been poorly managed or because the soil was exhausted. In his environmental history of Concord, Massachusetts, Brian Donahue persuasively argues that English settlers of the 1600s and 1700s farmed the land sustainably.[31] Their farms used a mixed husbandry system that had been developed in England over centuries. Soil fertility on cultivated land was maintained by application of manure, either by allowing livestock to graze on cropland after the harvest or by hauling manure from animal pens. Organic garbage was added to manure piles to provide more fertilizer. Many of the nutrients in the manure ultimately came from floodplain meadows, which were hayed to provide animal feed in the winter. The nutrient levels of these meadows were maintained by sediments from seasonal flooding, which was often controlled by farmers using canals. Wood-

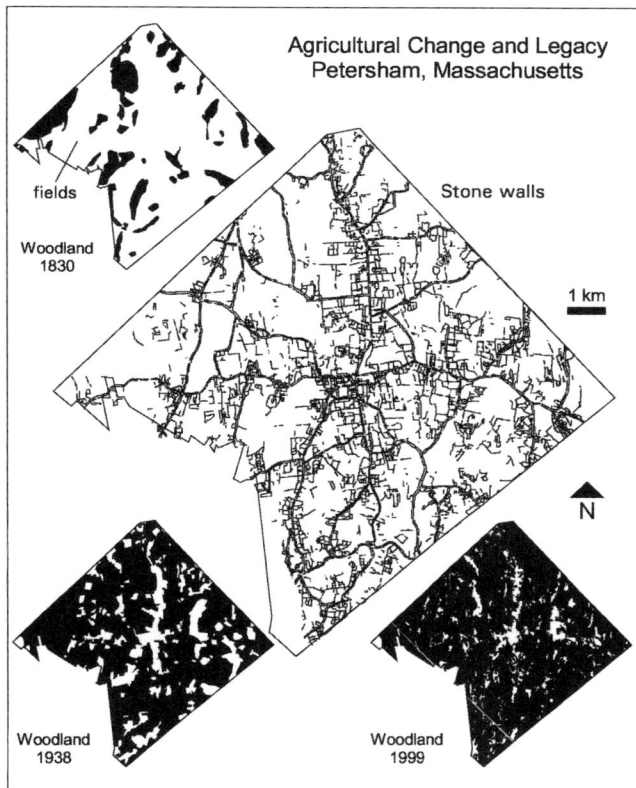

Agricultural Change and Legacy
Petersham, Massachusetts

fields

Woodland
1830

Stone walls

1 km

N

Woodland
1938

Woodland
1999

Figure 10. Location of stone walls, and changes in woodland cover between 1833 and 1999, in the town of Petersham, Massachusetts. Stone walls demarcated the boundaries of fields and pastures in the 1800s. These maps illustrate the shift from open fields to mature forest that has occurred in much of southern and central New England. (Foster and Aber, 2004)

lots were a final component of this balanced system. The trees provided building materials, and the understory was grazed by livestock.

As Donahue emphasized, this system did not maximize productivity, but instead ensured "security through diversification."[32] Each farmer raised a wide variety of crops and livestock, so no farming family or farm community was dangerously vulnerable to the failure of a single crop. Placing a priority on security was necessary when farms and villages were largely self-sufficient, but it became less important in the 1800s as farming increasingly became part of the market economy. Many farms in New England and the Appalachians were abandoned not because the soil was exhausted, but because they could not compete with farms that had richer soil and flat, less rocky terrain in producing commodities for a profit. In contrast, farmland in the southeastern deciduous forest region was often abandoned because of a decline in soil fertility. This was particularly true for large tobacco plantations, where land eventually became unproductive even though it was "rested" for twenty years after producing tobacco for three years.[33] Large fields of cash crops (tobacco and cotton) could not be sustained with the traditional mixed husbandry farming tradition.

Farms reverted to forest not only on the rocky soils of New England or the exhausted soils of Maryland tobacco plantations, but throughout the eastern deciduous forest region.[34] Forests in eastern North America usually regenerated with no systematic efforts at forest restoration. Farmland was simply abandoned, often after a transition from cropland to grazing. By the time half of the forests of Ohio had been cleared, forests in New England were already growing back. At the end of the nineteenth century, as the hardwood forests of southern Minnesota and Wisconsin were being cleared, forests in eastern Ohio were growing back. Similarly, abandoned tobacco and cotton farms in the southern United States reverted to forest, although in many cases this was pine woodland rather than deciduous forest. The overall rate of clearing was greater than the rate of forest regrowth for eastern North America until about 1872, after which the total amount of forest steadily increased.[35] Even when clearing for agriculture and timber reached a peak in the 1870s, about 50 percent of eastern North America was covered with forests. Much of this forest was young second growth, but these young forests helped preserve the biological diversity of the deciduous forest ecosystem.

Until the mid 1800s, much of the lumber in North America was produced as forest was cleared for farmland, but as farming moved on to the prairies, commercial logging companies began to supply the wood needed for construction and fuel. After the 1860s, the main source of lumber shifted from the northeastern states to the Great Lakes region, and in the 1890s it shifted to southern floodplain forests and the coniferous forests of the Pacific Northwest.[36] Forest clearing became more efficient as logs were hauled out of the woods on spur railroad lines and transported to steam-powered lumber mills. Entire regions were stripped of trees, leaving a desolate cutover landscape with abandoned train tracks and mills.

The rapid deforestation of entire regions of the continent resulted in growing alarm about depletion of natural resources and an impending "timber famine." Maps showing the shrinking amount of "virgin forest" (forest that had never been cleared) confirmed the fear that the nation was drawing on its last forest frontiers and would soon run out of new sources of wood. These analyses ignored the expanding area of second-growth forest and the

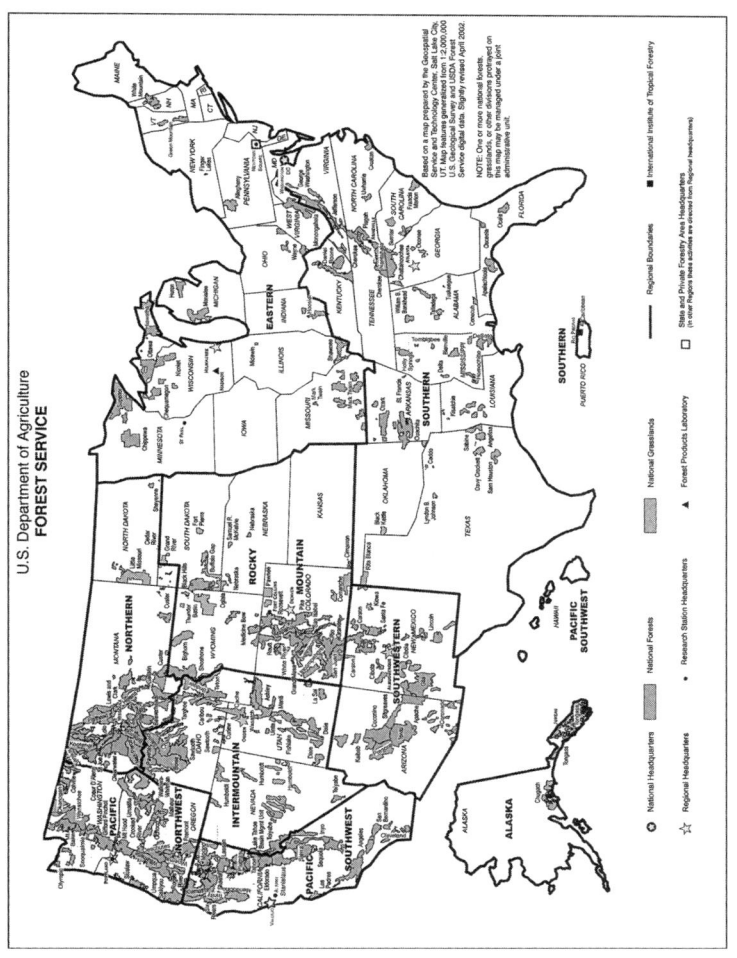

Figure 11. Location of national forests and national grasslands administered by the U.S. Forest Service. (Reprinted with permission from the U.S. Department of Agriculture, Forest Service)

productivity (in terms of amount of wood produced per year) of the nation's forests. Concern about future timber supplies and protection of the nation's watersheds led to the establishment of forest reserves (which later became national forests) in the 1890s. Because the focus was on uncut "virgin forest," the early national forests were all in the western half of the United States.[37] Only in the 1930s were large areas of deciduous forest incorporated into the national forest system.[38] These were often in areas where the forest (which was often mature second growth on long-abandoned farmland) had been recently harvested.

National forests were managed by the U.S. Forest Service, which adopted the general practices and goals of German forestry. The primary goal was sustainable production of wood facilitated by either replanting or natural regeneration of forests following timber harvests, with a secondary goal of protecting watersheds. Private lumber companies also shifted to sustainable management of their lands.[39] "Tree farming" became economically viable when there were few uncut stands left to exploit, and as tax laws were changed to reduce the burden of owning "standing timber." Finally, technological changes that reduced the need for wood as a building material and a fuel source helped avert a timber shortage.

Although the national forests were established primarily to provide a stable wood supply, other forests were preserved primarily because of their esthetic and recreational value. Setting aside land to preserve natural beauty and serenity was initiated with the preservation of Yosemite Valley in 1864 and Yellowstone National Park in 1872.[40] In both cases, the main focus of preservation was unusual and spectacular natural scenery. Adirondack Forest Preserve in upstate New York was the first extensive area of deciduous forest in the United States set aside to protect an intact, forested landscape. Although proponents of a park emphasized the importance of the Adirondack forests for supplying water to New York City and to New York State's navigable waterways, the main motivation for many was to protect an area of natural beauty for outdoor recreation.[41] A 289,000-hectare (715,000-acre) forest preserve was established in 1885, and expanded to more than 3 million acres to create Adirondack State Park in 1892. In 1894 a new constitution for the state of New York stipulated that Adirondack State Park should remain "forever wild," and commercial logging was banned within the preserve. Later many other parks and nature reserves were set aside to protect deciduous forest, including Great Smoky Mountains National Park, which covers more than 2,000 square kilometers.

By the end of the twentieth century, farms and cities had replaced much of the world's deciduous forest. The regions where great expanses of deciduous forest originally grew in Europe, East Asia, and eastern North America were centers of intensive agriculture and concentrated industrialization. They supported some of the largest metropolitan areas in the world. Despite this, deciduous forest ecosystems survived, and in some regions, such as eastern North America and Japan, they recovered and expanded. Many nations have placed increasing emphasis on forest protection and restoration, using variations on the scientific forestry approach developed originally in German universities or the natural areas preservation approach that was pioneered in U.S. national parks. Traditionally these two approaches have been in tension because of distinctly different goals. As recently as the 1960s, for example, some of the ancient deciduous forests of England were converted to conifer plantations rather than preserved as valuable reservoirs of natural diversity.[42] Plantations with one or two species of trees can produce forest products and protect watersheds sustainably, but do little to preserve biological diversity. Foresters have developed methods for harvesting trees sustainably from diverse natural forests, however, opening the way to restoration of deciduous forests outside of national parks and nature reserves on land that is managed for timber production.

Implications for Conservation

The trees and other plants of mid-latitude deciduous forests are remarkably resilient. They survived drastic shifts in climate during the Pleistocene and then massive environmental changes caused by people. Just as forests reassembled on barren ground left by melting glaciers, they reassemble on land abandoned after farming, logging, charcoal burning, or other intense human activities. These new forests are never the same as the forests that were originally cut down, but this is nothing new. The assemblage of species that constituted deciduous forests probably differed during each of the approximately 19 Pleistocene interglacial periods. The resilience of the temperate-zone plants is cause for cautious optimism. With careful planning and management, we should be able to restore these forests and sustain them even in the face of rapid environmental change caused by human activities.

CHAPTER 5

Giant Trees and Forest Openings

W hen land is no longer used for farming in the humid temperate zone, forests will eventually return. Unless the soil has been removed by strip mining, paved over, or poisoned by industrial toxins, the regrowth of the forest is remarkably swift, measured in decades rather than centuries. As a result, second-growth forests now cover large portions of the northeastern United States and the interior of Japan, and regenerating forest is beginning to appear on abandoned farmland in Europe. Does this represent a phoenix-like resurrection of temperate deciduous forests? A more pessimistic view is that the intricate and diverse primeval forests that were destroyed by agricultural clearing have been replaced with simplified and degraded woodlands. How do these new forests differ from the original forests?

To answer these questions, we need to determine the characteristics of the forest before agriculture spread throughout the temperate zone. As discussed in the previous chapters, we can create an imperfect picture of the original forest from historical accounts and from the pollen record in lakes and ponds. Another key source of information is ancient forests that were never cleared. Most patches of "virgin" or primeval forest in the temperate zone are small and isolated, but there are a few regions where landscapes are still dominated by ancient forests. Although never completely cleared, these forests inevitably have been affected by humans in various ways and to various degrees. Hunting has occurred in all of these forests, and in most cases large mammals have been extirpated, often leading to profound changes in the forest ecosystem. Also, relict patches of old forest were often subjected to periods of livestock grazing or selective timber harvesting. And particular species of trees have declined or disappeared in some ancient forests because of the spread of introduced insects or pathogens. Even though none of these forests have remained completely untouched by people, however, they are still our best source of information about how forest ecosystems functioned before widespread clearcut logging and farming.

A Remarkable Primeval Forest in Poland

To enter the Strict Reserve at Białowieża Forest in Poland, visitors walk along a narrow road through a large, sunny meadow filled with whinchats, common quail, and other grassland birds. At the edge of the forest they pass through an imposing wooden gate, entering the shade of gigantic oaks and limes. This is the heart of one of the largest expanses of forest in Europe, extending from Poland into Belarus. Parts of this region were never cleared for agriculture and have been heavily forested for thousands of years.

The aspect of the forest is distinctly different from the heavily managed forests in most parts of Europe. Instead of carefully managed stands of young trees that have been pruned, thinned, and cleaned up, the ancient oak-lime-hazel forest of Białowieża is a jumble of fallen logs, dead trees, and live trees of different sizes. The guided tours and guidebooks for Białowieża are geared to a European audience, so there is a heavy emphasis on the ecological importance of dead wood in all stages of decomposition, from standing dead trees to crumbling logs to mounds of organic soil that were once logs. Dead wood and its ecological importance are more familiar to visitors from forested parts of the United States and Canada, where virtually every nature trail has signs extolling the ecological value of snags (standing dead trees) and decaying logs. The role of dead wood is especially clear at Białowieża, however. Uprooted trees and fallen branches cover 12 to 15 percent of the forest floor.[1] Woodpeckers are not only common, but also remarkably diverse, with seven species regularly nesting in this forest.[2] They forage on dead wood, and some species build their nest cavities in snags. Standing and fallen trees also support a remarkable diversity of fungi, which is obvious from the wide variety of mushrooms sprouting out of logs. As logs decompose, they create a well-fertilized and protected site for the germination of herbaceous plants and tree seedlings.[3]

For visitors expecting an ancient forest to look like Muir Woods in California, with its clean columns of massive coast redwoods, the Białowieża forest is surprisingly complex and cluttered. Large oaks—300 to 400 years old—are scattered among younger trees with a wide variety of heights and trunk diameters. Enormous Norway spruces occasionally tower above the canopy of hardwood trees. The understory is open, with only scattered shrubs and saplings, but the ground is covered by a rich diversity of herbs and ferns. When one of the ancient spruces or oaks collapses, it leaves a large gap in the forest canopy. This results in a sunny area on the forest floor covered with a dense tangle of saplings and herbaceous plants. This mix of young and old trees is typical of ancient deciduous forests. It reflects a history of relative stability in which trees live for hundreds of years, then die and fall, opening the way for young trees to grow up into the canopy. Theoretically, this could lead to an almost unchanging forest dominated by a few tree species that are best adapted to growing in the shade on the forest floor and then racing upward when a gap opens in the canopy. Most deciduous forests are still recovering from massive disturbance by humans, however, so we would not necessarily expect them to show this type of dynamic equilibrium in which old trees are consistently replaced by younger trees from the same set of dominant species, in the same proportions. Restoring this equilibrium may require hundreds of years. The best place to search for evidence of a steady-state system is in an ancient forest such as Białowieża.

It is difficult to believe that an extensive area in central Europe, in a region where empires have collided for centuries, could still have a large, intact forest that was never cleared. The pollen records from two peat bogs in the southeastern part of Białowieża National Park, however, confirm that some sections of the park have been covered with forest for thousands of years, apparently without interruption.[4] At the end of the last glacial period the region was an open parkland or savanna with scattered pines. It became more tundralike during the Younger Dryas cold period, but after that it was continuously covered with forest. There is no evidence in the sediment record of frequent fires or of agricultural clearing in the region near the bogs. The composition of the forest changed slowly over thousands of years as the bogs, which had originally been lakes, filled in. Initially the landscape was mostly covered with pine with a few scattered birch trees. Later birch became more frequent and pine declined. Eventually a mixed deciduous forest dominated, but the species composition of the forest continued to shift as new types of trees spread into the region. Oaks, for example, arrived later than most other deciduous tree species, and common ash appeared even later. According to the pollen record, European beech never reached Białowieża, and it is not found there today. The composition of the forest has constantly changed, but the region has been covered with a deciduous forest with a diversity of tree species for 5,000 to 8,000 years.

Some other sections of Białowieża National Park are covered with coniferous forests (a mix of Norway spruce and Scots pine) rather than deciduous forest. These areas were covered with an open pine savanna in the 1600s and 1700s.[5] Analysis of tree rings of living and dead trees reveals that during this period fires swept through these sites an average of every six years. Frequent low-intensity fires singed the trees (fire scars can still be seen in tree ring samples), but usually didn't kill them. The fires may have been ignited by human activities such as collecting honey from bee colonies in natural cavities in trees (which involves the use of fire and smoke to ward off the bees) and charcoal burning. After 1781 the frequency of fires declined precipitously, and since then the pines have progressively been replaced with spruce, which is less tolerant of fire but more tolerant of shade (and thus able to grow up under a pine canopy). It is uncertain whether the open pine savannas were purely a result of fires caused by humans or were originally maintained by natural fires.

In contrast to these conifer forests, the deciduous forests in Białowieża National Park apparently were never subjected to frequent fires. Despite this, the vegetation of deciduous forests is also changing. This may reflect changes in the way these forests have been managed by people over the past two or three centuries.[6] From the early 1400s until 1798 Białowieża Forest was protected as a hunting reserve for Polish royalty. Limited harvesting of timber and other forest products occurred during this period and continued after 1798, when Poland was partitioned and the forest became part of the Russian Empire. From 1888 until 1914 the forest was once again managed as a hunting reserve, this time for the Russian tsars. Intensive game management during this period had a lasting impact on the forest. Populations of red deer and bison were boosted by exterminating predators and providing hay to reduce winter mortality. During World War I, however, most of these large, herbivorous mammals were shot by soldiers and local people to provide food, and timber was harvested intensively in some parts of the forest.

After World War I the best remaining ancient forest was protected as a national park. In 1921 the park covered 4,700 hectares, but subsequently it was enlarged to 10,500 hectares.[7] The Strict Reserve within the national park has been carefully protected from human disturbance, and the vegetation and large mammal populations have been monitored since the 1920s. In contrast to many old-growth forests, the forests in the Strict Reserve are embedded in an almost continuous complex of natural forest reserves and

managed forest that covers 1,250 square kilometers, more than half of
which is in Belarus (where there is another large national park that protects
ancient forest). This complex of forests supports an almost complete set of
extant large European mammals, including red deer, roe deer, European
bison, and moose (called elk in Europe), as well as predators of these large
herbivores (gray wolf and lynx) and an active population of dam-building
European beavers. As a result, Białowieża Forest represents our best opportunity to learn how European forest ecosystems worked before forests were
heavily managed by people.

The Strict Reserve was protected because it lay deep within the forest
where relatively little logging or other direct destruction of the forest occurred. This forest has all of the characteristics of a vibrant old-growth forest, with ancient trees nearing the end of their maximum lifespan and
vigorous seedlings and saplings in the understory and in forest openings. Is
this really a steady-state system, however, in which each tree species is successfully producing the next generation?

In 1936 five vegetation transects were established within ancient forest in Białowieża National Park.[8] Along each transect, every tree was identified and the diameter of its trunk was measured. These measurements were
repeated approximately once every decade from 1936 to 1992. The overall
structure of the forest did not change; the number of trees per hectare for
different size classes (from small saplings to old trees that are more than
one meter in diameter) was similar in 1936 and 1992. The species composition of the trees changed substantially, however, with major declines in
Norway spruces, pedunculate oaks, Norway maples, birches, poplars, and
willows. The decline in early successional species—birches, poplars, and
willows—is not surprising, because these species only grow in relatively
open, sunny areas. Large natural disturbances such as storms had not disrupted the canopy to create habitat for these species along any of the transects
(although this has happened in other parts of the national park). Spruces,
oaks, and maples are characteristic of mature forest, however, so it isn't as
obvious why populations of these species declined. The situation with Norway spruce (the species that suffered the most substantial decline) is particularly perplexing because this is a shade-tolerant species that should be
able to reproduce in forest conditions. Despite this, Norway spruce was
replaced as the dominant species by three deciduous trees: European hornbeam, small-leaved lime (a relative of the American basswood), and common
ash.

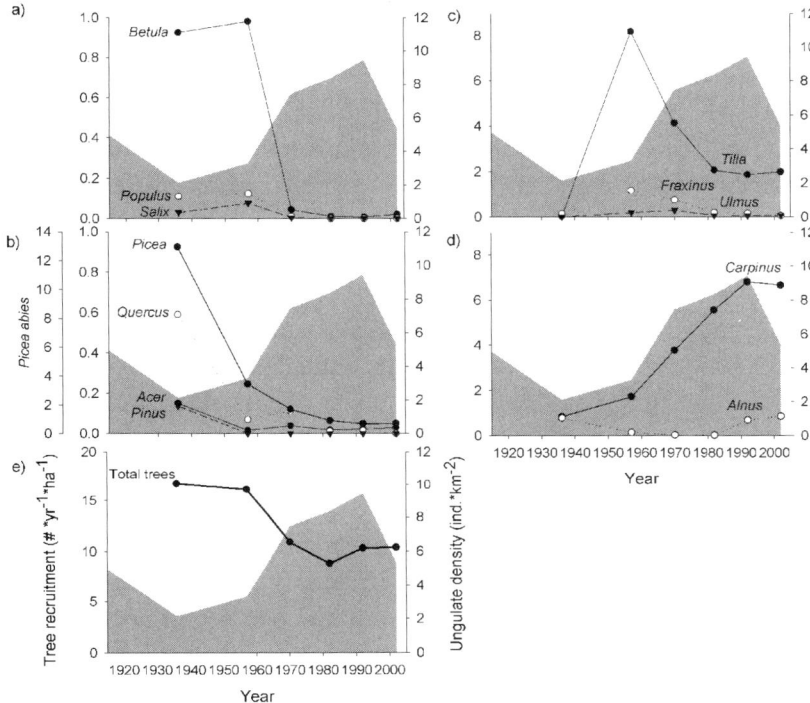

Figure 12. History of the abundance of ungulates (deer and bison) and of the recruitment rate (number of tree saplings per hectare per year) for different species of trees in Białowieża National Park, Poland. Note that spruces (*Picea*), oaks (*Quercus*), maples (*Acer*), pines (*Pinus*), and birches (*Betula*) showed high recruitment rates when ungulate densities were low. Only a few species, such as hornbeam (*Carpinus*) and limes (*Tilia*), reproduced successfully after ungulate density increased. (Kuijper et al., 2010a; reproduced with permission of John Wiley and Sons, Inc.)

The best approach to understanding these changes is to investigate the survival of seedlings and saplings of various species of trees. Analysis of recruitment of new trees on the permanent vegetation transects in Białowieża Forest demonstrated that spruce, oak, and maple seedlings had high survival rates between 1916 and 1936, when the density of deer and other large ungulates was extremely low following intensive hunting during the First World War.[9] In contrast, hornbeams showed the highest recruitment rates as the density of ungulates increased after 1936.

Hornbeams, in fact, thrived during this period of high ungulate densities. Ironically, hornbeams are a preferred food species for deer, but the saplings are especially tolerant of browsing because they quickly grow lateral branches when their tops are cropped off. They appear to survive despite high deer densities and may benefit from deer removing saplings of other species that compete with hornbeams for light and nutrients. Fenced plots that prevented browsing by deer and bison supported a diversity of saplings, including maples, elms, birches, and spruces.[10] In contrast, nearby unfenced plots were dominated by hornbeams. The implication is that browsing deer reduce the diversity of tree saplings by removing virtually all saplings except hornbeams.

The ungulate-exclosure experiments also indicated that tree regeneration is heavily influenced by other factors that affect young trees at different ages. The success of small tree seedlings is mostly determined by the density of herbs and ferns on the forest floor.[11] The maximum number of seedlings occurred in sites with intermediate herbaceous cover. For small saplings (bigger than seedlings but less then 50 centimeters tall), the greatest densities are found at sites with high soil fertility. It is only taller saplings (greater than 50 centimeters) that are heavily affected by the presence of browsing mammals. Thus, a tree seedling must survive a gauntlet of environmental conditions, each of which may favor one species over another. The combination of factors will determine which species of trees eventually reach the canopy.[12] For example, wych elm seedlings are greatly outnumbered by hornbeam and lime seedlings in most parts of the forest, but are virtually the only tree seedlings that successfully grow in sunny forest openings covered with dense stands of stinging nettles. In contrast, hornbeam and lime seedlings are particularly common in areas where wild pigs have plowed up the forest floor while rooting for food, removing most of the herbaceous cover. Other species of tree seedlings such as Norway spruce and European mountain ash grow especially well on top of decomposing logs.[13]

Although hornbeam is becoming increasingly common in the forest canopy, it may eventually be replaced by lime, which grows taller and lives longer.[14] The hornbeam canopy could slowly be overtopped and shaded by the many young lime trees that are now present in the understory. Lime saplings were almost absent from the forest in the 1920s following a period of intense browsing by deer and bison, and a large age cohort of limes (trees with a diameter of 1–2 meters) is missing from the forest today because of a long period more than a century ago when there was little successful reproduction. This is the same period when many of the tall Norway spruces in the forest germinated. Thus, the composition of the forest may still reflect a period of several decades when bison and deer populations were artificially boosted for hunting by the Russian royal household.

Although gigantic old pedunculate oaks are a prominent component of the ancient forest, oak saplings are infrequent in the understory because they do not grow well in shade. Some oak seedlings may successfully grow in the small openings caused by tree falls, but they grow best in much larger openings with full, direct sunlight.[15] Areas adjacent to the ancient forest where farming was abandoned 40 years ago already support a developing oak woodland with a mix of mature trees and saplings. Numerous oak saplings survived in fields despite browsing by deer and other native ungulates. The immense oaks of the ancient forest may have originated in similar large openings that were caused by either natural or human disturbances.[16]

Frans Vera argued that open woodland dominated by oaks was widespread in Białowieża long before agricultural clearings were made in the forest.[17] This is confirmed by pollen records from sites in a depression on the forest floor and in a bog in the northern part of the forest.[18] Although the concentration of oak pollen increased during the past few hundred years, oak pollen is present in the pollen record for the past 1,200 to 1,500 years. Vera hypothesized that large grazers (especially aurochs, the ancestor of domestic cattle, and tarpans, a type of wild horse) maintained parklike woodland similar to the traditional wood pastures of Europe (such as those in the New Forest in England). According to Vera's hypothesis, the presence of grazers allowed young oaks to grow successfully in competition with faster growing, more shade-tolerant limes and hornbeams that would normally out-compete them. All tree seedlings suffered high mortality because of grazing, but oaks were more likely to survive because they have an exceptionally large taproot with stored food that they can use to grow back after their shoot is nipped off.[19] Equally important, acorns are efficiently transported into favorable growing spots in open woodland by Eurasian jays, which bury them in order to have a stored food supply for the following winter, spring, and early summer.[20]

Aurochs and tarpans survived in Białowieża Forest long after they were extirpated from most of Europe, but they finally disappeared in the seventeenth century.[21] After their disappearance, the wood-pasture landscape may have been sustained by domestic livestock, especially domestic cattle and horses, because some grazing was permitted in the hunting reserve. Oak regenerated during this period, but stopped regenerating as grazing of domestic livestock was increasingly restricted by forest regulations. According to Vera, the decline of domestic grazers initiated a transition from open, heavily grazed wood pasture to closed-canopy forest, which in turn led to the decline in young oaks. Vera contends that this explanation applies not only to Białowieża Forest, but also to most oak-dominated lowland forests of central and western Europe.

Although it was never completely cleared to create extensive farmland or converted to tree plantations, the structure of Białowieża Forest has still been heavily influenced by human activities. After a longer period of protection from human disturbance, perhaps the mix of tree species in the forest will return to some ancient steady state. Another possibility is that the mix of trees has always been in flux as a result of natural disturbances (windstorms and fires) and changes in climate and animal populations.[22] If there is a steady state, it may take the form of more general structural characteristics of the forest, not the particular tree species that make up the forest canopy. The steady state may be characterized by a range of trees of different sizes, from seedlings to ancient giants, with a complex forest floor interrupted by decomposing logs, massive roots of tipped up trees, and open, sunlit patches where trees have fallen. Tall trees emerge above the canopy, and there are several layers of vegetation below the canopy. The mix of species making up the forest layers may shift over time while the overall structure changes little and consistently provides the full range of microhabitats needed by the rich diversity of lichens, mosses, insects, birds, and other forest organisms. Perhaps some areas within a large expanse of forest support pine savanna and open oak woodland. It may be the overall structure of the forest—not the dominant tree species at any particular time—that best defines the steady-state conditions of an old-growth forest.

This hypothesis is supported by research on old-growth forests in eastern North America, many of which were never subjected to major human disturbances. Between 1976 and 1991, for example, there was relatively little change in the overall structure of numerous old-growth forests in eastern North America.[23] The total diameter of trees and the density of trees changed very little. Larger trees showed a higher mortality rate than small trees, but their death was compensated for by the growth of smaller trees. Trees died at an average rate of 1 percent per year, providing a constant source of snags for animals that require dead trees. Although the basic structure of these forests did not change much at these sites, there was often a substantial shift in the species composition of canopy trees. Most notably, the proportion of American beech either increased or decreased at most sites. In some sites, American beeches were replaced with sugar maples. These changes were not due to pathogens killing the beeches or other obvious causes, and may reflect basic environmental shifts (such as climate change) that favored one species over another. Although these subtle changes might be caused indirectly by human activities, they might also be natural changes of the sort that occurred long before the advent of agriculture or industrial pollution.

Young Trees in Old Forests

Old-growth forests are characterized not only by giant trees and large amounts of decomposing dead wood, but also by openings in the tree canopy that support tree saplings and early successional species. These canopy gaps are more frequent in old-growth forests than in younger forests because old-growth forests have numerous old trees. Windstorms frequently blow down giant trees that have been weakened by age. Also, when a 300- or 400-year-old oak dies, it leaves a large hole in the canopy, particularly if it takes other trees down as it collapses. In old growth hemlock-hardwood forests in the Great Lakes region, the size of canopy gaps and the total percentage of the forest covered by canopy gaps increase steadily with the age of the forest as the average size of dead trees increases.[24]

In old-growth forest in Białowieża National Park, trees fall at an annual rate of 200–450 per 100 hectares.[25] The resulting openings drive a cyclical change in the forest.[26] During the first 60 years canopy gaps are dominated by saplings that grow quickly, competing for space in the forest canopy 35 meters above the ground. From 60 to 200 years the restored canopy remains closed as trees slowly grow larger, but after 200 years the canopy thins and new gaps begin to open up, allowing another set of tree seedlings to grow on the forest floor. Eventually this initiates a new cycle.

Andrzej Bobiec and his colleagues mapped the distribution of these three stages in the cycle at sites in ancient forest in Białowieża National Park.[27] They found that 40–48 percent of the forest was covered with open canopy gaps or stands of young trees less than 60 years old; 32–40 percent was closed-canopy mature forest, and 20 percent was old forest where trees were dying and the canopy was beginning to open up. This mosaic of patches is a key feature of old-growth forests. Thus, the biological diversity of old-growth forests depends not only on the vertical complexity created by multiple layers of vegetation and dead organic matter (from leaf litter to tall snags), but also on the horizontal complexity encountered as one moves from canopy gaps to stands of old trees to younger mature forest. The exceptionally high diversity of birds nesting in Białowieża Forest, for example, is partially due to this mosaic of habitat patches.[28]

Natural disturbances produce canopy gaps of different sizes, resulting in both small and large patches of low vegetation. In canopy gaps in 14 old-growth forests in North Carolina, Tennessee, Ohio, Pennsylvania, and New York, the percentage of land immediately under forest openings ranged from 3 percent to 24 percent.[29] The rate of gap formation could be estimated based on the approximate age of the gaps. Gaps formed in 0.5–2 percent of the forest canopy each year, so any particular point in the forest would be subject to a canopy gap every 50 to 200 years. This is shorter than the maximum age of many of the trees in the forest, but gaps can occur repeatedly at some areas while other areas remain undisturbed for long periods, allowing some trees to survive for hundreds of years. The picture that emerges is not of a timeless, unchanged forest of ancient trees but of a constantly shifting mosaic of young and old forest patches.

The forests that grow on abandoned farmland are "even-aged," meaning that most of the trees started growing at the same time. Even when these forests become mature, they are more homogeneous than an ancient forest. Most of the trees are about the same height, and there is relatively little dead wood and only a few vertical layers of vegetation. Because the trees are still young and vigorous, canopy gaps are infrequent and small. If a tree dies, the gap is usually filled by the horizontal growth of branches of neighboring trees. It is well known that these forests do not support species that require large snags or logs. In addition, they may be missing species that depend on forest openings caused by tree falls.

Canopy Gap Specialists

One of the most elegant sights in the deciduous forests of North America is a male hooded warbler fluttering through the understory. It moves gracefully from branch to branch, repeatedly fanning its tail to reveal bright white patches. When an insect is startled into flight by these rapid movements, the warbler flies after it and deftly captures it with an audible snap of the bill. As the warbler moves into a patch of sunlight, its yellow body and face glow in striking contrast to the velvety black head and throat.

Hooded warblers build their nests and search for insects in the under-story, so they are most common in forests with a dense layer of shrubs and small trees. They are common in young, regenerating forests where the canopy is still open and shrubs have not been shaded out.[30] As the tree canopy becomes denser and more continuous, however, hooded warblers decline and disappear. The closed canopy shades out shrubs and saplings, causing many to die. After the forest has grown for 150 or 200 years, however, canopy gaps are large enough to permit the growth of dense clusters of shrubs and saplings, providing new habitat for hooded warblers.[31] Some European bird species may display a similar pattern. In a comparison of 50 canopy gaps and 50 sites within closed-canopy forest in Białowieża National Park, Robert Fuller found that several species of forest birds (particularly dunnock, blackcap, and chiffchaff) were concentrated in canopy gaps.[32]

Much of the landscape of eastern North America is now covered with young, closed-canopy forest that provides little appropriate habitat for species like the hooded warbler. As these forests age, canopy gaps will become more frequent, providing sufficient habitat. This transition will require decades if not centuries, however. In the meantime, selective harvesting of trees can be used to create artificial openings filled with shrubs and saplings that attract hooded warblers, boosting their breeding populations.[33] In regions dominated by young forest, selective harvesting of trees or small groups of trees may be a reasonable way to sustain habitat of canopy-gap specialists until the forest is old enough to produce a sufficient number of natural canopy gaps.

Thus, the loss of old forest threatens not only species that require big trees and complex, multilayered forests, but also species that thrive in forest openings. The clearing of most of the bottomland hardwood forests in the southeastern United States in the late 1800s and early 1900s affected both types of species. The ivory-billed woodpecker disappeared as the forest was logged because it needed extensive old-growth forests with giant dead trees for feeding and constructing nest cavities. In contrast, the Bachman's warbler probably became extinct because it nested in forest openings caused by tree falls.[34] Although Bachman's warbler may have benefited from the early stages of logging when selective removal of the most valuable trees created many openings, it declined along with the ivory-billed woodpecker when the forest was completely cleared.

Canopy gaps are also important for foraging bats, which often concentrate along the edges of forest openings.[35] In South Carolina, bats were more active in these openings than in the interior of the closed-canopy forest.[36] Canopy gaps may have a higher insect density or they may simply be a more efficient place to catch insects while flying because of the lack of obstacles. When automatic recording devices that detect the ultrasonic calls of bats were placed in a large number of natural openings in deciduous forest at Tomakomai Experimental Forest in Japan, bat species with short, rounded wings were detected in small canopy gaps more frequently than in large forest openings.[37] Unlike these agile, short-winged species, bat species with long, pointed wings foraged in the center of large openings in the forest or above the forest canopy.

The importance of canopy gaps is widely recognized for old-growth tropical forests.[38] They are also important for deciduous forests in the temperate zone, but their ecological role has not been as widely recognized. Most research on forest ecology in the temperate zone takes place in highly managed forests or in fairly young second-growth forest rather than in old-growth forests, so it isn't surprising that the key role of canopy openings is often overlooked.[39] They play a central role in forest regeneration and enhancing biological diversity only after the forest is hundreds of years old.

Forest Catastrophes and the Growth of New Forests

Cathedral Pines, a nature reserve in Connecticut managed by the Nature Conservancy, was established to preserve an old stand of white pines and American hemlocks, many of which were more than 50 meters (150 feet) tall and more than 300 years old.[40] This stand of old trees was small (only 17 hectares, or 42 acres), however, making it intrinsically vulnerable to damage or destruction. Nearly all of the old trees at the site were blown down when the area was hit by an unusual New England tornado in 1989. Within minutes this ancient woods was converted to a pile of broken branches and uprooted trees. The storm initiated the process of forest regeneration, but the effect was much different from the limited impact of a small forest opening. The structure and species composition of the site have been almost completely transformed by a single catastrophic event.

The Cathedral Pines case illustrates why it is important to protect large areas of ancient forest. Small, isolated stands of old trees can be eliminated in a single windstorm or fire. In contrast, when a cluster of intense tornadoes hit the Tionesta Natural Area in Allegheny National Forest in Pennsylvania in 1985 and leveled 1,295 hectares (3,200 acres) of old-growth forest, thousands of acres of ancient forest escaped damage.[41] The result is a complex landscape of young forest and old-growth forest that supports a high diversity of animals and plants. The large openings created by the storm are an alternative route to forest regeneration that differs from the type of regeneration that occurs in canopy gaps. Many plants and animals that are not found in canopy gaps quickly colonize large openings created by storms, fires, or timber harvests.

The recovery of the forest in an area where the tree canopy has been destroyed by a blowdown can occur in two basic ways. If numerous tree seedlings and saplings were already present on the forest floor before the blowdown, then these will grow quickly and become the dominant trees in the new canopy.[42] Thus the new canopy can form from shade-tolerant tree species that were already established in the ancient forest before the disturbance. The result is similar when young trees sprout from the stumps of trees that have been snapped by the wind. Alternatively, if there are few seedlings in the forest before the disturbance (or if these seedlings were destroyed by an intense fire that burns through the fallen dead wood on the windthrow site), then the area may be colonized by fast-growing trees such as aspens (poplars) and cherries that germinate from seeds and that grow well only in open, sunny conditions. These early successional species grow more quickly than shade-tolerant forest trees and become the dominant species in the initial closed-canopy forest. Early successional species tend to be short-lived, however. In the absence of another major disturbance, they are eventually replaced by longer-lived trees that can grow in the shade of the pioneer species.

The recovery of vegetation in a 900-meter-wide windthrow that covered 400 hectares was monitored after the tornado ripped up the canopy at Tionesta Natural Area.[43] During the six years of the study, numerous tree seedlings grew up within a thick growth of burnweed and Allegheny blackberry. The tree seedlings eventually overtopped the dense layer of shrubs and herbs, which then began to thin out. Before the storm only American beech seedlings were present on the forest floor in good numbers, so this species was well represented among the rapidly growing young trees in the large openings created by the tornado. However, most of the tree seedlings were yellow birch, a species that does relatively well in open areas. Yellow birch seeds were common in the soil of the mature forest, and the crop of new seedlings was probably derived from this "seed bank." Yellow birch and American beech are on their way to becoming the dominant species in the new forest canopy. Although there were few seedlings of eastern hemlocks at the time of the tornado, numerous hemlock seedlings grew from seeds. These seedlings were later decimated by a combination of severe drought and heavy browsing by deer. In areas that were fenced off by the interlocking trunks and branches of fallen trees, however, young hemlocks escaped deer browsing and were growing quickly. If they grow tall enough so that their needles are out of reach of browsing deer, hemlocks may become part of the new forest canopy.

The details of how trees became established and survived in the massive Tionesta windthrow show that there is no simple, predictable model for forest succession following a major disturbance. The eventual composition of the recovered forest will depend on which tree seedlings and saplings are already growing on the forest floor; how many saplings are produced from sprouts growing out of damaged trees; which types of tree seeds are available from nearby trees or in the soil; and the subsequent survival of the tree seedlings and saplings.[44] Survival of young trees depends not only on competition with other plants for sunlight and nutrients, but also on weather events such as droughts and on the depredations of herbivores. One broad generalization about forest succession, however, is that large disturbances can favor plant species, including trees that may eventually become part of the mature-forest canopy, that almost never survive in a small canopy gap. Thus, natural disturbances such as blowdowns or fires can increase the overall diversity of a forest.

Of course when large-scale disturbances are too frequent, forests are unlikely to ever become old enough to acquire typical old-growth characteristics such as giant trees and large, decomposing logs. Instead the forest is a mosaic composed of patches of vegetation that are in different stages of recovery from fires, insect infestations, or windstorms.[45] Each patch consists primarily of trees of about the same age that started growing together soon after the last disturbance. Only a small proportion of the landscape escapes a major disturbance long enough to support old forest with trees of many different ages. This pattern is common in many regions of the boreal forest of Canada, northern Europe, and Siberia where large, intense forest fires have historically been frequent.[46] In contrast, many regions in the deciduous forests south of these fire-prone boreal forests are not subject to frequent natural disturbances.[47] Deciduous forests often grow in relatively wet areas where fires are infrequent and small. Most natural disturbances are due to the death of a single large tree or a small group of trees, so the forest primarily regenerates because of trees growing in canopy gaps. The result is large areas of old-growth forest of the sort seen at Tionesta Natural Area. This type of forest must have been widespread in eastern North America before logging and agricultural clearing, especially in regions with high enough rainfall to make natural wildfires infrequent and with a low incidence of windstorms large enough to destroy the forest canopy. Ancient forests with massive hardwood trees grew in the Ohio River valley and the coves (deep valleys) of the Great Smoky Mountains, for example, because these regions were buffered from large-scale disturbances. Based on studies of numerous old-growth forests, James Runkle concluded that this pattern characterized the "central mesophytic forest" of eastern North America, the exceptionally diverse deciduous forest stretching from the southern Appalachian Mountains to the southern Great Lakes and the Mississippi River.[48] This region is too far inland for major hurricane damage, too wet for frequent fires, and too far east or north for frequent tornadoes and downbursts.

Some regions of North America's deciduous forest are more prone to large-scale disturbances, however. This occurs at the borders of the eastern deciduous forest, where deciduous forest blends into more fire-prone ecosystems such as the boreal coniferous forest in the north, the tallgrass prairie on the west, and longleaf pine savannas in the Southeast. In contrast, the interior parts of the eastern deciduous forest have been described as "asbestos forests" because fire is so infrequent.[49] Near the coast, however, these forests can be blown down by major hurricanes. Severe hurricanes occur infrequently on the scale of human life spans, so an ecologist may study a forest throughout his or her career without ever experiencing a major blowdown. During the long life of an oak or a maple, however, major hurricanes are almost inevitable near the Atlantic Coast and even far inland unless the site is protected by deep valleys or mountains.[50]

In 1938 New England was hit by a major hurricane that caused massive damage to forests in a 150-kilometer-wide swath from Long Island Sound inland into the mountains of Vermont and New Hampshire.[51] Winds exceeded 200 kilometers per hour. Many of the even-aged forest stands found in New England today started growing immediately after this storm in areas where trees were leveled. In Harvard Forest in New Hampshire, the 1938 hurricane created numerous small patches with different levels of damage, from largely intact forest canopy with a few downed trees and branches to small openings to extensive areas where most trees were blown down. As a result, the current forest is a complex mosaic of different types of patches.

There are small areas with homogeneous trees of the same age; other areas with trees of a wide range of ages, including old trees that survived the 1938 hurricane; and areas that have a mix of even-aged and multiple-aged groups of trees.

Although the 1938 hurricane was an unusual event, forests in New England were also heavily damaged by severe storms in 1635, 1788, 1804, 1815, 1821, and twice in 1869, so a 300-year-old oak growing in this region has survived numerous major storms.[52] All of these storms were classified as F3 hurricanes, with winds strong enough to blow down numerous trees and destroy houses. In addition to these eight major hurricanes, 48 less severe hurricanes hit New England between 1620 and 1997, causing minor or moderate damage to trees and buildings. Thus, hurricanes are frequent enough to ensure that most forests in New England are in a state of constant flux caused by sudden, severe disturbances.

The effect of disturbances at a particular site can be worked out by carefully analyzing living trees and the layers of dead wood on the forest floor. Using methods akin to an archaeological dig, investigators can unravel the history of the forest. Living and fallen trees are dated from the pattern of their growth rings. Major windstorms result in numerous fallen trees that are all pointing in approximately the same direction. The trees knocked down by a more recent storm overlay trees knocked down in earlier storms, so the sequence of natural disturbances can be worked out. One of the most revealing studies of this sort was completed in Pisgah Forest near Ashuelot, New Hampshire.[53] Because this forest has never been cleared or plowed, the historical record revealed by dead wood is complete for almost 400 years. Nearly all of the trees that grew on the plot at the time of the surveys had started growing soon after a major fire or windstorm, so relatively large disturbances, not small canopy gaps, were the main source of new canopy trees at this site. This suggests that relatively large openings dominated by young vegetation occurred regularly in New England forests.

The islands of Japan are also subjected to hurricanes (which are called typhoons in the Pacific Ocean), so the dynamics of forest regeneration may be similar to the pattern on the East Coast of North America. An analysis of tree rings for a mixed forest of fir and various hardwood species (including *Acer mono*, *Magnolia obovata*, *Quercus mongolica*, and *Kalopanax pictus*) in Shiretoko National Park in northern Hokkaido revealed the pattern of canopy gap formation between 1890 and 2000.[54] Sudden increases in growth at the seedling or sapling stage show when a canopy gap formed and young trees grew quickly to reach the canopy. The number of gaps per year was constant for most of the 110-year period, but there was a substantial spike in canopy gaps during the period around 1979–1980 when two typhoons passed over the forest. Also, more large gaps caused by the death of several adjacent trees occurred during this period, but there is no evidence for the sort of large-scale blowdowns that resulted from the 1938 hurricane in New England.

Farther south in Japan, deciduous forests at mid elevations in the mountains are dominated by Japanese beech. Regeneration of trees occurs primarily in small canopy gaps, many of which are caused by moderate typhoons that regularly hit the islands.[55] Occasionally severe typhoons cause larger blowdowns, however. In 1991, for example, a severe typhoon blew down a large area within an old-growth beech forest (which has numerous trees more than 250 years old) in the Daisen Forest Reserve on Honshu. In the aftermath, most of the saplings growing in small canopy gaps were beeches, but a large opening (1.7 hectares) created by the storm was dominated by Viburnum furcatum and four species of maple. Also, Japanese cherry birch seedlings grew where the typhoon had uprooted trees and laid the soil bare. Although these large openings are infrequent, they appear to be important for maintaining a diversity of trees in the forest canopy.

Fire and Oaks

Fire may have been another important disturbance in some lowland forests in northeastern North America before European settlement. This was clearly the case where dry, sandy soils supported pitch pine and pine-oak "barrens."[56] The abundance of oak pollen in lake sediments indicates that fire was also important in determining the composition of deciduous forests on better soil. Just as in the old-growth forests of Poland, in New England oaks generally don't grow successfully in canopy gaps. Oak saplings do very well, however, where a fire has removed not only the forest canopy, but also the seedlings of competing tree species that are more sensitive to fire.[57] Oaks have been common in the forests in the coastal and lowland areas of southern New England for the past thousand years, suggesting that fires have been frequent enough to permit oaks to become an important component of the forest canopy.[58] Descriptions of oak forests along the northeastern coast of the United States at the time of European settlement indicate that they were open and parklike, with grass rather than shrubs

and saplings in the understory. This is consistent with the occurrence of
regular ground fires. Also, fire scars on ancient trees in Maryland and
Ontario indicate that fires occurred in oak woodlands on average every
6–20 years. Surprisingly, this rate did not change after European settlement.
There was also no substantial change in the concentration of charcoal in
lake sediments from oak-pine sites in coastal Massachusetts. The charcoal
particles reveal that fires were frequent both before and after European
settlement. Fires may have occurred before Europeans arrived because of
burning of agricultural fields and woodland by Indians, but it is difficult to
demonstrate this connection.

Originally the forests of the mountains of northern New England were
distinctly different from the oak-chestnut forests in southern New England.[59] They were dominated by species such as American beech, sugar
maple, yellow birch, and eastern hemlock, all of which are sensitive to fire
and consequently don't grow in areas with frequent fires. These species
were virtually absent from southern New England, which is not true today
after years of fire suppression. Later, the pollen record shows that these
trees declined in northern New England and that the forests began to resemble the oak-dominated forests of coastal areas.[60] Much of this change
occurred after European settlement and may be due to frequent fires during the period of forest clearing. The situation may be similar to that in
Białowieża Forest in Poland, where occasional fires may have favored
long-lived oaks, changing the character of the forest canopy for centuries.
Now that fires are suppressed, however, oaks are steadily being replaced
by maples, beeches, and other shade-tolerant trees in both southern and
northern New England, as well as in many other regions in eastern North
America.

In contrast to New England, oaks were an important component of deciduous forests in many parts of Europe where wildfires were apparently
infrequent. Frans Vera attributed the occurrence of oaks in these forests not
to frequent fires, but to large grazers such as aurochs, tarpans, and European bison, and to the domesticated grazers that later replaced them. Vera
hypothesized that parklike woodlands maintained by grazers composed a
large proportion of the lowland forest of western and central Europe. This
radically different view of prehistoric European landscapes would explain
the large number of species that depend on farmland and other artificially
created open habitats today, such as the many species of crows (carrion crows,
hooded crows, jackdaws, and rooks) that thrive in rural areas of Europe.

Vera's hypothesis doesn't seem to account, however, for the large number of species that do best in closed-canopy old-growth forest.

To determine whether open woodlands were widespread, Jens-Christian Svenning analyzed information about the pollen record as well as the fossil record for plant leaves, insects, and terrestrial mollusks in lake and bog sediments for numerous localities in northwestern Europe. He analyzed the pattern not only for the current interglacial period, but also for previous interglacial periods when the activities of people presumably had much less influence on woodland ecosystems. His basic conclusion was that during previous warm interglacial periods and before the spread of agriculture during the current interglacial period, most of the landscape in upland sites was covered with closed-canopy forest with occasional openings (probably forest glades). At upland sites that have relatively high soil fertility and no evidence of river flooding, pollen from grass or herbaceous plants is scarce. Along floodplains, however, pollen from grass and herbaceous plants is much more abundant, indicating extensive open areas. Also, beetles from floodplain areas are usually dung beetles that depend on animal dung as a food source, and grassland beetles, which fit the profile for a relatively open landscape with numerous large mammals. Periodic flooding along rivers probably knocked down trees, creating glades that would then be maintained by large mammals such as horses, aurochs, and other large herbivores (including hippopotami and Murr water buffaloes during previous interglacial periods). Pollen and macrofossil analysis also show that oak parkland and open grassland occurred in sites with dry sandy or calcareous soil, and that grassland occurred continuously in a British chalkland since the last glacial period. Just as in prehistoric New England, the ancient landscape of Europe may have had a mix of closed-canopy forests dominated by shade-tolerant trees, and more open habitats with low herbaceous vegetation and trees and shrubs that grow in more sunny conditions.[61]

The Ecological Importance of Ancient Forests

Because deciduous forests are frequently subjected to severe disturbances that destroy the forest canopy and initiate the growth of a young forest, it is tempting to conclude that ancient forests are not important for preserving biological diversity. This clearly is not the case, however. Comparisons of ancient forests with second-growth forests that are recovering from agricultural clearing or clearcut logging demonstrate that many species of plants and animals are substantially more abundant in ancient forests. This is particularly true for organisms that are not very mobile and so do not quickly colonize second-growth forests. Forest herbs (which include many of the characteristic wildflowers that bloom on the forest floor in early spring) have substantially lower abundance and diversity in second-growth forests than in old-growth forests. Their diversity is severely reduced by clearcut logging, and remains low even after decades of forest recovery.[62] Also, many species of woodland wildflowers were substantially more abundant in the forests that had never been cleared than in forests that had grown up on farmland in Massachusetts.[63] Wildflower species that have seeds dispersed by wind or vertebrates are usually found in post-agricultural forests, but species that are dispersed by ants or by gravity (by simply dropping to the ground) were absent from many of these forests, presumably because they are not very effective at dispersing their seeds over long distances to colonize an isolated new forest.

A particularly rigorous comparison of ancient and more recent woodlands was completed in Lincolnshire, England.[64] Historical records show that these ancient woodlands have been continually covered with trees (including highly managed coppice woodland) since at least 1600, and many of these sites have probably been wooded for millennia. In contrast, most of the recent woodlands were planted on former farmland in the eighteenth and nineteenth centuries. Ancient forests had a higher diversity of herbaceous woodland plants, and many plant species grow primarily in ancient woodland. These species have rarely colonized "recent forests" even though some of these are centuries old. Recent woodlands that are adjacent to ancient forest have a higher diversity of these plants than do recent woodlands that are more isolated, but they still are not as diverse as ancient forests. Either the environment of managed, planted forests is not suitable for these species or many species of small woodland plants are not effective at colonizing new forests, or both.

Lichens are even better indicators of old-growth forest than are wildflowers. Many species of lichens have very specific habitat requirements, and a large number of species are restricted to ancient forests that are centuries old, with different species in hardwood forests and conifer forests.[65] Francis Rose successfully used lichens as indicators of ancient open woodlands ("wood pastures") in England.[66] Thirty species of lichens are associated with woodlands with very old trees and a history of continuous woodland cover. Steven Selva used a similar approach to analyze the distribution of

lichens in old forests in northeastern North America to successfully distinguish ancient, relatively undisturbed forests from forests that were not as
old.[67] Some species of lichens do not colonize a forest until it has acquired
ancient tree trunks with acidic bark and damp, shady conditions, and this
may require centuries.

Salamanders are another group of organisms that recover slowly after
forest removal. Clearcutting of mature forests in Pisgah National Forest in
North Carolina resulted in an estimated 80 percent reduction of salamander
populations as well as a significant drop in the number of salamander species.[68] Twelve species of salamanders were recorded in mature forests, but
some species of *Plethodon* salamanders are almost completely missing from
clearcuts. Comparison of sites that had been logged at different times, from
19 to 120 years before the study, demonstrated that it requires about 70 years
before the density and number of species of salamanders fully recover. In
the Ozark Mountains of Missouri, *Plethodon* salamanders were 5 times more
abundant in old-growth forest than in second-growth forest that had been
harvested 70 years previously, and 20 times more abundant than in regenerating clearcuts that had been harvested 5 years previously.[69] *Plethodon* salamanders breathe almost entirely through their moist skin, so they require a
moist environment. The large, heavily decayed logs and dead branches on
the floor of an old-growth forest provide a much better habitat than either
recent clearcuts or young second-growth forests.

Even organisms that are more mobile than woodland wildflowers and
salamanders may be more abundant in old-growth forests than in second-
growth forests. These species could colonize new forests relatively quickly,
but they tend to be more abundant in old-growth forests because of their
habitat requirements. Although there are no bird species that are restricted
to old-growth forest in northeastern North America, the birds encountered
in old-growth sites and second-growth sites are surprisingly different.[70] The
overall density of birds is higher in old-growth forests, probably due to the
greater complexity created by multiple layers of vegetation. Many bird species
are substantially more abundant in old-growth forests than in younger forests: brown creepers are twice as abundant and magnolia warblers are 40
times more abundant. Some species such as Blackburnian warbler and Swainson's thrush are considerably more abundant in old-growth forests because
hemlocks make up a large part of the canopy and these species are associated with coniferous forest. They also occur in young coniferous forests, but
most second-growth forests are dominated by deciduous hardwood trees.

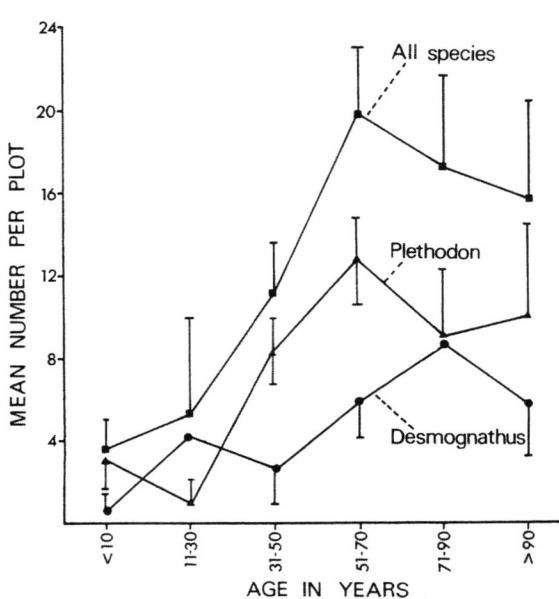

Figure 13. Relationship between the age of the forest and salamander density (average number per 50-meter-by-50-meter plot) for two important genera of salamanders and for all salamander species in Pisgah National Forest, North Carolina. (Petranka et al., 1993; reproduced with permission of John Wiley and Sons, Inc.)

Other species depend on structural features of old-growth forests that are not commonly found in second-growth forests. Barred owls and brown creepers nest in large snags; hooded warblers and American redstarts use the dense, low vegetation in large canopy gaps; and winter wrens forage and nest among the branches and root masses of large, fallen trees. Also, old hemlocks and white pines produce many more cones than young trees, so old-growth forests attract many more red crossbills and other seed-eating finches.

The main implication of these studies is that ancient forests serve as a source of sensitive species that are largely missing from second-growth forests. These species may decline or disappear in heavily disturbed forests, but are eventually able to recolonize if populations are sustained in surviving patches of older forests. In the landscapes that preceded intensive agriculture and forestry, disturbances such as windstorms and fires might decimate these species in large swaths of forest, but they would eventually return as seeds or young animals dispersing from adjacent undisturbed forests or from islands of surviving forest. This link between ancient forests and regenerating forest has been broken in most parts of the world, however. Much of the new second-growth deciduous forest grows on land that

was once cultivated and is a great distance away from the nearest old-growth
forest. Also, many forests are managed on a relatively short rotation to maximize wood production. If the entire forest is managed in this way, then there
will be no refuge for species that would naturally arrive in a regenerating
forest late during forest succession. The result will be a steady loss of biological diversity, particularly in sensitive groups such as wildflowers and
salamanders. Protecting some areas of older forest as a source of late-
successional species will sustain biological diversity and the complexity of
the forest ecosystem. This is important because loss of late-successional
species is likely to have unanticipated effects on the stability and sustain-
ability of the forest. Loss of terrestrial salamanders, for example, which are
major predators of insects and spiders on the forest floor in many deciduous forests, could influence everything from rates of decomposition of dead
plant material to growth rates of plants.

To sustain forest diversity we should protect remaining old-growth forests, as well as areas of mature forest that can eventually become old growth.
But how do we recognize old-growth forests and forests that can potentially
become old growth? There are numerous definitions of "old growth."[71]
Some include the stipulation that the forest has never been disturbed, or at
least has never been completely cleared. By this criterion, the ancient forest
at Tikal National Park in Guatemala would not qualify as old-growth forest
because it was once the site of an ancient Mayan city. Forest has been growing continuously on the ruins since the city was abandoned more than a
thousand years ago, however, and the current forest has ancient trees and
multilayered vegetation. Other definitions focus on the time since the last
major disturbance. For deciduous forests the minimum age is typically set
at 250 to 300 years. If this is used as the sole criterion, however, then sparse
and stunted bonsai-like trees growing on cliff ledges or on very poor soils
qualify as old growth because they consist of old trees. Identifying these old
forests is useful for botanists who analyze growth rings of trees to investigate climate change or to date other events, but ecologically these stunted
forests may have little in common with an old forest growing on rich, moist
soil. They are better referred to as "ancient forests" rather than old-growth
forests.[72] I use the term "old-growth" to refer to forests that share some key
ecological characteristics. These forests have grown under favorable conditions without a major disturbance (either natural or artificial) for so long
that new generations of trees have grown up in canopy gaps. They have acquired a set of recognizable characteristics: large, old trees; large fallen trees

in an advanced state of decomposition; tip-up mounds created when the root mass is exposed after a large tree falls; relatively large canopy gaps; a relatively open canopy; multiple layers of vegetation; and an even mix of trees of many different heights, diameters, and ages. Comparison of old hemlock-hardwood forests of different ages in northern Wisconsin and Minnesota (177–374 years old) shows that all of these characteristics come together when a stand reaches an age of 275 to 300 years since the last major fire or blowdown.[73] After this threshold is reached, stands have many more trees that are more than 70 centimeters in diameter and many more large logs in an advanced state of decay. Canopy gaps account for 10 percent of the area of the forest, and some canopy gaps are larger than 200 square meters. Even before the 275-year mark is reached, old forests have canopy gaps, snags, and fallen logs and branches. The number of logs and snags and the size of canopy gaps progressively increase with the age of the forest. Mature forests acquire these characteristics gradually through time, but they cannot be considered old growth until all of these structural characteristics are in place. Using criteria such as these, the tropical moist forest at Tikal would be considered old growth even though it was once the site of a densely populated urban center.

Species that Require Young Forest and Large Forest Openings

Historical changes in deciduous forest have been intensively studied in only a few regions, but these studies consistently reveal forests that are in flux in complex and not entirely predictable ways. Although forests generally change slowly through the replacement of individual trees or small groups of trees that die, larger disturbances also play an important role by shifting a forest into a new trajectory of change that may determine the mix of tree species for centuries. Large disturbances also have an equally important (but more ephemeral) impact by creating large areas of low, scrubby vegetation and, later, young forest that support numerous species that are absent from the interior of mature forests. Many of these "early successional species" specialize on a particular early stage of forest succession, and they are absent from both the deep shade of the forest floor and the canopy gaps in mature and old-growth forests. Although large openings are rare and temporary events at any one locality, they may appear at a constant rate across the regional landscape, dependably providing adequate habitat for early successional species. As older openings grow into mature forest, new openings are created by fires, floods, windstorms, or other natural disturbances. Where people have reduced or eliminated natural disturbances (with suppression of wildfires or damming of rivers, for example), early successional species begin to disappear. As a result, even carefully protected natural forests may lose much of their biological diversity if something is not done to create forest openings. Thus forest diversity is threatened not only by the loss of ancient forest, but also by the loss of the youngest stages of regenerating forest.

The New England cottontail, a species of rabbit that is restricted to New England and a small area along the border in neighboring New York, has declined by 80 percent since the 1960s and is now a candidate for listing under the U.S. Endangered Species Act.[74] It has disappeared from a large proportion of its former range, including all of Vermont and most of New Hampshire.[75] New England cottontails have declined as their preferred habitat—low woody vegetation—has dwindled. They are especially common in thickets growing on abandoned farmland, but are also found along powerline corridors, clearcuts, shrubby swamps, and other areas with low, scrubby vegetation that provides dense cover.[76] Some of the old fields and thickets once used by these rabbits became housing developments, but most of the habitat loss was simply due to forest succession. Saplings grow into tall trees and then shrubs are shaded out, converting a scrubby habitat into a young forest that is unsuitable for cottontails.

Not only has there been a decline in the total amount of low woody vegetation (often called "scrub shrub" or "shrubland" even though it may be dominated by tree seedlings and saplings rather than shrubs), but also the remaining habitat is often restricted to small, isolated patches. Intensive monitoring of rabbits equipped with collars housing radio transmitters demonstrates that those living in small shrubland patches have low survival rates. On average, rabbits living in patches smaller than 2.5 hectares die at twice as high a rate as those living in larger areas of continuous scrub shrub. This is due to a high rate of predation during the winter, when generalized predators such as coyotes and red foxes move into small shrubland patches.[77] These predators thrive in lightly developed rural habitats and may overwhelm the rabbits in islandlike patches of shrubland. As a result the future of New England cottontails depends on larger areas of scrub shrub. These could serve as source populations from which rabbits disperse to recolonize smaller scrub-shrub patches where rabbits have disappeared. The average distance between separate populations of New England cottontails was 17 kilometers in northern New England and 50 kilometers in southern New

England, however, and movement across such great distances is probably unlikely for a species that does not stray far from cover.[78] Corridors connecting the remaining populations, perhaps along powerlines or railroad tracks where low vegetation is maintained, could be used to help sustain the regional cottontail population. Alternatively, rabbits could be captured and moved to establish populations in isolated patches of good habitat.

If New England was almost completely forested before European settlement, then it is difficult to envision how New England cottontails would have found a sufficient amount of low woody vegetation. They must have moved from windthrow to windthrow, escaping to a new opening when the tree canopy in an older opening began to close. But hurricanes, downbursts, and tornadoes large enough to create extensive forest openings were probably too infrequent and unpredictable to sustain a population that must constantly disperse between isolated patches of habitat by hopping. As John Litvaitis pointed out, however, the forest landscape before European settlement had sources of scrub-shrub habitat that have subsequently become rare, such as scrub-oak barrens and abandoned beaver ponds that had dried up and become "beaver meadows" covered with alder thickets.[79] Scrub-oak barrens may have been major sources of cottontails that dispersed into recent windthrows, using beaver meadows and floodplain shrub swamps as stepping stones. Many of these patches of shrubland were small, so they would not be favorable cottontail habitat in the modern landscape. Instead of being surrounded by roads, farmland, and residential areas with inflated densities of medium-sized predators, however, scrub-shrub patches in the pre-agricultural landscape were generally surrounded by mature forest where the predator populations were very different. Large predators such as gray wolves and mountain lions probably kept down the density of smaller predators that were more likely to prey on cottontails. The main predator of cottontails was probably the bobcat, a species that has declined along with the New England cottontail since the 1960s.[80]

The New England cottontail is almost identical to the eastern cottontail, a more adaptable species that became widespread in gardens, farmland, and other open areas after it was introduced into New England. The two species are difficult to distinguish, and positive identification depends on examination of the skull or on mitochondrial DNA analysis.[81] Despite their close similarity in appearance, however, the eastern cottontail does not appear to be an ecological replacement for the more specialized New England cottontail. Eastern cottontails gravitate to more open, grassy habitats. The response of the two species when they are captured reflects deep behavioral differences: New England cottontails usually squeal and wiggle to escape while eastern cottontails are rigid and docile.[82] When they encounter a predator, the first response of New England cottontails is to dive into the nearest thicket (which is usually not far away for this species) and dodge through the tangle to escape. In contrast, eastern cottontails often graze in open areas with no dense cover close at hand. Their first response to a predator is to freeze and depend on camouflage to escape detection. Thus the loss of New England cottontails from early successional forests would not be offset by the expansion of the introduced eastern cottontail, which evolved behavior suited to more open habitats.

Even in the pre-agricultural landscape New England cottontail populations must have been relatively small and scattered except—briefly—in areas where the forest had not yet recovered from a major fire or hurricane. It isn't surprising that most other species of mammals in New England have broader habitat requirements and so don't depend on patches of ephemeral, early successional habitat.[83] Isolation of patches of good habitat is potentially much less of a challenge for birds, however, which can fly long distances in search of favorable breeding areas. Many North American bird species are specialized for using low vegetation dominated by shrubs and tree saplings, and they quickly colonize favorable habitat created by natural or human disturbances. Six years after a tornado ripped through Allegany State Park in New York, for example, the low thicket growing on the large windthrow area supported a dense population of chestnut-sided warblers and other specialized scrub-shrub bird species.[84] Similarly, prairie warblers, blue-winged warblers, yellow-breasted chats, and other scrub-shrub birds quickly colonized a deciduous forest that was heavily damaged by a tornado in Ozark National Forest in Arkansas.[85]

Many of the bird species that use scrub-shrub vegetation in North America have habitat requirements that are more specific than New England cottontails. Some species are primarily found in low thickets with a dense tangle of vines and shrubs, while others are most abundant in areas dominated by numerous young trees. As the vegetation becomes taller, some species decline while others increase. The result is that early successional forests in eastern North America have a distinctive assemblage of birds dominated by species that are not found in closed-canopy forests or in small canopy gaps.

My students and I have studied these "scrub-shrub specialists" in two of their major habitats in eastern Connecticut: recent clearcuts where all or nearly all trees have been logged, and powerline corridors where trees are selectively removed to create low vegetation dominated by shrubs to prevent tree branches from contacting electrical powerlines. Both of these artificial habitats have high diversity and density of scrub-shrub birds, including many species that have declined substantially during the past few decades.

To reach clearcuts we entered state forests early in the morning and walked for a mile or two along a woodland trail while listening to the songs of wood thrushes, red-eyed vireos, and worm-eating warblers and other forest birds. The border of most clearcuts was abrupt. We stepped out of the forest shade into a sunny sea of tree saplings, many of which had sprouted from the stumps of harvested trees. Seedlings and saplings of canopy trees (maples, oaks, beeches, and black birches) covered most of the clearcuts, with only a small component of maple-leaved viburnum, blueberry, and other shrubs. The bird songs also changed as we neared the opening. The most abundant species in the clearcut were birds associated with shrubby habitats: gray catbirds, common yellowthroats, blue-winged warblers, chestnut-sided warblers, and eastern towhees.[86] These species are absent or substantially less common in mature forest. All of these species appear to thrive in clearcuts and so appear to benefit from timber harvests. Several studies demonstrate that they not only reach high densities in clearcuts, but that they also reproduce successfully.[87]

Ironically, widespread efforts to enhance biological diversity in North American forests by shifting from clearcutting to selectively logging single trees or small clusters of trees will not produce appropriate habitat for these scrub-shrub birds, and consequently may lead to a loss of biological diversity. In northern hardwood forests in New Hampshire, for example, chestnut-sided warblers and common yellowthroats are substantially more frequent in clearcut plots (clearings of 8–12 hectares) than in "group selection" plots (clearings smaller than 0.65 hectare where small groups of trees have been harvested).[88] In oak-dominated forests in Missouri, shrubland species such as blue-winged warblers and eastern towhees were much more frequent in clearcuts than in group-selection or single-tree selection cuts.[89]

In Connecticut we found that ten or eleven years after the forest was harvested, the trees on clearcuts had grown tall enough to attract some of the typical birds of closed-canopy forest, and that scrub-shrub birds had declined. Thus, clearcuts can provide appropriate habitat for scrub-shrub specialists, but this habitat is transient. More stable scrub-shrub habitat is found along electrical powerline corridors. In Connecticut these have been

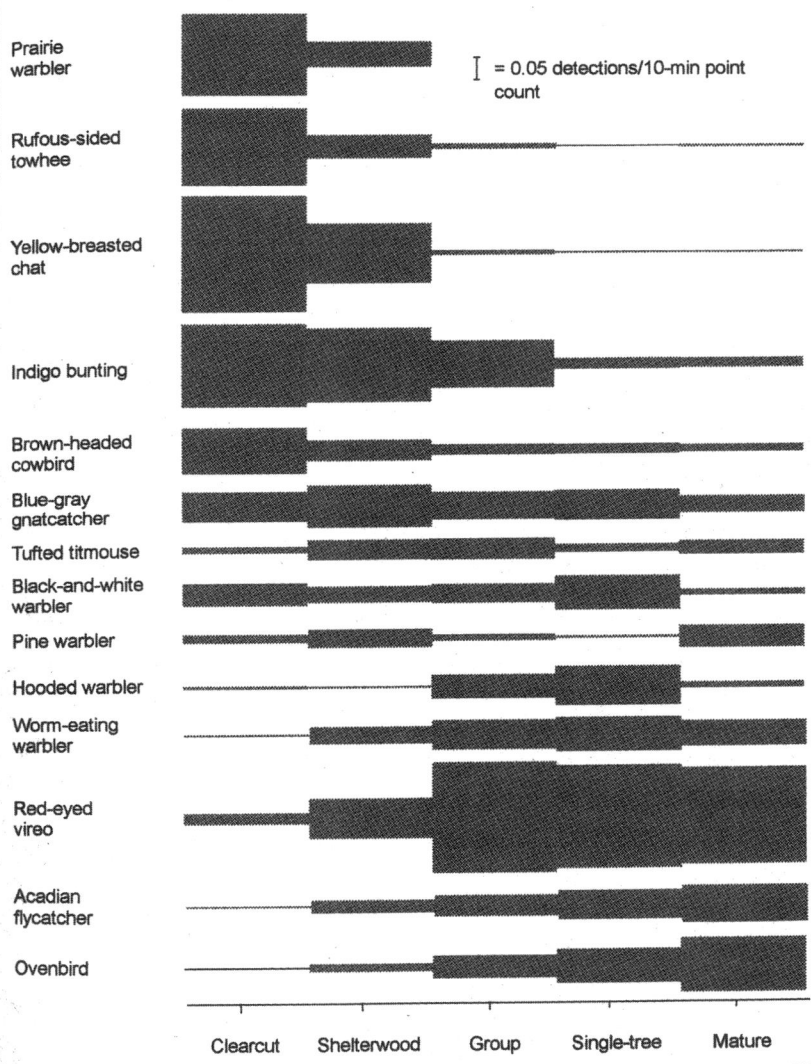

Prairie warbler

Rufous-sided towhee

Yellow-breasted chat

Indigo bunting

Brown-headed cowbird

Blue-gray gnatcatcher

Tufted titmouse

Black-and-white warbler

Pine warbler

Hooded warbler

Worm-eating warbler

Red-eyed vireo

Acadian flycatcher

Ovenbird

I = 0.05 detections/10-min point count

Clearcut   Shelterwood   Group   Single-tree   Mature

Figure 14. Relative abundance of common bird species in areas managed with different forestry methods: clearcut (most mature trees removed over a large area), shelterwood (many large trees left as residual cover after most of mature forest was harvested), group selection (groups of trees removed to create openings smaller than 4 hectares), single-tree selection (individual trees selectively removed from mature forest), and mature (trees not harvested during the past 50 years). Timber harvesting at the recently harvested sites had occurred within the previous six years. (Annand and Thompson, 1997; reproduced with permission of John Wiley and Sons, Inc.)

managed by selective removal of trees since the 1950s, when most electrical utilities ended the practice of indiscriminate herbicide spraying of all vegetation along powerlines. This more selective approach is now used in most of the northeastern United States. The most abundant species in this habitat are gray catbird, prairie warbler, blue-winged warbler, eastern towhee, and field sparrow. Some species that specialize in dense thickets, such as white-eyed vireos, yellow-breasted chats, and brown thrashers, are also present in small numbers. We never recorded these three species or field sparrow in the many clearcuts we surveyed in the same region. The difference between the bird assemblages in clearcuts and powerlines illustrates a basic difference in vegetation structure between two types of low, woody vegetation: young forests are dominated by tree seedlings and saplings, while early successional habitats are characterized by a complex tangle of shrubs, vines, young trees, and herbaceous plants. "Young forests" grow where the canopy trees have been destroyed by wind, fire, or logging and young seedlings and stump sprouts immediately begin to grow. "Early successional habitat," in this context, refers to areas where there are no tree seedlings or stump sprouts available to immediately start a new forest, so the site is colonized by pioneer plant species with seeds that can disperse effectively over great distances. This pattern is seen in abandoned farm fields and beaver meadows, and in burns where exceptionally hot forest fires have killed seedlings, roots, and seeds. The selective removal of all tall-growing trees on powerline corridors produces somewhat similar vegetation.

We initially tried to find populations of scrub-shrub birds in more natural habitats such as major windthrows or recent burns where a forest fire had killed the canopy trees, but we could not find enough natural sites to complete a statistically valid comparison with clearcuts and powerline corridors. Natural disturbances still create new habitat for these species, but people have ensured that these disturbances are infrequent and have a limited effect. People have suppressed fires, built dams to control river flooding, and removed beavers from much of the landscape. Blowdowns caused by storms still occur, but the landscape is dominated by relatively young forest with few old trees, so the openings are smaller and less frequent than in older forests. Without artificial habitats—abandoned farm fields, clearcuts, and powerline corridors—scrub-shrub specialists would have a difficult time finding enough breeding habitat each year.

Scrub-shrub species reached exceptionally high densities during the period between 1850 and 1960 when farms on marginal land were gradually abandoned in eastern North America.[90] This resulted in a continuous supply of "old fields," newly abandoned farmland covered with shrubs, vines, herbs, and tree saplings. These old fields eventually became woodland, however, and when the supply of old fields dwindled, nearly all species of scrub-shrub specialists declined. Initially this trend caused little alarm among conservationists because the populations of these species had clearly been inflated by the abundance of artificial habitat, but the declines have been so rapid and sustained that there is now concern about the future of many scrub-shrub species.

Scrub-shrub specialists contribute greatly to the diversity of the bird fauna in forested regions.[91] There are 16 species of scrub-shrub specialists— species that are largely restricted to low, woody vegetation—in eastern North America.[92] This list does not include species such as gray catbird, common yellowthroat, and eastern towhee that are common in this habitat but are also frequently found in other habitats. Ten of these species occur in southern New England, two of which (the yellow-breasted chat and golden-winged warbler) are on the verge of regional extirpation even though they were once common breeding birds. Most of these species have declined in eastern North America since 1966, some at a rate of more than 2 percent per year.[93] An entire assemblage of species is threatened by the steady loss of scrub-shrub habitat due not only to the scarcity of newly abandoned farmland, but also to the scarcity of shrublands caused by natural disturbances.

Scrub-shrub birds are also found in early successional woody habitats in Europe and East Asia. A list of bird species in the low scrub and tall scrub stage of forest succession at Białowieża, for example, includes red-backed shrike, yellowhammer, and garden warbler, all of which are largely restricted to open habitats dominated by shrubs.[94] In England and other countries that have been intensively cultivated for millennia, coherent sets of bird species associated with different stages of deciduous forest succession are often difficult to identify because forests seldom undergo natural succession. Most forests are harvested and replanted as tree plantations or they are maintained as low coppice by frequently harvesting stems. Early successional bird species have adjusted to different habitats in this heavily managed landscape, with different species gravitating to young conifer plantations, coppice, hedgerows, heath, or forest edge. A successional sequence of bird species occurred in areas of southern England where scrub replaced grassland because of a decline of sheep grazing, however. As shrubs invaded and created low, open scrub, common whitethroats, linnets, and yellowhammers replaced the skylarks and meadow pipits of open grassland.[95] These species were in turn replaced by dunnocks, garden warblers, and lesser whitethroats as the thicket became taller and denser. Similar changes involving many of the same species occur during forest succession following timber harvesting in deciduous broad-leaved woodlands in France.[96] In the Ardennes region of Belgium, clearcut openings in Norway spruce plantations supported bird assemblages that were distinctly different from bird assemblages in farmland, mature broad-leaved forest, or conifer plantations.[97] These artificial openings have become a major refuge for species that previously were found in heathland, moorland, or coppice, habitats that have largely disappeared from the region. Species that now largely depend on scrubby clearcuts in conifer plantations include great gray shrike, melodious warbler, and European stonechat.

Three scrub-shrub species (common whitethroat, lesser whitethroat, and yellowhammer) declined significantly in British woodlands between 1967 and 1999.[98] These declines appear to be due to maturation of forest plantations and abandoned coppice, although changes in farming practices and (for the two whitethroats) in the African wintering areas may also be factors.[99] These declines have occurred despite efforts to restore traditional coppicing, which maintains low, dense woody vegetation used by many shrubland birds, in nature reserves in Great Britain.

Japan apparently does not have a great diversity of scrub-shrub specialists. When Reiko Kurosawa studied 50 sites in southern Hokkaido that had a wide range of open, nonforested types of habitats, including agricultural meadows, pastures, dune grassland, volcanic barrens, and alder and willow thickets, she only found one species, meadow bunting, that specialized in scrub-shrub habitat.[100] Other possible scrub-shrub specialists in Japan are Japanese bush-warbler, bull-headed shrike, brown shrike, thick-billed shrike, and Gray's grasshopper-warbler.[101] Two of these species, brown and thick-billed shrikes, have shown severe population declines and range contractions in the past few decades, and they are now so infrequent that they are unlikely to be detected frequently enough on standardized surveys for analysis of habitat use.[102] Since the 1970s forests in Japan have matured due to the end of coppicing and a decline in rotational harvesting of tree plantations and second-growth forests.[103] As a result, the amount of scrub-shrub habitat has declined, and this has caused range contractions in many early successional forest bird species. The declines have occurred among both long-distance migrants and non-migratory birds, but have been more severe for species that spend the winter in the tropics, suggesting that early successional species are negatively affected by habitat changes in both the northern breeding areas and tropical wintering areas.

Implications for Conservation

Ancient deciduous forests should have a high priority for conservation because so few have survived. They harbor species that are not found in younger forests or other habitats, and they provide our best source of information about how deciduous forests functioned before agricultural clearing. Because of the direct and indirect impact of people, all remaining old-growth forests are changed and incomplete to various degrees. Despite this, they perpetuate some of the key ecological processes that are missing or muted in young second-growth forests, such as canopy gap formation and decomposition of massive dead trees. These processes are often the key to understanding the adaptations of particular species of microbes, plants, and animals, leading to insights that would be difficult to develop in the absence of information about their ancestral habitats.

Many ancient forests are small groves surrounded by cleared land. These relict patches are vulnerable to invasion by species from the surrounding open landscape, including predators and introduced plants, which often modify the structure of the forest and the way in which the forest ecosystem functions. Equally important, small patches of ancient forest, whether isolated by clearings or embedded in younger, second-growth forest, are susceptible to sudden catastrophic destruction. Consequently the most valuable ancient forests are extensive areas of old growth in heavily forested regions. These invaluable deciduous forest ecosystems are found in only a few places, such as Białowieża Forest in Poland and Belarus; Adirondack Park, Tionesta Natural Area, and Great Smoky Mountains National Park in the United States; and Shiretoko National Park in Japan. These parks and forest reserves are large enough to absorb large-scale natural disturbances (such as a swarm of tornadoes that hit Tionesta Natural Area) without complete destruction of old-growth forest. Instead disturbances enhance diversity by creating a mosaic of young and old forest patches that support a wide range of species with different habitat needs.

Before agriculture the importance of old-growth forests varied among regions. In regions prone to frequent fires or severe hurricanes or typhoons, old-growth forest may have been restricted to a few protected sites, such as uplands protected from fire by swamps or deep valleys that are shielded from the wind. Moreover, openings created by disturbances may have been sustained for long periods by the activities of large grazing and browsing mammals. Most of the landscape may have been more open than modern second-growth forests, with large areas of scrub shrub and young forest, or open parklike woodland dominated by tree species that do not grow in the shade of the forest canopy. If nature reserves in these landscapes are managed solely for closed-canopy forest, then biological diversity will be severely reduced. For this reason, there are increasing efforts to understand the nature of disturbances in landscapes that have not been heavily modified by people. These disturbances could then be reintroduced or simulated in nature reserves to maintain biological diversity. A major goal should be to reverse the severe declines in early successional species that have occurred in deciduous forests on all three continents.

Restoration of early successional and open woodland habitats is easier than restoring old-growth forest, which requires hundreds of years to develop its distinctive features. Some features of old-growth forest can be created in mature second-growth forests by active management, however. For example, selective cutting of trees can speed up the growth rate of remaining trees and provide large snags and woody debris. This approach might be particularly useful in second-growth forests adjacent to old-growth stands because it may boost populations of old-growth species and provide a refuge if the old-growth stand is blown down. The best long-term solution, however, is to set aside some second-growth forest as future old growth. This goal will take decades or even centuries to completely achieve, but these forests will become progressively more valuable for preserving distinctive species and ecological processes as they mature. A good model for managing a large forest is Larry Harris' proposal of a central core that is never harvested surrounded by a larger peripheral ring where slices of forest are harvested in rotation around the core.[104] This approach will ensure that each stage in forest succession, from low scrub shrub to ancient forest, is always represented in the landscape. The central old-growth core will serve as a repository of biological diversity for reseeding forests on the periphery, while the extensive working forests around the core will serve as a protective buffer for the old-growth forest.

CHAPTER 6

Forest Islands and the Decline of Forest Birds

I stopped to listen in the slanting, early-morning light. The flutelike notes of a wood thrush resonated among the tall black oaks and tuliptrees. The challenge was to pinpoint this sound and the songs of other birds so that I could mark their locations on a detailed map of the study site in the Connecticut College Arboretum. Red-eyed vireos sang their monotonous songs—short, choppy phrases separated by brief pauses—from different directions. I worked to disentangle the sources for three different males. Two ovenbirds alternated their loud, penetrating songs, countersinging to defend territories in distinctly different locations. A downy woodpecker rattled and drummed in the distance. This complex soundscape represented a rich diversity of forest birds that I had monitored since 1982, when I picked up the reins of a much older bird population study initiated in 1953.[1] The bird study was coordinated with intensive surveys of the vegetation of the study area in order to simultaneously track changes in the forest vegetation and the bird populations that depend on it.

The heart of the study area is a patch of old forest straddling a deep granite ravine.[2] The forest was badly damaged in 1938 when a major hurricane swept across Connecticut. Most of the 300-year-old hemlocks were toppled or broken off at the trunk, but many of the old oaks survived. In 1953, the forest canopy was still under repair, with new trees growing into the gaps created by the storm. The plan was to follow this process over the next few decades as the forest slowly recovered. The bird study would throw light on how forest bird populations respond to these changes. The logical prediction was that birds of mature forest would increase in abundance and diversity as the canopy filled in with large trees.

Over the next two decades the forest vegetation changed in predictable ways.[3] The average diameter of trees increased. The shrub layer thinned as the tree canopy closed to create deep shade. Instead of increasing in response to the growth of trees, however, populations of many forest bird species declined precipitously, and some previously common species disappeared entirely.[4] The decline of hooded warblers wasn't surprising, because this species gravitates to the dense shrub layer of forest openings. As canopy gaps disappeared, their habitat shrank. Most of the other declining species typically live in closed-canopy forests, however. Why should they decline as the forest recovered?

It turned out that this was not merely a local mystery involving one particular storm-damaged forest in coastal Connecticut. A similar pattern emerged in numerous bird censuses in wooded parks and nature reserves at widely scattered locations across the eastern deciduous forest, from Virginia to Wisconsin.[5] The same species—ovenbirds, black-and-white warblers, red-eyed vireos, and wood thrushes—declined at different sites, and all of these sites experienced a loss of diversity of forest birds. In the 1960s and 1970s these census sites provided the best overview of long-term population changes in woodland birds across eastern North America. Densities of particular species were determined by mapping their breeding territories using a standardized protocol developed by the National Audubon Society in the late 1930s.[6] The method is time-consuming, but it provides a reasonable measure of the abundance of territorial males. It is less useful for monitoring female songbirds, which are quieter and more secretive. The population declines were particularly surprising because each of these study sites was in a well-protected conservation area. The habitat of these species had not been destroyed or heavily modified in any apparent way. The implications were alarming. Forest bird populations were plummeting in widely scattered locations across the eastern deciduous forest for no obvious reason.[7]

Searching for the Cause of Bird Population Declines in Eastern North America

The declines were particularly well documented for several parks with natural woodland in the Washington, D.C., metropolitan area.[8] A graph illustrating these declines was published in 1976 in the newsletter of the Audubon Naturalist Society of Washington and, soon thereafter, in a scientific paper.[9] Researchers immediately began searching for the cause of these declines. Because the breeding habitat of these birds appeared to be intact, they looked to problems during the nonbreeding season. Nearly all of the declining species were Neotropical migrants, birds that nest in the temperate zone and spend the winter in the New World tropics. Many of these species spend more time each year in their tropical habitats than in their northern breeding areas. Extensive clearing of forests in some of the most important wintering areas—in southern Mexico, northern Central America, and the Greater Antilles—may have deprived them of critically important winter habitat. Thus the decline of wood thrush populations in Washington's Rock Creek Park or in the Connecticut College Arboretum may have resulted from tropical forest destruction in Veracruz or Guatemala rather than from environmental changes in the northern study sites.

Initial impressions were deceiving, however. Forest birds had not suffered a general decline across the eastern deciduous forest, as would be expected if population changes were driven by massive loss of tropical winter habitat. The severe problems with forest songbirds were documented in Audubon census sites, and these were not a representative sample of the eastern deciduous forest. An Audubon census requires eight to ten early-morning visits each year during the early part of the breeding season, so it entails one or two visits each week in May, June, and early July. Most censuses were run by unpaid volunteers, so only censuses in convenient locations were likely to be sustained for 20 or 30 years to determine long-term population trends. As a result, most censuses that were not discontinued after a few years were located in relatively small parks or forest reserves near cities or university campuses. Nearly all were surrounded by an urban or suburban landscape. The partial collapse of the forest songbird community—with steep population declines for some species and complete extirpation of others—was restricted to these islandlike patches of forest surrounded by suburbs or farmland.

Comparable collapses did not occur in extensive forests far from cities and suburbs. Populations of forest songbirds changed over several decades in the extensive forests of Great Smoky Mountains National Park, White Mountain National Forest, New York's Allegany State Park, and Hiram Fox Wildlife Management Area in Massachusetts, but these changes did not involve consistent or catastrophic declines.[10] In most cases the population changes of particular species in large forests were readily explained by changes in the local environment (such as succession from young forest to old forest, or the decline in the amount of food following a major outbreak of forest caterpillars). Consistent, simultaneous declines in many species of forest birds were restricted to smaller forests in more developed areas. If population declines were due to loss of tropical winter habitat, then one would expect them to be more widespread.

Another potential explanation for these declines came from studies of the diversity of plants and animals on oceanic islands. Small, isolated islands have relatively low numbers of species for particular groups of organisms such as trees, birds, lizards, or dragonflies. Extinction rates are high on small islands because the populations are small (and thus more vulnerable to natural disasters and periods of severe weather). If the island is far from the mainland, species that go extinct are not readily replaced by new species immigrating across the ocean. The interplay of high extinction rates for species already on the island and low colonization rates for species new to the island results in low diversity. Perhaps forested nature reserves had become like these oceanic islands.[11] They are wooded islands in a suburban sea, so they have small populations of forest specialists that may be prone to local extinction. If dispersal rates from other forests are low, then these forest islands may steadily lose species after they become isolated.

This explanation may apply to species of animals and plants that are not very mobile and so are truly marooned when a suburban nature reserve becomes a forest island. Understanding the "island effect" may be extremely important for sustaining the diversity of salamanders or ant-dispersed woodland wildflowers such as violets and trilliums, for example, since these groups have limited abilities for dispersing across large areas where the forest has been cleared. The island hypothesis doesn't seem to apply to highly mobile organisms such as songbirds, however, because they can easily cross the developed spaces between patches of forest. It particularly doesn't seem to be applicable to Neotropical migrants, which fly hundreds or thousands of miles southward for the winter. Although the adult birds typically use pinpoint navigation to return to the same areas where they nested the year before, the young of the year disperse widely across the landscape. They probably find every patch of good breeding habitat as they move northward across the continent, so it is difficult to believe that urban woodlands are truly isolated for these species. Urban woodlands such as those in New York's Central Park often have an exceptionally high density of Neotropical migrants during spring migration when migrants briefly stop to feed and rest, so migratory birds clearly find these woodland patches. Why do so few stay to nest in what appears to be appropriate habitat?

Another possibility is that these species declined not because of loss of winter habitat and not because the sites became too geographically isolated, but because the habitat conditions inside small woodland reserves are not as favorable for nesting as the conditions within extensive forests. In small forests, the interior is more exposed to wind and sun from the edge of the forest, and drier conditions may reduce the density of insects, which are the main summer food for most forest songbirds. The density of insects is lower in leaf litter on the forest floor near the edge of the forest than deep within the forest, for example, and this is correlated with a lower density of ovenbirds, which feed by walking around the forest floor searching for insects.[12]

Although subtle, fine-scale changes in the climate and habitat may be important for some species, the most important environmental problem for songbirds nesting in small patches of forest in eastern North America appears to be low rates of reproduction due to predators and parasites. Small forest islands are invaded by large numbers of medium-sized predators (such as raccoons, domestic cats, and crows) that are associated with the suburban houses or farmland in the surrounding landscape. These predators generally are not a major threat to fast-moving, agile adult songbirds, but they are effective predators on eggs, nestlings, and newly fledged young. In some small forests the rate of predation is so high that few young are produced.[13] After successive seasons of nest failure, adults may disperse to another breeding area, resulting in a steady decline in abundance of particular species of songbirds in the forest.

Small forests may also be invaded by another species that is particularly common in open habitats, the brown-headed cowbird.[14] The cowbird first moved into eastern North America from the Great Plains during the nineteenth century after forests were cleared for agriculture. They are now common wherever there is open habitat with feedlots, cornfields, bird feeders, or other artificial sources of grain or seeds. Cowbirds are "brood parasites" that never construct their own nests or raise their own young. Instead, female cowbirds search for the nests of other species, often observing other birds closely to discover where nests are located and when they are left unprotected by the parents. The cowbird then quickly lays an egg in the host's nest, often after removing a host egg. Cowbird nestlings are larger and more aggressive than the nestlings of many of their hosts. They eat a disproportionate amount of food, resulting in the starvation of host nestlings. Host parents sometimes raise their own young along with a cowbird chick, but the overall effect on reproductive success can be severe. Even if some host nestlings survive, the number of surviving offspring is greatly reduced. Also, the consequences are more severe than for predation because songbirds quickly build a new nest and try again after their eggs or nestlings are taken by a predator. Most forest songbirds do not respond to cowbird parasitism by renesting, however, so they do not have a second chance for a successful breeding season.

Even if small patches of forest have favorable breeding habitat and low densities of predators and parasitic cowbirds, they may not be as attractive to nesting birds as large forests. Forest birds live not only in an environmental landscape containing food and nest sites, but also in a social landscape containing competitors and potential mates. This is a relatively new way to view forest songbirds. In contrast, it has long been known that birds that nest in dense colonies are attracted to other birds during the breeding season. The colony provides protection from predators because there is safety in numbers; predators can only prey on a small proportion of a large population. A lone pair is unlikely to be successful at raising young. Arctic terns were lured back to an island off the coast of Maine, where they had been extirpated by the 1930s, by setting up numerous tern decoys with recordings of nonaggressive tern calls to create the illusion of a dense colony.[15] Clearly the social setting is as important to terns as the availability of good sites for constructing nests or the abundance of fish in the water surrounding an island. This approach seemed to be irrelevant to forest birds, which live in spacious, apparently self-contained territories, each of which supports a breeding pair. If this description of the behavior of forest birds were complete and accurate, then the minimum area of forest occupied by particular species should be set by the size required for a single breeding territory.

Our view of bird territories derives from intensive observations of particular individual birds. Birds are marked in some way (usually with brightly colored bands or rings on the legs, with different color combinations for different individuals). By mapping the movements and activities of individuals, a researcher can work out the boundaries of each breeding territory and determine which male is mated to which female. Intensive studies of this sort confirmed the view that most forest bird species have self-contained territories that support a single, monogamous breeding pair and their young.

These studies occasionally revealed birds leaving their territories and mating with individuals other than their mates. And once molecular genetics methods could be used to test paternity, it became obvious that many species were not as monogamous as previously assumed.[16] Many songbird nests contained nestlings fathered by males on adjacent or even distant territories rather than by the female's mate. Another technological advance revealed what was happening. Tiny transmitters with short antennae were attached to the backs of birds so that their movements could be tracked with a radio receiver throughout the day. This showed that songbirds often range beyond the boundaries of their territories. In some cases they appeared to be searching for food, but often they were mating with individuals other than their mates. Both males and females leave their territories for this reason. Also, radio transmitters revealed that after the young fledge, they typically move well beyond their parent's territory and may end up hiding and feeding in habitats such as forest openings.[17] They also may join highly mobile flocks with birds of the same and other species. All of these observations suggest that breeding territories are not as self-contained as previously assumed. If songbirds partly depend on a larger area of habitat than the vigorously defended breeding territory, and if they potentially benefit from mating with more than one individual, then they may be attracted to a cluster of territories of the same species. Small forest remnants may not be large enough to support these clusters and permit wide-ranging movements of birds beyond the territory boundaries. The most conclusive and direct evidence for why birds decline in small forests comes from research on predation and cowbird parasitism, but the social neighborhood and the availability of food and other resources outside the territory boundary may also be important.

A study of black-throated blue warblers in the White Mountains of New Hampshire provides new insights into how birds select breeding habitats.[18] Late in the breeding season, males that had successfully fledged young sang at a rate more than five times greater than did males who had not produced young. Successful males may sing frequently to model songs for their young (which may be learning the song pattern of their species) or to reaffirm possession of a territory for the next breeding season. Also, males that do not produce fledglings are known to leave their breeding territories, so the territories become silent during the late breeding season. The distinct difference in singing activity between successful and unsuccessful territories occurs during a period when young birds are moving out of their parents' territories for the first time. They may be searching for potential nesting sites where they can settle after they return from their first winter in the West Indies or Central America. Because successful males tend to sing, an area of forest with lots of singing males represents good breeding habitat.

To test whether young birds tap into this information about the quality of breeding habitat, the researchers set up CD players that broadcast the sounds of singing males or (in other plots) the calls and songs of successful parents and their nestlings. These sounds were played throughout the day for four to six days at each of 18 sites. Hand-painted decoys of male black-throated blue warblers were also added to these experimental plots to enhance the illusion of occupied territories. These simulations of groups of apparently successful birds were set up in areas where no black-throated blue warblers were nesting during the year of the study.

The results were surprisingly clear. Young birds that had recently left their nests were found more frequently at sites with recorded songs and calls than at control sites with no recorded sounds, suggesting that young birds were drawn to these areas. Even more remarkably, during the next breeding season many birds established breeding territories in the areas where recorded songs were played, even though the recordings had not been played since the previous summer. Very few birds established territories in the control sites that had no recordings; four times as many birds settled in the sites where recordings had been played. The study sites did not fit the profile of preferred habitat for black-throated blue warbler, with a sparser shrub layer than one normally finds in breeding territories of this species, so, to paraphrase the title from the paper on this study, social information trumped habitat information in the selection of breeding territories.[19] As the authors point out, this means that birds can respond quickly to changes that affect nesting success in different habitats. If a change in climate or in insect densities results in a shift in the best type of vegetation for nesting, then the black-throated blue warbler population will quickly track this shift to a new habitat. On the other hand, if high parasitism and predation rates or other problems reduce nest success in small patches of forest, then young birds will not be attracted to these sites because few males will be singing after the nesting season. Consequently, the population will decline and eventually disappear.

This may be what happened as the small Audubon census sites were increasingly affected by suburban development in the surrounding landscape. Also, remnant patches of forests have small population sizes, so even if resident birds nest successfully, the resulting post-breeding song chorus will be muted compared with the many singing birds in the interior of a large forest where many territories are clustered together. In addition, forest remnants may be too isolated for young birds to easily find in their forays away from their parents' territory. Although songbirds can travel great distances, young birds may not be prone to fly over large expanses of open habitat where they are vulnerable to predation by hawks. If they stay close to

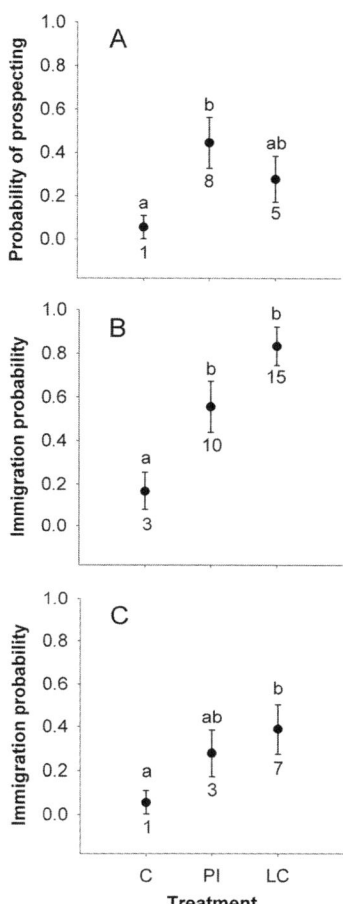

Figure 15. Responses of black-throated blue warblers to forest sites with location cues (LC, playback of songs and presence of models of male black-throated blue warblers), public information cues (PI, recordings of the begging calls of fledglings and the songs of adult males, along with models of females and fledglings), and no playbacks or models (C, controls). All playbacks occurred late in the breeding season. Graphs show the proportion of sites (a) visited by black-throated blue warblers of either sex during the late summer when playbacks occurred, (b) occupied by males during the subsequent spring, and (c) occupied by females during the subsequent spring. Values below error bars indicate number of sites visited or occupied. Error bars with the same letters are not significantly different. (Betts et al., 2008; reprinted by permission of the Royal Society)

cover, then they would be unlikely to discover isolated islands of forest surrounded by farmland or suburbs.

Importance of Large Forest for Bird Conservation in Eastern North America

Whatever causes forest birds to avoid or suffer population declines in small
forest reserves, the result is a distinct difference in bird communities in
small and large forests in eastern North America. Compared with the interior of extensive forests, small, isolated patches of forest generally have a
low diversity of bird species that are restricted to mature forest. Small forests have high densities of more generalized species such as chickadees
and titmice that are found in a wide range of habitats (including gardens in
wooded residential areas, for example), but fewer forest specialists. A number of studies from different regions in the eastern deciduous forest reveal
a clear relationship between forest area (the amount of continuous woodland in a forest) and the diversity and abundance of specialized forest
birds. This pattern holds up even if one compares similar areas (such as
100-meter-radius circular plots) in large and small forests. The plots in large
forests have more species of forest specialists, and the abundance of particular species is higher. When my students and I surveyed plots in 46 forests
in southeastern Connecticut, for example, we heard few or no forest specialists in the smallest forest patches.[20] Species like wood thrush, red-eyed
vireo, and ovenbird were missing or infrequent in small forests, but were
consistently present—typically at high densities—in larger forests. Some
species, such as cerulean warbler, were present only in the largest forests.

A similar relationship between the density of forest birds and the size of
forest patches has been found in many parts of the eastern deciduous forest,
from Wisconsin and Ontario to Maryland.[21] The most thorough study was
completed by Chandler Robbins and his colleagues, who counted birds in
271 forests in western Maryland and adjacent states.[22] They used a standard
method in which an observer stands at a survey point for 20 minutes and
records all of the individual birds that are seen or heard. They found that
numerous species of forest specialists consistently increased in density (as
measured by the average number per survey point) as forest area increased.
Scarlet tanagers, for example, were present in only 10 percent of the survey
points in forests smaller than 2 hectares, but at almost 90 percent of points
in the forests larger than 1,000 hectares. Consistent results of this sort across

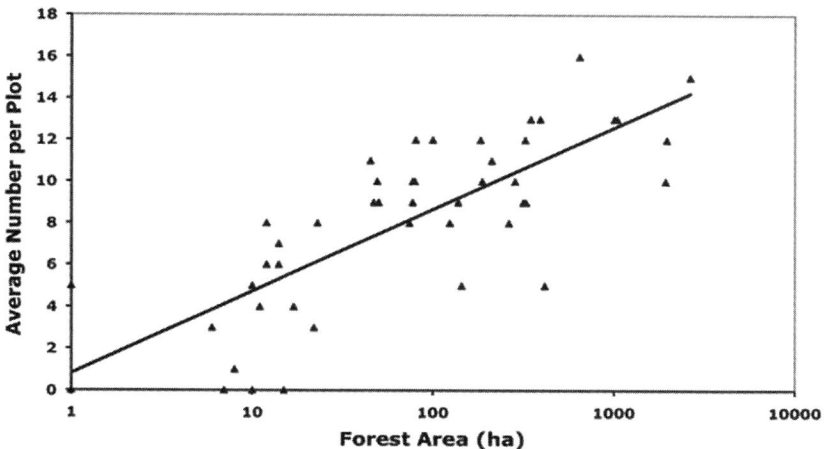

Figure 16. Relationship between the area of forest (in hectares) around a bird survey point and the average number of mature-forest birds that were detected during a 20-minute survey. Only species that are mature-forest specialists were included. (Reprinted from Askins et al., 1987, with permission from Elsevier)

numerous studies indicate that small forests simply do not provide good habitat for many species of migratory forest birds.

Large forests not only support diverse communities of specialized forest birds, but may also produce the young birds that support regional populations of particular species, including the populations in small forests. Nest success is so low in many small forests that adults probably do not replace themselves during their lifetimes. The persistence of forest birds at some of these sites suggests that they are maintained by immigration from sites with better nesting success. The best evidence for this (albeit indirect) comes from coordinated studies by five sets of researchers who worked in widely dispersed regions across the American Midwest.[23] Study areas in northern Wisconsin and southern Missouri were in regions with more than 90 percent forest cover, while the other study areas were in predominately agricultural areas. Forest birds in the heavily forested regions generally produced more young than were necessary to offset the estimated adult mortality each year, while those in forests surrounded by farmland did not produce a sufficient number of young birds to compensate for annual mortality. It appears that heavily forested regions not only sustain local populations of forest birds, but also help sustain populations in small woodlots and forest reserves throughout the Midwest. The extensive forests of the Appalachian Mountains and foothills may play a similar role along the eastern coast of North America. Thus, areas with extensive, unbroken forest may support "source populations" that export young birds to other forests with less favorable conditions for reproduction. Small forests may have "sink populations" that depend on immigrants because they don't produce enough offspring to be self-supporting. If this is the case, then large forests in heavily forested regions are critical for conservation of forest birds, and they should be protected from fragmentation caused by roads, powerlines, and residential development.[24]

There is a general perception that wildlife and forest managers respond only slowly to new insights from ecological research because of a lack of communication between researchers and the people who manage conservation areas.[25] This clearly has not been the case with songbird conservation in North America, however. Especially after the 1970s, research results have been rapidly incorporated into new approaches to conservation. This change was largely due to the Partners in Flight Program, an international effort that brought together researchers and conservationists from throughout the Western Hemisphere to protect all of the links in the migratory systems of songbirds—breeding areas in the north, winter areas in the south, and migratory stopover areas in between.[26] In the eastern deciduous forest the main focus of this collaborative effort among federal and state conservation agencies, nonprofit conservation organizations, and researchers has been protection of breeding habitat for forest songbirds, and large blocks of unfragmented habitat have become a priority. This is reflected in the "matrix forest" program of the Nature Conservancy, where the goal is to focus conservation efforts on protecting the remaining areas of extensive, unfragmented forests in the northeastern United States.[27] This not only reflects research on the distribution of forest birds, but also takes into account insights about ecological processes that take place across regional landscapes. Although a storm or fire may completely level a small forest, in an extensive forest it will merely create openings that increase habitat diversity. Also, extensive forests can encompass the home ranges required by large animals such as bobcats and black bears. The Nature Conservancy program is unusual in placing such a heavy emphasis on the importance of sustaining unfragmented forests, but conservation plans of many state and federal agencies have been influenced by these landscape considerations as well as by more traditional goals of timber management, game management, and protection of endangered species.

Protecting large areas of breeding habitat will not be sufficient for preserving populations of migratory forest birds, however, because they also depend on tropical habitats for much of the year. Some of the species that require mature forests for nesting use a much wider range of habitats in the winter, including gardens, coffee and citrus plantations, second-growth forest, and even recently abandoned fields and cattle pastures.[28] Some of these species are nomadic during the winter, opportunistically moving between different sites and habitats in search of food.[29] Other species remain at sites in specific types of natural habitat during the winter and so are more vulnerable to habitat destruction in the tropics. Wood thrushes, for example, primarily spend the winter in mature tropical rainforest or moist forest in southern Mexico and Central America, regions that have suffered severe deforestation during the past half century.[30] Northern waterthrushes are primarily found in coastal mangrove swamps in the West Indies and coastal Central and South America, and cerulean warblers are largely restricted to mid-elevation tropical forests in the northern Andes.[31] The preferred habitats of these three species have been extensively cleared for agriculture or (in the case of the northern waterthrush) shrimp aquaculture.

Wood thrushes and cerulean warblers experienced substantial and statistically significant population declines between 1966 and 2007 on Breeding Bird Survey routes in Canada and the United States.[32] This survey is a system of nearly 4,000 roadside bird survey routes that provide our best picture of population changes for particular species of birds across temperate North America. Most woodland bird species have not declined across the eastern United States and Canada along these routes, perhaps because abandoned farmland has grown up into forest in many regions.[33] The decline of two species with relatively specialized winter habitats is therefore notable. These species are also sensitive to forest fragmentation in their breeding ranges, however, so they may have declined because of a combination of factors.

Shifting the Perspective to Japanese Forest Birds

Soon after completing an analysis of the bird communities in forests of different sizes in eastern North America, I had the chance to study forest birds in and around the city of Kyoto, Japan. Kyoto was an ideal place to do this study. Large areas of second-growth deciduous forest are protected on the hills and mountains in and around the city to preserve the setting for ancient Buddhist temples, Shinto shrines, and imperial palaces. As a result, I could survey birds in isolated patches of forest with a wide range of sizes. I used the same bird and vegetation survey methods that I had used to study forests of different sizes in Connecticut to compare 19 forests in the Kyoto area.[34] I coordinated this study with two Japanese colleagues, Hiroyoshi Higuchi and Hidenori Murai, who used the same methods to survey birds and vegetation in 23 forests in Tokyo. Hiroyoshi Higuchi had already found preliminary evidence that several species of forest birds were missing from forests smaller than 100 hectares in central Honshu.[35] In our joint study of forest patches in Kyoto and Tokyo, we tested this conclusion using standardized methods. We counted the number of birds we heard or saw during two 20-minute observation periods at survey points in each forest.

The results for Kyoto, where forests ranged in size from 15 to 2,300 hectares, were similar to the results that my students and I obtained in Connecticut in eastern North America.[36] Forest specialists were significantly more abundant in large forests than in small forests. In contrast, there was no significant relationship between forest area and total abundance of generalized bird species (species that are frequently found in both open and wooded habitats) or forest edge species (species that are more abundant on the edge of the forest than in the forest interior).[37] Several forest species were more abundant in large forests than in small forests, including Eurasian jay, Japanese green woodpecker, pygmy woodpecker, and masked grosbeak. In addition, two forest specialists (blue-and-white flycatcher and coal tit) were completely missing from small forests. Thus, several species were less frequent or absent in small, isolated patches of forest. As in North America, most of these "area-sensitive" species were forest specialists. One difference, however, was that only one area-sensitive species (the blue-and-white flycatcher) in Kyoto migrates to the tropics for the winter, whereas most of the area-sensitive species in eastern North America are tropical migrants.

Interestingly, there was no evidence that forest birds were more common in larger woodlands than in smaller woodlands in Tokyo. The different results for Kyoto and Tokyo may be due to differences in the size and isolation of forests in and around the two cities. The Kyoto region includes many large, protected forests, while most of the woodlands in the Tokyo region are in relatively small parks. Also, the Tokyo sites were more isolated from

Figure 17. View across Kyoto, Japan, showing one of the smaller sites (the wooded park in the middle distance of the photograph) and one of the largest sites (Mount Hiei in the distance) included in a study of the abundance of forest birds in forests of different sizes. (Askins et al., 2000)

other forests. On average, 47 percent of the area within two kilometers of the Kyoto sites was covered with forest, compared to an average for Tokyo sites of only 8 percent. The Tokyo sites may display little variability in the number of forest birds because they are nearly all small and isolated. All except one of the Tokyo sites is smaller than 480 hectares, and when we analyzed the distribution of birds in the sites smaller than 480 hectares in Kyoto, we found no relationship between the abundance of mature-forest birds and the area of forests. Thus, forest fragmentation may have had such a pervasive negative effect on the abundance of forest birds in the wooded parks of Tokyo that it is now difficult to detect differences among forest patches of different sizes.

To understand the effect of forest fragmentation on bird communities in Japan, it is important to get away from the densely settled metropolitan areas and complete studies in areas that have larger forests that are less disturbed. A study in southern Hokkaido provides additional evidence that some species of Japanese birds need large areas of continuous forest.[38] While working on her master's degree with me at Connecticut College, Reiko Kurosawa completed standardized surveys of birds on circular plots with a 50-meter radius in sites with broad-leaved deciduous forests. The 53 sites showed an impressive range in area, from 1 to 5,045 hectares. The abundance of forest birds detected from survey points increased with forest area. Eleven species were absent or less abundant in smaller forests. As in eastern North America, many of these area-sensitive species in southern Hokkaido are forest specialists and tropical migrants.

Although there is compelling evidence that some Japanese birds require large, continuous areas of forest, the reason that they are infrequent or absent in apparently good habitat in small forests has not been determined. In contrast to eastern North America, in Japan brood parasites are not primarily associated with the forest edge or the interior of small forests. In place of the edge-loving cowbird of North America, the major brood parasites of forest songbirds are three species of forest-dwelling cuckoos, the oriental cuckoo, northern hawk-cuckoo, and little cuckoo. Each of these species has specific hosts that have evolved defenses against their brood parasites. In central Honshu, for example, the primary hosts are eastern crowned leaf-warblers for oriental cuckoos; Siberian blue robins and blue-and-white flycatchers for northern hawk-cuckoos; and Japanese bush-warblers for little cuckoos.[39] In most cases the egg of the cuckoo mimics the egg of its primary host; little cuckoos lay solid reddish brown eggs like those of Japanese bush-warblers, and northern hawk-cuckoos lay blue eggs like those of their primary hosts. Despite these adaptations, the hosts tend to reject cuckoo eggs, so the density of cuckoos and the proportion of host nests that they parasitize are relatively low. Also, in forests in southern Hokkaido the most abundant brood parasite, the oriental cuckoo, is a forest specialist that is found more frequently in large forests than in small forests.[40] It is unlikely, therefore, that brood parasitism accounts for the low densities of forest birds in small forests.

A more likely problem in smaller forests is a high density of predators that prey on the eggs and nestlings of forest birds. Jungle crows are an important nest predator in Japan, and in southern Hokkaido jungle crows are significantly more abundant on the forest edge than in the forest interior, and in small forests than in large forests.[41] Thus nest predation by jungle crows might make small forests unfavorable breeding sites for many species of birds. Jungle crow populations have increased dramatically in Japan because of the availability of artificial food sources such as garbage, so their impact on populations of forest songbirds may be increasingly severe.[42]

It is not known whether other native Japanese predators such as red foxes, Japanese martens, raccoon dogs, mice, and snakes have higher densities in small forest patches than in large forests. It is important to determine whether they are more active along the forest edge than in the forest interior and whether they cause a higher frequency of nest failures in smaller forests. Like the oriental cuckoo, some of these species may be forest specialists that have coexisted with forest songbirds for a long period and do not threaten their populations. Predators are more likely to present a problem if their densities have been boosted because of human activities. Researchers should especially focus on predators that thrive in agricultural or residential areas because they benefit from new food sources such as garbage or grain left in fields after harvesting, or because larger predators such as wolves have been extirpated. Inflated populations of small native predators could result in population declines in vulnerable bird species, especially species that nest close to the ground and have open cup nests (rather than nests that are covered with a domed roof or nests located in cavities in trees).

Another threat to forest songbirds in Japan comes from introduced predators. In North America, raccoons and domestic cats are two of the most serious predators of eggs and young birds. Raccoons, in particular, are believed to be one of the major causes of the decline of forest birds in regions where the forest has been fragmented. Domestic and feral cats are also common in Japan, and unfortunately raccoons have been introduced to several regions in Japan.[43] In 1962 raccoons escaped from a zoo in Aichi Prefecture and established a wild population. Raccoons are known to have escaped or have been abandoned in 17 prefectures. This is a potentially serious situation because raccoons not only cause serious crop losses for farmers, but they may also have a major impact on populations of birds and other native animals.

It is also important to determine if there are other problems for forest birds living in smaller forests, such as a relatively low density of food, a lack of appropriate nesting sites, more frequent disturbance by people, or the absence of sufficiently large populations to sustain important social interactions within a breeding population of a particular species. Discovering the reason that some species of birds are less abundant in small forests might reveal how forest reserves in Japan should be managed to provide good habitat for forest birds.

Many species of forest birds in Japan may also face a threat from destruction of habitat that they need during the winter and during their migratory journey to the tropics. Since the 1970s forest harvesting has declined in many heavily forested parts of Japan because of competition with cheaper timber products from North America and Southeast Asia. Along with the abandonment of farmland and the decline of coppice cutting for fuel and building materials, the decline in timber harvesting has led to an expansion of mature forest, which should provide additional habitat for forest birds. Partly because of demand for wood from Japan, however, there has been extensive deforestation in Southeast Asia.[44] This is a major overwintering area for many long-distance migratory birds (the "Indomalayan migrants") that nest in Japan. Yuichi Yamaura and colleagues predicted that migratory forest migrants might have declined in Japan despite the recovery of forests in their breeding areas because of loss of tropical forests used as winter habitat.[45] In contrast, nonmigratory birds (which remain in their breeding areas during the winter) and short-distance migrants (which migrate farther south or to lower elevations in Japan during winter) may have increased because their habitat has improved at all seasons. They tested this prediction by using data from nationwide bird surveys that were completed in 1978 and 1997–2002. The analysis revealed that the average range size of forest bird species that are residents or short-distance migrants increased, while the ranges of Indomalayan migrants shrank.[46] Several other analyses of long-term trends in Japanese bird populations also indicate that this group of species has declined in recent decades.[47] These results suggest that loss or degradation of winter habitat may be a major factor in the decline of Indomalayan migrants, but the causes of these declines need to be investigated more thoroughly to ensure that the most effective conservation actions will be taken. Even without further research, however, it would be judicious to prevent fragmentation of large areas of continuous forest in Japan by roads and development, and to help with efforts to establish national parks and develop sustainable forestry in the wintering areas for migratory birds that nest in Japan. These efforts would have many environmental benefits in addition to bird conservation.

Searching for General Patterns in the Response of Birds to Forest Fragmentation

The goal in ecology, as in any branch of science, is to develop general theories that can be applied in a wide range of situations. It would be much easier to understand and manage natural systems if research on ecological patterns in one region could be applied to other regions. For example, can the extensive research on the effects of habitat fragmentation on forest birds in the eastern United States and Canada be applied to birds in similar deciduous forests in East Asia and Europe? Are large, continuous forests needed to protect specialized forest birds in all of these regions? Are smaller forests consistently less valuable as breeding habitat for these species for the same underlying reasons, such as higher predation and parasitism rates in small forests or avoidance of sites without a cluster of territories of birds of the same species? If the patterns are consistent across regions and between continents, then we could develop a convenient set of guidelines for conservation of forest birds in temperate deciduous forests.

Unfortunately, ecological patterns are seldom this straightforward and predictable, and responses to habitat fragmentation by birds in different regions and habitats are highly variable.[48] The density of forest birds and the number of species per survey plot increase with forest size in southern Hokkaido and Maryland, but not in Tokyo or southwestern Ontario. Even in North America, forest specialists nesting in small forests do not always have low nest success rates. In some cases birds in small forests have similar nest success rates to the interior of large forests; this is true, for example, for worm-eating warblers in western Connecticut and wood thrushes in southwestern Ontario.[49] As Frank Thompson and his colleagues argue in their review of the effects of forest fragmentation on birds in different parts of eastern North America, the nature of forest islands is determined by the landscape in which they are embedded and by even larger scale patterns in the distribution of species.[50] Cowbird parasitism, for example, will not be a problem in landscapes that don't have farmland or open fields where cowbirds feed. The parasitism rates for one well-studied species, the wood thrush, reflect regional differences in cowbird abundance. Rates of cowbird parasitism vary from 80 percent of nests in the agricultural Midwest to 20 percent in many parts of the more heavily forested East Coast.[51] Consequently, forest birds face different levels of risk in different geographical regions.[52] In landscapes with more than 90 percent forest cover and very little open habitat, the rate of cowbird parasitism may be low in both small and large forests. In regions with a highly fragmented mosaic of farmland and small woodlots with a total forest cover of less than 15 percent, cowbird parasitism may be frequent in all forests, even in the center of the largest blocks of forest. The proportion of nests parasitized by cowbirds is most likely to show a clear relationship with the size of forests in intermediate situations with about 50 percent forest cover and some cowbird feeding areas. In these landscapes, cowbirds tend to lay their eggs in nests of birds with territories on the forest edge, close to preferred feeding areas. Birds in the interior of large forests are largely buffered from cowbird parasitism because the relatively small number of cowbirds supported by this landscape can find a sufficient number of host nests without commuting deep into the forest interior where they are far from their feeding areas.

Predation on nests may vary in a similar way across landscapes, with higher predation rates in landscapes that have more open habitat and less forest cover. In extremely open or highly developed landscapes predators may be so abundant that they penetrate into the interior of even the largest forests, so there is little difference in the bird populations of large and small forests. Where their regional density is low, they have little impact on birds nesting in small, isolated forests. This may be the reason that specialized forest birds could survive in the northeastern United States in the eighteenth and nineteenth centuries, when much of the region was cleared for farming and woodland was restricted to relatively small, islandlike patches. Cowbirds had not yet spread east into this region from the prairies, and populations of predators (such as raccoons and crows) were suppressed by hunting and trapping by a dense population of farmers who protected their crops and livestock. Today the landscapes in this region generally support high densities of cowbirds and medium-sized predators. Hence, there is no reason to necessarily expect that the effects of habitat fragmentation on forest birds will be similar at different times or in different places. The effects of forest fragmentation in Asia and Europe may be distinctly different from the typical pattern in eastern North America. This may be a predicable result of differences in the abundance and distribution of predators, which may largely depend on how human activities reduce or boost predator populations.

Forest Fragmentation and Woodland Birds in Europe

Some of the key studies of the effects of forest fragmentation on birds in western Europe were completed in regions of England and the Netherlands with less than 2 percent forest cover.[53] Based on the examples from North America and Japan, one would not expect a strong difference in bird populations in small woodlands and large woodlands in a landscape with this much deforestation. Moreover, in some of these studies even the largest woodland patches were small compared to woodland patches in North American and Japanese studies. For example, in a study of the relationship between woodland size and abundance of particular species of birds in 160 woods in southwestern England, the smallest woodland patch was 0.02 hectare and the largest was 10.3 hectares.[54] In contrast, the largest forest patches in most North American and Japanese studies cover hundreds or thousands of hectares.[55] Analyses of bird distributions in woodlands that are 10 hectares (25 acres) or smaller necessarily focus on distinctly different ecological processes than analyses that include much larger forests. A tiny patch of woods may be too small to support a single breeding territory for a particular type of forest songbird. Also, a tiny patch will have little or no forest interior habitat that isn't influenced by a habitat edge with surrounding farmland or residential areas, so it may primarily support bird species that thrive in edge habitats. In the range of 0.02 to 10 hectares, the comparison is effectively between pure edge habitat and small woodlands dominated by edge habitat rather than a comparison between the interiors of small and large forests that characterizes most North American and Japanese studies of forest fragmentation. Not surprisingly, these European studies generally show only weak relationships between woodland size and the abundance of particular species of woodland birds.

A few European studies have compared small woodlands with much larger forests that cover hundreds of hectares, but even these studies indicate that the bird communities in small and large forests are similar, suggesting that most European bird species do not require extensive, continuous forests. As Hugh Ford pointed out, Britain appears to be lacking "big forest" bird species.[56] Moreover, European ecologists usually do not classify birds as "forest specialists" or "forest-interior species" because so few species of birds would qualify for this designation. Most of the birds in populated parts of western Europe would be classified as generalists in North America, so it isn't surprising that they are not restricted to large patches of forests. They are flexible enough to use the forest edge and the gardens and hedgerows outside the forest. For generalists, woodland may merely be one of many favorable habitats in a complex mosaic of farmland, woods, and towns. These generalized species may be equally abundant in forests of different sizes or more abundant in small forests than in large forests. For true forest specialists, however, a forest reserve is an island of favorable habitat surrounded by unsuitable and even dangerous habitats, so the area of the forest is a potentially important factor. Forest specialists such as northern goshawk, capercaillie, Eurasian golden oriole, and pied flycatcher are now so uncommon and localized in Great Britain and other parts of northwestern Europe that they usually are not detected in regional surveys of forests of different sizes and so are not included in analyses of the effects of forest area on birds.[57] Perhaps the specialized bird species that are most sensitive to forest fragmentation have already disappeared from much of western Europe.

A key question for conservationists is whether it is better to protect a single large reserve that encompasses an extensive area of continuous natural habitat or a set of small reserves that occupy somewhat different habitats. In England and the Netherlands, the total number of bird species may be smaller in a single large forest reserve than in a set of small forest reserves of the same combined area.[58] A set of small woodlands in scattered locations may have greater habitat variability than one finds within a single large forest, and they may have more edge-loving and generalized species. The pattern is similar in North America if all bird species, including both generalists and forest specialists, are considered together. Survey plots in small forests have an equal or greater number of species compared with survey plots in the center of large forests.[59] This broad-brush analysis of the total number of species obscures the distinct difference between the bird communities of small and large forests in North America, however, because small woodland patches are dominated by generalized species that are common in suburban gardens and they are missing forest specialists that are restricted to larger forests. In contrast, in England and the Netherlands, forests of different sizes have similar types of birds, and most species do not appear to respond to the size of the forest, so large forests aren't as important for maintaining biological diversity.

There are a few species of European birds that have lower frequencies in small woodlots than in extensive forests, however. For example, when Hugh Ford counted birds in 20 woods ranging in size from 0.14 to 18 hectares near Oxford, England, he found great spotted woodpecker, Eurasian jay, and coal tit only in larger woods.[60] Interestingly, the ranges of all three of these species extend across Eurasia to Japan, where each of these species is more frequent in truly large forests than in smaller forests.[61] Similar surveys in the Netherlands indicated that a number of species of "mature forest birds" was greater in large forests than in small forests, and that particular species (such as great spotted woodpecker, short-toed treecreeper, and Eurasian nuthatch) occur less frequently in smaller forests than in large forests.[62] Again, the largest forests were relatively small (20 or 39 hectares), and smaller woodland patches may have been too small to accommodate a single breeding territory for some species. Great spotted woodpeckers, for example, were absent from woodland patches smaller than 1 hectare, which is considerably smaller than the typical home range size of 4–60 hectares for a breeding pair.[63] Research near Stockholm, Sweden, where it was possible to compare forests of a greater range of sizes, from 2 to 700 hectares, provides better evidence that some species are restricted to larger forests than would be expected on the basis of the size of their breeding territories.[64] Willow tits, crested tits, and coal tits were found only in coniferous forests larger than 25 hectares, while marsh tits were found only in deciduous forests larger than 10 hectares.

These studies all analyzed the frequency of occurrence of particular types of birds in forest patches. A limitation of this approach is that sampling was not standardized in woodland patches of different sizes. Researchers spent more time in larger patches to ensure that the bird censuses were thorough. The result is that more species would have been detected in larger forests purely by chance because observers were covering more ground and spending more time in these sites.[65] Less common species would be especially prone to being detected only in larger sites even if they don't actually prefer large forests. As any birder knows, the chance of seeing unusual species increases as one spends more time in the field. This makes the results of these studies difficult to interpret. This problem was avoided in North American and Japanese studies by comparing counts from the same type of survey plots in forests of different sizes. The occurrence of birds in these standard plots (not in the forest as a whole) was then compared for forests of different sizes. If the frequency of occurrence of a particular species is greater for plots in larger forests, then we have conclusive evidence for the importance of large forests for this species. Many of the European studies are not standardized in this way, so, although it appears that some species require larger forests, the results are not conclusive. We can't rule out the possibility that the species was only detected in larger forests because larger forests were sampled more intensively.

Another way to address this problem is to determine if the density of particular types of birds differs in small and large forests. This is a more intensive and dependable assessment of bird distributions than surveying standardized plots that sample one or a few points in each patch of forest. Using this approach in the Netherlands, Erik Matthysen found that the average density of Eurasian nuthatches was 0.34 per hectare in forests larger than 100 hectares, but ranged from 0.13 per hectare to 0.27 per hectare (depending on the type of woodland) in forest fragments of 1–30 hectares.[66] The lower densities in fragments are not caused by low survival or reproductive rates at these sites, however. This is not surprising because Eurasian nuthatches nest in cavities in trees, so their nests are much better protected from predators than the open cup nests of many forest birds.[67] Instead, forest fragments have relatively low nuthatch densities because few young birds establish territories at these sites. The low recruitment of young birds in small, isolated woodlots may result from high mortality when they travel away from the woods where their parents nested. They must cross long stretches of open farmland and urban areas before finding a potential territory of their own. The result may be high death rates for these relatively inexperienced birds. This hypothesis is supported by evidence from color-marked birds. Young birds that were raised in nests within the set of small forest patches were labeled with a set of colored leg bands (rings), with a different color combination for each individual. Marked first-year birds dispersing from territories within the set of small wooded patches in late summer accounted for only part of the new recruits that sustained the population and compensated for mortality of adults each year. Many new recruits did not have color bands and may have dispersed during the autumn or winter from larger forests to the north.

The Eurasian nuthatch study suggests that forest area is a factor that should be considered by people trying to protect woodland bird populations in Europe, but the effects of forest area are more subtle and mild than they are in North America and Japan. In North America, in particular, conservation organizations and land managers work with the knowledge that an entire group of specialized forest birds depends on maintenance of extensive, unfragmented forests that cover hundreds or thousands of hectares. This clearly is not the case for deciduous forests in western Europe. Why should European birds differ in this respect?

Why Are European Birds Tolerant of Forest Fragmentation?

There are several reasons that woodland birds can survive in regions with only tiny remnants of deciduous forest, but the most compelling explanation comes from a consideration of the history of European forests.[68] As we have seen, European forests have been severely disrupted during long spans of geological time and during the much shorter span of human history. During the repeated advances of continental glaciers across northern Europe, deciduous forests were restricted to small, relict patches on the southern edge of the continent. Highly specialized birds that required extensive areas of continuous forest probably became extinct during this period or evolved more generalized behavior.[69] Severe forest destruction again occurred as intensive agriculture spread across Europe during the last few millennia, leaving few large, continuous forests. It is unknown whether some specialized forest birds became extinct because of the spread of farming, but the patchy distribution of species such as white-backed woodpecker and black woodpecker suggest that at the least regional disappearances occurred.[70]

Most of the birds recorded in archaeological and paleontological sites in Great Britain during the Mesolithic (the part of the current interglacial period before farming was introduced) are still familiar species of the British countryside.[71] Although researchers have speculated that forest specialists such as white-backed woodpecker and black woodpecker became extinct in Britain after forest clearing, there is no direct evidence that these species were ever present. Woodpeckers are not well represented in archaeological sites, however, so the lack of evidence is not conclusive. Among the Mesolithic British birds that live in forests, only the Eurasian eagle-owl and capercaillie are known to have disappeared from Britain, and the latter was subsequently reintroduced.

In contrast to western Europe, relatively large areas of deciduous forest survived in North America and Japan even when continental glaciers reached their maximum extent, and even in the wake of extensive agricultural clearing. Specialized forest birds had refuges in both of these regions until recently, when they became threatened as networks of paved roads brought fingers of intensive development deep into the remaining forest.

Because of Europe's history of frequent and severe forest destruction, the forest-dwelling birds that survived there are mostly generalized species that use a variety of habitats and have wide distributions.[72] In contrast there are a larger number of specialized forest species in North America and Japan. Many of these are related to species that inhabit tropical forests to the south, which also tend to be highly specialized to particular habitats.[73] Also, the surviving European birds may have slowly adapted to a landscape of patchy forests and extensive open areas.[74] Species that require large forests in Japan are found in surprisingly small forests in Europe. The smallest forest in which black woodpeckers were detected, for example, was 579 hectares in Japan and only 11 hectares in Poland.[75]

Whether birds require large forests is determined not only by the adaptations of particular bird species, but also by the nature of the landscape surrounding forests. Agricultural areas in the Midwest of North America and urban areas in the Tokyo metropolitan region may represent "hostile landscapes" for breeding birds, with high densities of nest predators and (in North America) parasitic cowbirds. Birds that attempt to nest in small forests may have very little success. In landscapes with few predators or parasites, however, small forests may be much better places to nest. In Europe, rates of nest predation are generally higher in forests than in areas settled by humans, so the agricultural or urban landscape is not a source of invading predators.[76] Long periods of intensive human land management apparently have reduced the density of predators, and this may open the way for forest birds not only to tolerate adjacent farms and villages, but also to move out of the forest into these artificial landscapes.

Winter Wrens Around the World

North of Kyoto, Japan, a footbridge crosses a clear mountain creek in the village of Kibune, leading to a trail that climbs through tall forest to a series of Buddhist shrines and temples along a mountain ridge. Crossing this bridge on a sunny spring morning, I heard the familiar trills and warbles of a winter wren coming from a dense stand of streamside shrubs. I have heard similar winter wren songs along a stream in a 2,000-year-old redwood forest in California, in an ancient deciduous forest in Poland, and in hemlock ravines in Connecticut, as well as in the distinctly different setting of an English garden. Winter wrens are the most widely distributed type of songbird in the Northern Hemisphere, found from Newfoundland and Alaska to Georgia, and from Kamchatka and Iceland to North Africa.[77] "Winter wren," which is the North American common name, fits for the region between southern New England and the Gulf Coast, where this species spends the winter after migrating south from the boreal forests of Canada, but it is a misnomer for regions where these wrens are summer or year-round residents.

At the time I first sketched out this chapter, the winter wren was considered a single species, with distinctive subspecies in different parts of North America and Eurasia, as well as on various islands off the coasts of both continents.[78] All populations are similar in appearance and they have similar, intricate songs. In 2010, however, the American Ornithologists' Union decided that these populations should be split into three species, with different species in eastern North America, western North America (west of the Great Plains), and Eurasia.[79] These new species are called winter wren, Pacific wren, and Eurasian wren, respectively. Despite this reclassification, the three different populations are closely related and are clearly descended from a recent common ancestor, so to prevent confusion I refer to all three types collectively as "winter wrens."

Judging by their present distribution, wrens probably evolved in the Western Hemisphere, which has numerous species in addition to the two types of winter wren. Wrens reach their maximum diversity in the mountains of Central and South America, where more than 20 species may be found in a single region.[80] Only one species of wren (called simply "wren" in British bird books) is found in Europe and Asia. It appears that the ancestors of Eurasian wrens invaded Eurasia from North America, probably by crossing the Bering Strait from Alaska.[81]

The various populations of winter wrens now inhabit a wide variety of landscapes, from heavily forested areas that have never been cleared to largely agricultural landscapes with only relict patches of woods. Their response to forest fragmentation is dramatically different in different regions, however, indicating that they have adapted to human activities in landscapes where forest clearing has occurred for thousands of years. In England they are not only found in woods and on remote sea cliffs, but are also common in gardens, particularly large rural gardens with dense shrub cover.[82] They reach their highest density in damp woods with thick cover, especially along streams. Wrens are by no means restricted to woodlands in England, however, so it is not surprising that they are found in both large and small patches of forest.[83] In fact, studies of large numbers of woodland patches in southeastern England showed that wrens had higher densities in small woods than in larger woods.[84]

In continental Europe the same species (the Eurasian wren) displays a distinctly different pattern. In both Sweden and Italy, for example, wrens have higher densities in larger forests.[85] In aspen-birch forests in Sweden, wrens were completely missing from small forest fragments surrounded by farmland but were present in large, continuous forests with an area of more than 1,000 hectares. In oak-dominated deciduous forests in Italy, the density of wrens was significantly greater in large forests than in small patches of forests. Like the Swedish study, the Italian study included much larger woodlands than the studies in southeastern England, encompassing a range from 0.12 to 536 hectares (as opposed to 0.02–10 hectares in England). Hugh Ford, however, did not find a difference in the density of wrens in small patches of woods and in larger control forests such as Wytham Woods (415 hectares) in southern England, so there appears to be a real difference in the responses of wrens in England and continental Europe.[86]

Based on his research on wrens in Białowieża Forest in Poland, Tomas Wesołowski hypothesized that wrens in Great Britain have adapted to an agricultural landscape with only small, remnant patches of forest.[87] Wesołowski was able to study wrens in forests that are presumably much more similar to their original habitat than any garden or woodland in England. He found that Białowieża wrens are habitat specialists that are concentrated in particular types of mature forest, especially ash-alder and oak-hornbeam forests with numerous fallen, uprooted trees. Most wren nests were located in the tangled roots of uprooted trees. These wrens had substantially larger territories than do English wrens, and on average they raised fewer young during the breeding season. Lower nesting success was due to a high rate of predation on nestlings at the Polish site, which has a much higher density of predators than rural areas in England. Wesołowski hypothesized that the eradication of predators in agricultural areas in England permitted wrens to nest successfully in a wider range of locations and habitats. Individuals with more flexible nesting behavior would reproduce more successfully, especially after mature forests with numerous old, uprooted trees disappeared from the landscape, and this would ultimately lead to the evolution of more generalized nesting behavior. This process probably began millennia ago, slowly converting a bird of deep recesses in old forest into a bird of well-planted gardens and small woods.

In North America the two species of winter wrens are most abundant in mature and old-growth forests with a high density of fallen trees, and thus are more similar to those of Białowieża Forest than to the well-studied wrens of the English countryside.[88] As Wesołowski pointed out, a published description of North American breeding sites fit his Białowieża data so well that "it could be reprinted in my results section."[89] It's not surprising, therefore, that numerous studies of Pacific wrens in western North America indicate that they are mature-forest specialists that tend to be less common in small patches of forest than in extensive forests.[90] There are only a few studies of the distribution of winter wrens in forests of different sizes in eastern North America because most forest fragmentation studies were done south of their breeding range, but, surprisingly, a large-scale study of deciduous forest patches of different sizes in heavily agricultural regions of Ontario (with 11–38 percent forest cover) indicated that the density of winter wrens was not correlated with woodland area.

In Japan, Eurasian wrens are found in moist deciduous and coniferous forests.[91] They were restricted to large forests in Hokkaido, but were detected in only 3 out of 53 forests included in the study, so this pattern was not statistically significant.[92] Thus, the winter wrens of northern Japan, North America, and Poland are much more specialized than are the ubiquitous wrens of England. In contrast to England, in many landscapes in North America the density of nest-raiding predators is higher in areas with extensive forest fragmentation than in the interior of large forests. Unless the density of predators declines, North American winter wrens may never adapt to open landscapes. Conversely, if changes in woodland and game management in England result in substantial increases in predators, then wrens may decline.[93]

Population Declines of Woodland Birds in Europe

Although woodland birds in Europe are more tolerant of habitat destruction and habitat fragmentation than are their counterparts in North America, populations of many species have declined during the past few decades. Based on an analysis of large-scale bird surveys in a large number of European countries, it appears that woodland species suffered widespread declines between 1980 and 2003.[94] Population declines among forest birds appear to be particularly severe in Britain even though the amount of mature broad-leaved forest has increased since 1947.[95] Analysis of several different long-term censuses showed that 10 of the 19 species that primarily nest in forests declined in Britain.[96]

Analyses of population trends for European birds with different migratory strategies revealed a higher proportion of declining species among those that migrate to sub-Saharan Africa during the winter than among either nonmigratory species or those that migrate within Europe.[97] Although most of the declining migratory species are associated with open habitat rather than forest, there is strong evidence from Britain that migratory forest birds are also declining at a rapid rate. Between 1967 and 1999, all 12 woodland species that are trans-Saharan migrants declined in England, and most of these species declined by more than 50 percent.[98] Three of these species primarily spend the winter in the Sahel, the arid zone south of the Sahara desert. These species declined between 1967 and 1978 when there was a major drought in the Sahel, but two of these species subsequently increased. The pattern of population changes for species that spend the winter farther south, in the humid, tropical forest, and savannah areas of West Africa, is particularly alarming. Six of the seven species displayed substantial, long-term declines. Habitat use by these species in their wintering areas in Africa has not been intensively studied, but most of these species appear to use fallow fields, recent clearings in the forest, and farmland with residual trees. It is therefore unlikely that they have declined because of rainforest clearing. They may have declined as a result of agricultural intensification in Africa, which would reduce the number of woodland edges, scattered trees, and fallow fields.[99] The ecology of the trans-Saharan migrants in the winter clearly requires more intensive research to find out whether population changes are driven by land use changes in the tropics.

In Poland's Białowieża Forest, the largest remaining area of mature and old-growth forest in Europe, forest birds (including trans-Saharan migrants) did not decline between 1975 and 1994.[100] Most species experienced relatively little population change during the first decade of the study and then—for no obvious reason—increased during the second decade. These results suggest that the decline of woodland birds in more highly disturbed forests in western Europe may not be due solely to changes in conditions in the wintering areas in Africa. Otherwise, one would expect similar declines in Białowieża Forest.

Implications for Conservation

The general lesson from studies of the impact of forest fragmentation on birds in eastern North America is that it is important to preserve large areas of continuous forest to sustain biological diversity. This approach should be central to conservation planning in the deciduous forests of eastern North America and Japan, and perhaps in other parts of East Asia where this relationship has not been intensively studied. This approach is also important in heavily forested parts of Europe, particularly in eastern Europe, where at least a few bird species appear to require larger forests. It is less important in historically deforested regions of Europe such as England and the Netherlands, where most species appear to tolerate or even favor small forests.

Hence, although there is a general tendency for specialized forest birds to disappear from small remnants of deciduous woodland after extensive areas of forest are cleared, this is not true in every region. Small woodlands are not necessarily intrinsically poor habitats for forest specialists as long as they are large enough to accommodate breeding territories. Their suitability largely depends on characteristics of the surrounding landscape, particularly the patterns of human land use and the effect of this land use on nest predators. Where human activities boost populations of medium-sized predators, then the loss of eggs and nestlings in small forests may eventually lead to the decline and loss of forest specialists. This is the pattern in many parts of eastern North America, where both agricultural and residential landscapes have high densities of predators and parasitic cowbirds. In contrast, rural areas in England are so heavily managed that predator densities are generally lower than they would be in the interior of an extensive forest, and forest birds do well in small forests and may even achieve higher densities than in larger forests.

The history of deforestation is also important. Regions of Europe where nearly all forests have been cleared for millennia may have lost some specialized forest bird species. Other species, such as the Eurasian wren, may have adapted to an open landscape with small patches of trees. Perhaps some of the "big forest" bird species of North America and East Asia will eventually make similar adjustments if we protect their current habitats long enough to give them the opportunity to evolve. In the near term, however, large blocks of continuous forests are critical for their survival. In all of these regions, it is important to monitor the effects of predators on the nest success of forest birds and to protect habitats with higher rates of nest success. It is equally important to protect the winter and stopover habitats for migratory forest birds. Specialized forest birds are among the most important predators of wood-boring and leaf-eating insects, so a reduction in their abundance and diversity could have a major impact on forest ecosystems.

CHAPTER 7

Missing Wolves and the Decline of Forests

I f current conditions prevail, the ancient beech forests of Kinkazan Island, a small island off the coast of northern Japan, will slowly disappear as forest openings expand and coalesce. The impending decline of this forest isn't due to some introduced fungus or defoliating insect that is decimating the Japanese beech trees. The problem is more subtle and difficult to detect on the time scale perceived by a human observer. Few beech saplings grow tall enough to reach the forest canopy, so old trees are not replaced when they die.[1] Most saplings are killed because of browsing by numerous sika deer. Ironically, the slow decline in the forest results from an ancient tradition of protecting plants and animals because the island is the sacred site of an eighth-century Shinto shrine. There are no natural predators of deer on the island and hunting is not allowed, so the number of deer has reached an exceptionally high density (50 per square kilometer). In 1984, after a winter of deep snowfall, half of approximately 600 deer starved to death, but the population has rebounded since then. Sika deer are remarkably versatile, feeding not only on the foliage of broad-leaved trees and conifers, but also on grass, dwarf bamboo, and (during the winter) tree bark. In some parts of Japan fallen leaves from deciduous trees make up a large proportion of their yearly diet, so their populations remain high even after most of the edible plants in the understory have been destroyed.[2] Also, when forest openings are colonized by grass, a new and nutritious source of food becomes available, helping to sustain high numbers of deer. So even after tree saplings and many species of herbs and shrubs have been stripped out of the forest understory, the deer population will not necessarily decline.

The result has been a steady change in the forest ecosystem. The forest understory is dominated by *Leucothoe grayana* and other unpalatable shrub species.[3] One interesting consequence was that Japanese macaques, the indigenous monkeys of Japan, were forced to shift their diet to include more plants that are inedible to deer, such as the leaves of Japanese barberry and the bark of *Zanthoxylum piperitum*.[4] A potentially more serious long-term trend is the invasion of canopy gaps by blackberry, *Aralia elata*, and silver grass.[5] Because of the high mortality of tree saplings, these openings do not undergo succession to forest but instead remain as dense patches of low shrubs and grass. Bit by bit, the ancient forest on Kinkazan Island is becoming an open grassland.

If grazing by deer is particularly intensive, then the site may undergo further change to become an expanse of Japanese lawngrass, *Zoysia japonica*.[6] *Zoysia* is a short, dense grass that thrives with moderate grazing. When cropped by deer or other grazers, it produces vigorous new rhizomes, stems, and leaves. Consequently, grazing increases the productivity of this grass, triggering the growth of tender young leaves that have high nutritional value for deer. The result is a "grazing lawn," which is the type of vegetation found on the short-grass prairies of North America and the Serengeti Plains in East Africa.[7] In these grassland systems, frequent grazing favors low-stature species that produce a dense carpet of grass with highly nutritious foliage and a large amount of food per bite, which increases the efficiency of grazers. This pattern is expected in ecosystems dominated by numerous grazers such as the prairie-dogs and bison of North American short-grass prairies or wildebeest and other antelopes in East Africa grasslands, but is unexpected in a Japanese deciduous forest. Sika deer and *Zoysia* lawngrass form a self-perpetuating system, however, because deer favor this grass by eliminating competing plant species that would overtop it and shade it out, and *Zoysia* seeds are readily dispersed in the droppings of deer.[8] A high proportion of *Zoysia* grass seeds pass through the gut of deer without being destroyed, and these seeds have a higher germination rate than seeds that have not passed through a deer. Kinkazan Island already has a 17-hectare patch of *Zoysia* grassland that supports an amazingly dense concentration of deer (814 deer per square kilometer).[9]

The key role of deer in this process was demonstrated by setting up exclosure plots surrounded by deer-proof fences.[10] When grassland on Kinkazan Island was "protected" from deer with fences, *Zoysia* grass was replaced with taller silver grass within a few years, and the silver grass was eventually overtopped and replaced by woody shrubs and trees saplings. In contrast, control plots outside the fences were still covered with short *Zoysia* lawngrass.

Kinkazan Island is located in the Pacific Ocean close to the Oshika Peninsula of northern Japan. The climate is cool and wet, a situation that generally favors closed-canopy forest. A missing element in the forest ecosystem,

Figure 18. Sika deer grazing on "lawn" of Japanese lawngrass (*Zoysia*) on Kinkazan Island, northern Japan. (Photograph by Seiki Takatsuki)

however, is an effective predator of deer. Predators would not only reduce deer densities, but would also probably force deer to spend most of their time hidden in the forest rather than grazing in highly productive grasslands where they are more easily detected.[11] Gray wolves were once common in Japan, but may never have reached the island, which is separated from the main island of Honshu by 600 meters of open water. After human hunting ended on Kinkazan Island, the deer population was limited not by predation but by the food supply. Because the diet of sika deer is remarkably broad, however, many preferred food plants could be eliminated from the forest before deer faced serious food shortages, setting the stage for major changes in the vegetation.

Kinkazan Island may represent an extreme case because deer are confined to a small island (960 hectares), so they don't have the opportunity to disperse to better feeding areas. Similar effects of deer browsing have been detected in deciduous forests on the main islands of Japan, however.[12] In fact, growing deer populations are affecting the structure and functioning of deciduous forest ecosystems in many parts of Europe and North America

Figure 19. Contrast in vegetation on two sides of a deer-proof fence 14 years after the fence was erected on Kinkazan Island, northern Japan. (Photograph by Seiki Takatsuki)

as well, often reducing the diversity of plants, birds, and other organisms, and sometimes threatening the continual regeneration of the forest by killing young trees. Intensive browsing by deer and other ungulates has become a slow-moving but serious problem that threatens the diversity and regeneration of deciduous forests in more and more regions of the world.

Missing Wolves

The emerging problem of high deer densities results from low rates of predation. Since the extinction of most large mammals at the end of the last glacial period, the main predators of deer throughout the northern temperate zone have been humans, wolves, and large cats. Cougars were important deer predators in eastern North America, and Eurasian lynx played this role in Europe. Tigers were found in the deciduous forests of Central China in historic times, even though they are restricted to forests of far southern and northern border regions today.[13] Tigers were also found in

Japan until the late Pleistocene. Various species of bears contributed to reduced deer numbers in different deciduous forest regions, but bears generally are not effective hunters of adult deer. Wolves are fast, and they hunt in
well-coordinated packs, so they are particularly successful in capturing
deer. Wolves were eradicated from nearly every temperate deciduous forest,
however, surviving only in isolated pockets in Europe and China, and on
the fringes of the boreal spruce-fir forest in the Great Lakes states and in
southeastern Canada in North America. Many deciduous forests have lost
all of their large predators. In eastern North America the American black
bear is the only widespread large predator; both wolves and cougars were
almost completely wiped out by the nineteenth century.

At one time the nocturnal howling of wolves characterized deciduous
forests as much as spring displays of woodland wildflowers or the summer
songs of thrushes. Each wolf pack—an extended family consisting of a
breeding pair and their grown young—defended a large territory by scent-
marking the boundaries with urine.[14] Howling was another territorial sig-
nal that announced at a distance that the territory was occupied. The pack
howled in chorus, with even the cubs joining in with high-pitched yips,
and other wolves could detect the chorus from up to 11 kilometers away.
The joint howling advertised the size, strength, and cohesiveness of the
pack. The territorial system ensured that each pack had a dependable supply of deer and other prey while incidentally ensuring that the prey populations were not eliminated by overhunting. Because of constant culling by
wolves, deer populations probably never reached the densities found in
many modern deciduous forests. As a result, enough tree saplings escaped
from browsing to ensure that gaps in the tree canopy were filled and the
forest regenerated. It has become increasingly clear that wolves, along
with human hunters and large cats, were an exceptionally important component of deciduous forest ecosystems because they reduced ungulate
populations.

Wolves of Japan

In 794, Emperor Kanmu moved the capital of his empire from Nagaoka-kyō
to Kyoto, where it remained for hundreds of years. The new capital—laid
out in a regular north–south, east–west gridwork of streets in the style of
Chinese cities of that period—was surrounded by steep, heavily wooded
mountains. Although the city became a center of cultural refinement, it

continued to be surrounded by surprisingly wild country. In his engaging history of Japanese wolves, Brett Walker summarized numerous reports of encounters with wolves on the roads leading out of the capital and even on the streets and compounds of the city.[15] Sometimes wolves were killed when they entered the precincts of the city, and sometimes they were considered auspicious signs. There are occasional reports of attacks on people. As late as 1034 a wolf chased and killed a deer on the grounds of Kamigamo-jinga, one of the most important Shinto shrines in Kyoto. Wolves were apparently common in Japan during this period, and were found in forests even on the edge of its largest city.

Traditional Japanese images and depictions of wolves are more ambiguous and complex than the negative depictions in traditional European cultures, which probably reflects the relative importance of livestock for East Asian and European farmers. Mixed farming with a heavy reliance on livestock was important in most parts of Europe. People depended on cattle, horses, sheep, pigs, and goats for meat, dairy products, hauling, plowing, and transportation. Large herds of cattle and flocks of sheep often ranged across the countryside, where they were vulnerable to attacks by large predators such as wolves. In Japan horses, cattle, and water buffalo were important domesticated animals, but they were used primarily for transportation and plowing. Hence livestock generally were not kept in large herds that were difficult to protect from wolves. Wolves could be a nuisance when they killed or injured the occasional horse or dog, and they were sometimes dangerous to people, but they were also considered an ally in protecting crops from deer, wild pigs, and monkeys. Wolves were often depicted as helpful (if somewhat unpredictable) spirits who would escort or guide people out of the woods to safety. One common name for the wolf was ōkami, which is significant because kami refers to the spirits of rivers and forests and "o" indicates something that is particularly revered or important.[16] The implication is that wolves were especially powerful spirits. The Chinese characters for ōkami have another meaning, however: "good wild animal." From a European perspective, this is a surprising label for the wolf, but it may reflect an ancient view of rice farmers that wolves were allies against the deer and wild pigs that raided their fields. Other Japanese names for wolf are ōinu (venerable dog) and "large-mouthed pure god."[17] Along with traditional Japanese names, the folklore about wolves as well as shrines dedicated to them as the messengers of gods suggest that these predators were viewed as spiritually powerful and beneficent.[18]

A lumber raftsman shot the last recorded wolf in Japan in the mountains of the Kii Peninsula, southeast of Kyoto, in 1905.[19] The dead wolf was purchased by an American biologist and eventually ended up in the Natural History Museum in London. Remarkably, the raftsman paid a local priest to perform a memorial rite for the wolf, and this rite became a tradition that continues to be performed each year. Given the special status of wolves in Japan as a "good animal" and a "great spirit," why were they hunted down and destroyed in every part of the archipelago, even in the most remote mountains?

This paradox is perhaps easiest to understand for the northern island of Hokkaido, most of which was not settled by the Japanese until the late nineteenth and early twentieth centuries. At the time of settlement, most of Hokkaido was inhabited by the Ainu, indigenous people who depended on hunting, fishing, and low-intensity agriculture. After the Meiji Restoration, the Japanese government initiated a pervasive and aggressive program of Westernization, and in Hokkaido this mandate took the form of intensive exploitation of natural resources and the establishment of large horse and cattle ranches modeled after those in the western United States.[20] Wild deer were commercially hunted on a large scale. Venison canneries were set up to process meat, saltpeter, skins, and antlers from sika deer.[21] The main prey for wolves was depleted at the same time that ranches with large numbers of horses and cattle were established. Not surprisingly, the wolves gravitated to ranches to prey on livestock. Edwin Dun, an experienced American rancher who had been hired by the Japanese government to help set up the livestock industry, introduced highly effective wolf eradication methods that had been developed to protect livestock in the American West. Bounties were established to encourage shooting and trapping of wolves, and wolves were poisoned with bait laced with strychnine.[22] By the early twentieth century wolves had been extirpated from Hokkaido.

If this were the entire story, then the eradication of Japan's wolves could be blamed on uncritical acceptance of Western methods and attitudes (including a deep animosity toward wolves) in a drive to rapidly develop a modern economy. Wolf eradication began on Honshu and the other southern islands of Japan long before the Meiji Restoration or extensive Western influence, however. Japanese attitudes about wolves changed rapidly after rabies arrived on the archipelago in 1730.[23] In many parts of Japan, rabid wolves came out of the forest and attacked villagers, resulting in immediate death from the attack or lingering death from rabies. Wolves quickly acquired a reputation as dangerous killers that should be eradicated. Villagers organized massive wolf drives in which beaters with drums and fireworks chased wolves into the open where they could be killed.[24] Local governments offered bounties on wolves, inducing professional wolf hunters to trap them. The process was less systematic than in Hokkaido, but the final result was the same. By the early twentieth century both the small subspecies of wolf found in the southern islands of Japan and the larger subspecies of Hokkaido had been driven to extinction.

As in North America and Europe, the consequences of removing wolves from forest ecosystems were masked as long as a dense rural farming population actively hunted deer. After the 1950s, however, many areas of rural Japan became depopulated and the popularity of hunting declined.[25] With the extinction of wolves and the decline in hunting by humans, populations of deer, Japanese serow (a wild ungulate resembling a goat), wild boar, and Japanese macaques increased. Equally important, these animals are no longer as cautious about predators, making them less reticent about leaving the dense cover of the forest. As a result, the remaining farmers in the mountains of Japan are besieged by numerous large mammals that raid their fields, often causing severe crop losses.[26] Tree plantations also are often badly damaged by deer and serow that kill seedlings and chew off the bark of older trees.

In an effort to restore forest ecosystems in Japan and protect farmers from deer, wild boar, and monkeys, a Japanese nonprofit organization called the Japan Wolf Association argues that wolves should be reintroduced to Japan.[27] Wild wolves could be transported from Inner Mongolia in China, which has the nearest population of relatively small wolves that are similar to the original gray wolves of southern Japan. Although this proposal is controversial, it may be gaining traction. In 2011 the mayor of Bungo-Ono in Kyushu approved a plan to introduce wolves to the mountains surrounding his city to control the deer population.[28] Although it may seem unlikely that gray wolves can thrive in a heavily industrialized and densely populated country like Japan, much of its interior has a sparse and declining human population. There may be enough space and clearly seems to be enough food to support wolves in some parts of Japan.

The change in vegetation and loss of biological diversity in Japanese forests because of the loss of large predators exemplify a more general phenomenon called "ecological meltdown."[29] Islands created by a giant reservoir in Venezuela, for example, were too small to support large predators.

As a result, the populations of various species of herbivores (everything from howler monkeys to leaf-cutter ants) exploded, resulting in low survival rates for tree seedlings and saplings. These isolated patches of diverse tropical forest will probably be converted to "an odd collection of herbivore-resistant plants" because of the absence of large predators.[30]

White-Tailed Deer Modify North American Forests

Walter C. Tucker Preserve is a natural area in a metropolitan park in central Ohio. The preserve encompasses second-growth deciduous forest dominated by sugar maple and American beech in upland areas, and honey locust, American elm, and several species of ash in lower, wetter areas. In order to monitor tree seedlings in this preserve, individual seedlings were tagged so that their fate could be determined.[31] The number of seeds of a particular species that germinated and grew into seedlings varied greatly from year to year because trees occasionally produced massive crops of seeds, followed by years when they produced few seeds. Also, the seedlings of particular tree species became established in some habitats more frequently than in others; ash seedlings, for example, were more frequent in lowland sites and sugar maple seedlings in upland sites. All of these factors potentially could play roles in the composition of the forest canopy in the distant future, but it turned out that they made little or no difference. Only 2 out of 2,553 tree seedlings grew beyond the seedling stage to become saplings. The great majority of seedlings died during their first year, and few made it past their second year. The decimation of tree seedlings resulted from browsing by white-tailed deer, which were abundant in the preserve (with 60–70 deer per square kilometer). If this intensity of browsing continues, then mature trees will not be replaced, and the forest will slowly change into an entirely different type of low, shrubby vegetation.

Tucker Preserve is a suburban park with an exceptionally dense population of deer, so it is not typical of forests in eastern North America. There is growing evidence that deer have major impacts on forest regeneration even in more remote, heavily forested regions, however. U.S. Forest Service biologists completed one of the best-designed and most ambitious studies of the effect of deer on young trees in the extensive forests of northwestern Pennsylvania.[32] They set up fenced enclosures in second-growth forest plots. These large enclosures (65 hectares) were partitioned into smaller fenced compartments (either 13 or 26 hectares). Each compartment had been prepared in the same way, with trees removed by clearcutting in 10 percent of the area, while the trees had been thinned in 60 percent of the area. The remaining 30 percent remained uncut. Captive-raised deer were released into the enclosures. Separate enclosures had different numbers of deer, with densities ranging from 4 to 25 deer per square kilometer. Each deer was equipped with a radio transmitter so that it could be monitored. The entire experimental set-up was replicated at four separate sites, and the fences were maintained and the vegetation and deer were monitored for ten years.

The goal was to determine the fate of tree seedlings and other plants at different levels of shading by the tree canopy and with different deer densities. If the enclosures had contained unnaturally high densities, then they would not necessarily reflect what happens in wild deer populations. The highest deer density was similar to the highest densities documented in the forests surrounding the enclosures, however, and was well below the natural density in the Tucker Preserve in Ohio or in many other sites in eastern North America.[33]

After ten years the vegetation in the enclosures with high deer densities differed substantially from the enclosures with few deer.[34] The average height of tree seedlings declined progressively as deer density increased. The diversity of tree seedlings also declined at higher deer densities. One species of tree seedling, black cherry, dominated the enclosures with large numbers of deer, and only black cherry seedlings normally grew into large seedlings or saplings in these areas. In contrast, the compartments with low deer densities had large numbers of birch, maple, and pin cherry seedlings. High deer densities also affected the ground cover, particularly in clearcut openings. Where deer were abundant, blackberry was replaced with dense stands of ferns that tend to shade out tree seedlings. In addition, grasses and sedges became more abundant. High deer densities also had a negative impact on the deer. Deer in the high-density enclosures suffered much higher winter mortality than those in enclosures with low densities.

The implication of this study is that a high density of deer will eventually convert a diverse deciduous forest into an almost pure black cherry forest, reducing overall biological diversity and setting the stage for a catastrophic collapse if black cherry suffers from an outbreak such as of scallop-shell moth, which defoliates this species.[35] We can be confident in these results because they come from a well-controlled, long-term experiment. One weakness of the experiment, however, is that the deer were unable to move to find better feeding areas and more protected areas in winter. Much as on Kinkazan Island in Japan, the deer were not able to travel great distances to adjust to changing conditions. Despite this, it turns out that the vegetation changes in the enclosures and on Kinkazan Island are consistent with changes documented in regions with high densities of freely moving deer.

The effect of white-tailed deer on forest vegetation is most extensively documented from work with exclosures, fences that keep deer out of a study plot, rather than enclosures that keep them in the plot. In the 1950s and 1960s exclosures were set up at a number of sites in Allegheny National Forest in Pennsylvania, in the same region as the large-scale experiment with deer enclosures.[36] All of the exclosures were in areas where the trees had recently been harvested by clearcutting. The goal was to find out how deer affected the growth of young trees in these clearcuts. The composition of the vegetation was compared in exclosures and nearby control plots in the 1970s, 9 to 22 years after logging, to determine whether the deer-proof fences had an effect on forest succession. Different types of vegetation were growing in the exclosures and control plots. To the distress of foresters who were managing this forest for timber, the control plots had low densities of saplings of the most commercially important tree species. While black cherry and red maple were more abundant in the fenced plots, the unfenced plots were dominated by American beech and striped maple, which are less valuable as timber trees. Deer browsing also affected the ground cover. Blackberry was much less common in the unfenced areas, which were dominated more by ferns and grass. This dense ground cover of ferns and grass may inhibit the growth of tree seedlings. Thus, the trajectory at these sites was the development of a more open forest with fewer tree species. Also, the key species that support sustainable logging in the region will not be well represented in the regenerated forest stands.

Numerous other studies have also shown that tree saplings are more abundant in plots that are protected from deer than in unfenced plots.[37] The low density of saplings in heavily browsed plots may result from either low survival of young trees or growth rates that are so low that tree seedlings do not become saplings, either of which would prevent them from becoming canopy trees. The exclosure studies that showed this pattern were all completed in regions with more than 8.5 white-tailed deer per square kilometer.[38]

When deer densities are high, they can also affect the abundance of smaller plants such as woodland wildflowers and shrubs. These plants are often more vulnerable than trees because they never grow tall enough to reach a safe height where their leaves are above the browsing range of deer. The potential impact of browsing by dense populations of deer is clear from the loss of diversity of forest understory plants at Heart's Content Scenic Area, a 50-hectare patch of old-growth forest in Allegheny National Forest in northwestern Pennsylvania.[39] Researchers studied this forest intensively in the 1920s and early 1930s, so we have a detailed description of vegetation before deer populations increased to their current densities of 10–15 per square kilometer. Even though this forest has been carefully protected, the diversity of shrubs and wildflowers collapsed between 1929 and 1995. In areas dominated by eastern hemlock and American beech, the number of shrub and herb species detected on 118 small plots dropped from 41 to 8. Virtually none of the less common wildflower species found on plots in 1929 were detected in the 1995 survey. The only species that increased were hay-scented fern and common wood fern, species that are known to be unpalatable to deer because their leaves have high concentrations of toxins.

Increased deer density was not the only environmental change at Heart's Content that might cause understory herbs and shrubs to decline.[40] Hay-scented fern may have increased not only because of deer, but also because the spores may have spread into the old-growth forest from nearby second-growth forests where logging has opened up sunny areas favored by this fern. Moreover, acid deposition ("acid rain") reduced the pH of the soil. Even in this protected reserve, there have been complex environmental changes caused by human activities since the 1920s.

Research in other parts of Allegheny National Forest, however, demonstrates that deer were the key factor in reducing the diversity of small woodland plants.[41] Flat boulders taller than 1.5 meters support a soil layer with wildflowers, shrubs, and small trees that are out of the reach of deer. The vegetation on these tall boulders was compared with nearby plots of the same size on the forest floor. The supposition is that any difference would be due to the effects of browsing by deer. Of course, the top of a boulder is different from the forest floor in other ways; plants growing on a boulder don't need to compete for nutrients with the roots of large trees, for example. To account for this, short boulders (lower than one meter) were also included in the plant surveys. These share the general characteristics of the tall boulders, but they are accessible to deer (as indicated by deer droppings).

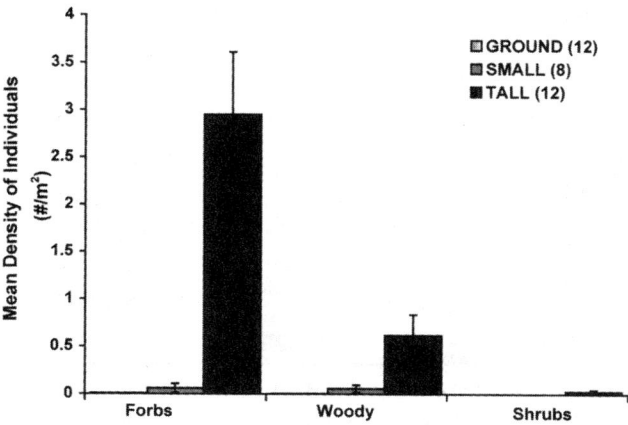

Figure 20. Average density of plants on the tops of tall boulders (greater than 1.5 meters tall), the tops of small boulders (less than 1 meter tall), and the forest floor. (Comisky et al., 2005; reprinted with permission from *American Midland Naturalist*, the University of Notre Dame)

The differences between tall boulders and the other two habitats (low boulders and the forest floor) were dramatic. The tall boulders had a much richer diversity of wildflowers. The surveys were completed in May when it could be determined whether herbs were large enough and healthy enough to flower. Although the areas sampled on tall boulders and on the forest floor were similar, 1,338 flowering herbs were found on tall boulders compared with only 6 on the forest floor. Canada mayflower, Solomon's seal, and red trillium were common on the tops of tall boulders. Short boulders had only a slightly higher diversity of wildflowers than the forest floor, indicating that accessibility to deer (and not special characteristics of boulder tops) was the key factor.

The authors of this study referred to tall boulders with the evocative name of rock refugia gardens.[42] These refuges may sustain wildflowers until deer populations decline so that they can recolonize the forest floor. If deer densities remain high, however, it is doubtful that these small, scattered refuges can sustain woodland wildflowers indefinitely.

A similar contrast was found for a fenced exclosure that was maintained in a wildlife management area in north-central Pennsylvania since the early 1940s.[43] When vegetation inside this small exclosure and on a nearby unfenced site were compared in 2007, ferns, grasses, and American beech seedlings dominated the unfenced plot instead of the rich diversity of wildflowers, shrubs, and young trees found inside the exclosure. This 60-year study thus provides important information about how protection from deer browsing over a long period results in high plant diversity in the forest understory, but the results are tenuous because they are based on a single exclosure and a single unfenced plot. The impact of deer browsing on composition of the forest understory was confirmed, however, with a comparison of 51 sites in mixed hardwood-coniferous forests in Wisconsin.[44] The diversity of herbaceous wildflowers declined as the browsing intensity (measured by the proportion of browsed and unbrowsed sugar maple twigs within the reach of deer) increased. In contrast, the density of ferns, grasses, and sedges increased with browsing intensity.

Numerous well-designed studies show that deer can potentially cause a major decline in the diversity of understory plants and the success of young trees. Studies of tree seedlings and saplings consistently show that few young trees will reach adult size wherever deer densities exceed 8.5 deer per square kilometer.[45] At high deer densities forest ecosystems are slowly being converted into what Thomas Rooney calls "fern parks" and "deer savannas." Fern parks are woodlands where the ground is almost completely covered with one or two species of ferns, and there are few wildflowers or tree saplings. If this situation persists, then the fern park may eventually become a deer savanna as many of the canopy trees die. The forest ecosystem is replaced by a very different type of ecosystem with much lower biological diversity.

Loss of Understory Birds

The loss of a diverse understory with numerous species of wildflowers and shrubs results in the decline of numerous species of animals, so intensive deer browsing has many unanticipated effects on forest diversity. Many species of forest birds depend on a dense and diverse understory for nesting and foraging, and these species often decline as the complex, shrubby understory changes to open expanses of fern or grass. David deCalesta found this pattern when he monitored birds in the large enclosures in northwestern Pennsylvania in which other Forest Service biologists had studied the effect of different deer densities on survival and growth of tree seedlings and saplings.[46] Bird species that primarily nest in the shrub layer and subcanopy declined by 37 percent as deer density increased from 4 to 25 deer per square kilometer. Four species were absent in plots with more than 8 deer per hectare. Interestingly, the number of species and total abundance of ground-nesting species and species that nest in the tree canopy were not significantly affected by deer density. Thus, increased deer density has the greatest impact on species that nest in shrubs or small trees.

The negative impact of dense deer populations on understory birds was confirmed by an intensive thirteen-year study of the Kentucky warbler, a forest understory specialist, at a Smithsonian Institution research area in Virginia.[47] During the period of the study, deer densities increased from 20 to 30 per square kilometer in most of the forest, but lower densities were maintained in one section of the forest that was surrounded with a fence. Deer drives and hunting reduced the density of deer in the fenced area to about one-fourth of the density in the surrounding forest. Kentucky warblers were found more frequently in this fenced exclosure, and they tended to shift into this area as deer densities increased in the rest of the forest.

What Is the Optimum Density of White-Tailed Deer?

Opponents of management of deer populations argue that if people do not interfere the deer population will eventually achieve a balance with their food supply. Long-term studies on the effect of high deer densities on forest vegetation indicate that this is not happening fast enough to prevent a collapse in biological diversity, however. In the absence of predators deer are able to sustain high densities even as the palatable species of the forest understory disappear. In oak forests, for example, production of acorns by mature oaks helps sustain deer through the winter.[48] Acorn production on tall oaks is not affected by deer density, so deer populations can remain high even as one plant species after another disappears from the understory. Also, white-tailed deer, like sika deer in Japan, can subsist on litterfall (leaves and twigs that have fallen from the canopy) in some situations. In balsam fir forests in Quebec fallen twigs covered with lichen are a major component of the diet of white-tailed deer during the winter.[49] The pattern is similar at a site in Maine, where conifer branches, hardwood leaves, and lichens in litterfall accounted for 85 percent of the winter diet.[50] When deer depend on food falling from canopy trees (in the form of nuts or dead leaves and branches), then the loss of palatable understory plants may not result in a decline in the deer population. The result may be a slow shift from one stable state (a diverse, closed-canopy forest) to a different stable state (an open savanna with much lower biological diversity).

Some regions in the eastern deciduous forests have lower densities of white-tailed deer, providing a good baseline for how forest ecosystems function with less intensive browsing of tree seedlings and other understory plants. Alejandro Royo and his colleagues studied forest vegetation in two forests in central West Virginia where deer densities are relatively low, ranging from 4.6 to 7.7 per square kilometer, which is not much higher than the estimate of 3–4 deer per square kilometer before European settlement.[51] In an ambitious, large-scale experiment, they compared the effects of ground fires, canopy gaps, and deer browsing on understory vegetation on 64 plots. Each plot covered 400 square meters. The understory in half of the plots was burned; the other plots were not burned. Researchers created relatively large canopy gaps (16 square meters) in some plots by girdling a group of trees. The size of these gaps matched the average for old-growth forests rather than the average for the second-growth forests where the study was completed. The researchers built tall fences to exclude deer around half of the plots. They ended up with plots that had different combinations of burned and not burned, fenced and not fenced, and canopy gaps present or absent. With replicates for each experimental treatment, they could objectively evaluate the effects of different types of natural disturbance on the diversity and density of the forest understory.

The results give fresh insight into the importance of natural disturbances for sustaining biological diversity in forests. Both fire and canopy gaps increased the diversity and density of understory plants. The combination of fire and canopy gaps resulted in even greater diversity than either disturbance alone. The maximum diversity of understory plants was found, however, in plots that had fire, canopy gaps, and deer browsing. In this situation deer *increased* the diversity of wildflowers and ferns, probably because browsing thinned out rapidly growing plants that would outcompete other species. This experiment shows that deer can be important in sustaining biological diversity if their density is not too high, suggesting that deer exclosures go too far by completely eliminating deer browsing. A low to moderate level of browsing enhanced the diversity of understory plants, particularly if combined with other forest disturbances such as fire and small openings.

Reducing White-Tailed Deer Populations with Hunting

Management of deer populations is a controversial topic in many parts of North America. In landscapes ranging from densely settled suburbs to extensive wilderness areas, deer are perceived as "overabundant" or "overpopulated," and there are debates about whether their numbers should be reduced. Although there is growing evidence that dense deer populations can diminish biological diversity in forests and prevent regeneration of tall trees, most of these controversies center on other, largely unrelated issues. In an elegant and incisive essay, ecologist Graeme Caughley pointed out that the term "overpopulated" is applied to particular species of animals for four distinctly different reasons, only one of which is relevant to our efforts to protect and sustain natural ecosystems.[52]

First, a species can be considered overpopulated if it interferes with human well-being. Most efforts to control deer in eastern North America focus on this interpretation of overpopulation. Deer are perceived as a problem because they feed on crops or ornamental plants; they run into roads and cause expensive and dangerous accidents with vehicles; or they help to spread the ticks that carry Lyme disease. Thus deer must be controlled because they create economic damage or pose a risk to human health and safety. Caughley's second definition of overpopulation applies to a species that reduces the numbers of another, more favored wild species. This logic is sometimes applied to predators that prey on game animals: wolves must be controlled because they reduce the number of elk. In the case of deer, this definition would apply, for example, when deer are preventing the growth of white oak trees that are favored by foresters. The third definition of overpopulation focuses on when a population is too dense for its own good. Deer may succumb to starvation or disease as their densities increase. As Caughley points out, wildlife managers strive to manage for populations of animals that are "larger, fatter, healthier and more fecund than those in an unharvested population," but "overpopulation" in this case may not differ from a typical population density before human management.[53]

These first three definitions all focus on the economic, health, and recreational concerns of people, and are not necessarily related to the stability and diversity of natural ecosystems. Caughley's fourth definition is the only one that is relevant to concerns about how deer are changing the structure of deciduous forests. In this case, overpopulation is defined as a population density that disrupts the ecosystem. Scientific research can objectively determine whether this is happening. We can answer questions about whether deer densities above 8.5 deer per square kilometer will ultimately change the structure of the vegetation and cause a collapse in biological diversity in the forest. In contrast, the first three questions are based on value judgments about the well-being of people and deer, so conflicts cannot be resolved with ecological research. As Caughley points out, as an ecologist his opinion on these issues is no more valuable than the opinions of his immediate neighbors, "a garage mechanic on one side and an Air Vice-marshal on the other."[54] People must apply values that have little to do with ecological research on whether the economic damage, health hazards, negative impact on favored species, and health of the deer herd warrant controlling the deer population. Ultimately, of course, one must also make a value judgment about whether killing individual deer (which is opposed in principle by animal rights organizations) or reducing the number of deer available for hunting each autumn (which is opposed by hunters) is warranted to prevent the deciduous forest from being converted to scrub or open savanna, with an inevitable collapse in the diversity of woodland organisms. In this case, however, ecological research can provide the critical information needed to make an informed decision.

When the decision is made to reduce deer populations and maintain them at a lower density to permit the recovery of the forest understory and the growth of young trees, the usual method is to open an area to deer hunting or to change hunting regulations to permit hunters to take more deer or to shoot does as well as bucks. Alan Rutberg argued that there is surprisingly little evidence that hunting actually reduces or stabilizes deer populations, however, because most studies do not compare areas with and without hunting, and because the number of deer that are harvested continually increases in many areas where hunting is permitted.[55] A long-term increase in deer harvest rates suggests that populations are increasing despite the harvesting of deer.

William Healy presents convincing evidence that hunting can reduce deer densities enough to sustain a diverse forest ecosystem, however.[56] He compared the vegetation in two sections of the watershed of Quabbin Reservoir in Massachusetts. More than 19,000 hectares of upland forest around the reservoir have been protected as a wildlife sanctuary since 1938, when the reservoir was created. Hunting was not permitted in this sanctuary, but deer are hunted in other parts of the watershed. Healy compared the vegetation in 8 areas inside the wildlife sanctuary with 16 areas outside the sanctuary.

Estimated deer densities were 10–17 deer per square kilometer inside the sanctuary but only 3–6 deer per square kilometer in the forest stands that are open to hunting. The density and diversity of tree seedlings was substantially higher in areas with hunting and relatively low deer densities. Black birch seedlings dominated high-density deer plots, whereas low-density plots supported numerous seedlings of many different species. Areas with high deer densities within the wildlife sanctuary had virtually no oak seedlings taller than 100 centimeters. Four species of oak seedlings were frequent in hunted sites (especially in more open forest stands that had been thinned), but were rarely found in the wildlife sanctuary. Another feature of this well-designed study was the erection of fenced deer exclosures in all of the high-deer-density areas. After six years these exclosures had substantially more tree seedlings than nearby unfenced plots.

Healy concluded that forests in the wildlife sanctuary at Quabbin Reservoir have steadily progressed in the direction of open savanna because so few young trees grow to maturity.[57] The prognosis is that the forest can no longer be managed effectively for timber production, and that areas managed to preserve biological diversity will never achieve the complex, multi-layered structure of an old-growth forest. Deer populations have fluctuated since 1938, but canopy trees support high deer densities even as the understory disappears. Old oaks that are reaching their peak years for production of acorns are common in the canopy, so mast crops of acorns sustain the deer through the winter.

The areas with low deer density around Quabbin Reservoir have been continuously hunted since the early twentieth century. Reducing deer density in a forest that has sustained heavy browsing for many decades may not result in the rapid recovery of the forest, however. A long-term study of the largest remaining expanse of eastern deciduous forest in Canada showed how a long period with dense deer populations can set the forest into a new trajectory that may continue even after deer populations are reduced. Rondeau Provincial Park supports a diverse deciduous forest with American beech, sugar maple, tuliptree, white ash, sassafras, and numerous other tree species. There are no wolves or mountain lions in the park. After hunting was banned in the park in 1974, the deer population increased from 27 to 55 per square kilometer. Between 1981 (when deer densities began to increase rapidly) and 1996, the density of young trees on permanent vegetation plots declined by 72 percent. In contrast, abundance of young trees remained high in two deer-proof exclosures that were set up in 1978.

In 1993 the deer population was culled by 67 percent to protect the deciduous forest, and the deer density has been maintained at 7 per square kilometer since then. After the deer herd was reduced, the density of young trees increased substantially but did not return to the 1981 levels. Canopy trees continued to die without replacement because of the absence of large saplings and small subcanopy trees. This age group of trees was largely missing because so few young trees matured between 1981 and 1993. One possibility is that the forest will become more open as more large trees die and are not replaced, creating forest openings that may expand as adjacent trees are blown down in storms. If fire were introduced, this site probably could be converted to oak savanna, but current efforts are directed at restoring "Carolinian" deciduous forest, which is a highly valued ecosystem in Canada. This effort may require a long period of recovery at low deer densities or additional active intervention to speed up the growth of young trees.

Would Natural Predators Reduce Deer Populations?

Wolves and mountain lions have been extirpated in nearly all eastern deciduous forests, and many forests do not permit hunting by people. Is this the main reason that deer populations have become so dense that they threaten the biological diversity and even the basic structure of the deciduous forest? Unfortunately this question is difficult to answer because wolves and mountain lions, both of which are efficient, year-round deer predators, are missing or very rare in most eastern forests. White-tailed deer have also increased because people have created favorable habitat in the form of forest edge, farmland, and suburban gardens, so their populations may have increased even in the presence of large predators.[58]

We can gain some insights from interactions between deer and large predators in other parts of North America. Most of these studies were done in the coniferous forests of western North America. The most dramatic example is the reintroduction of gray wolves to the high-elevation forests and grassland of Yellowstone National Park in 1995 and 1996. Wolves were exterminated in the park in the 1920s to protect elk and one of the few remaining herds of American bison from predation. Although other predators remained in the park or eventually returned, wolves were absent. Wolves are particularly effective predators of deer, including elk or wapiti (which is usually classified as the same species as the smaller red deer of Europe). Eventually elk numbers increased to such a high density that the National

Park Service began culling the herd each year so that they would not deplete their food supply. When this unpopular culling program ended, the elk population increased rapidly, which in turn led to a classic chain reaction of unintended ecological changes. Browsing by elk was so intense that aspen, cottonwood, and willow seedlings could no longer grow to maturity along streamsides.[59] As old trees died, they were not replaced, so an entire habitat and its dependent species were being lost. Beavers also declined because young streamside trees were their main source of food and building material for dams and lodges. The resulting decline in beaver dams led to a decline in many species that require open ponds.

Wolves were introduced with the hope that they could reverse the changes that had slowly eroded the biological diversity of the park. The reintroduction resulted in a remarkably quick ecological change that exceeded the expectations of the park managers. Not only did elk populations decline, but elk also became more wary, avoiding open streamside areas where they might be ambushed by wolves. They particularly avoided areas with poor visibility or a lack of escape routes.[60] Aspen groves in upland areas were still heavily browsed and failed to reproduce, but dense stands of aspen and willow saplings grew up in some streamside areas.[61] The streamside woodland is recovering along with populations of beavers, willow flycatchers, and other species that need this habitat.

Reintroducing mountain lions and wolves may be feasible in some of the more remote regions of the eastern deciduous forest, and the red wolf has already been successfully established in North Carolina.[62] This federally endangered species has been intensively bred in captivity to provide animals for reintroduction, and there are now small populations of these wolves in Alligator River National Wildlife Refuge and Great Smoky Mountains National Park. The return of an effective predator of deer may not depend on reintroduction efforts, however. Since the early 1900s coyotes have moved eastward from the Great Plains into the eastern forest. Coyotes are related to wolves (both are members of the genus *Canis*), but they are smaller and typically feed on smaller prey. Coyotes in the deciduous forests of the northeastern United States and southeastern Canada are considerably larger than the coyotes of western prairies, deserts, and sagebrush habitats, however. They also tend to live in larger groups than their western relatives, and they prey on both young and old deer. Deer may constitute a substantial part of the winter diet of these northeastern coyotes, which often kill healthy adult deer.[63] In many ways they act more like wolves than like western coyotes.

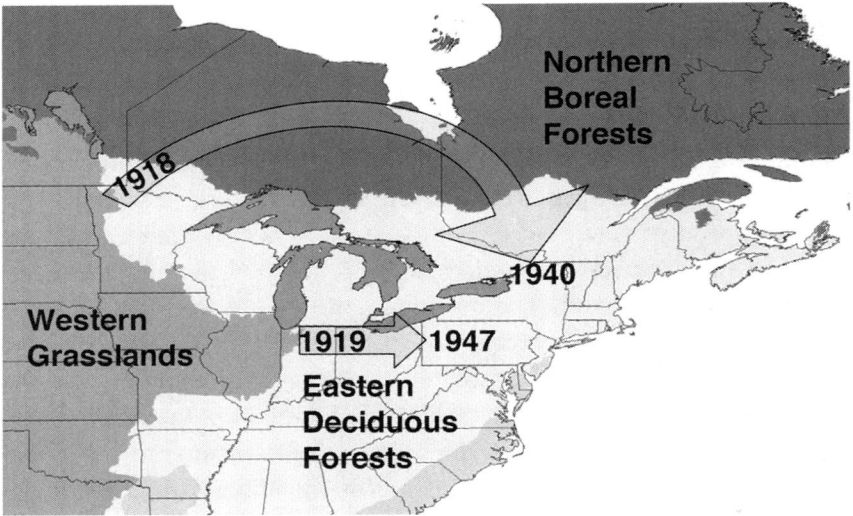

Figure 21. Colonization routes used by coyotes as they spread from the Great Plains into the eastern deciduous forest. (Kays et al., 2010; reprinted by permission of the Royal Society)

Their large size has often been attributed to a better diet in the moist, productive eastern forests, but recent genetic analyses show that there is much more to the story than better nutrition. Coyotes apparently hybridized with wolves as their populations spread across southern Ontario, and the result is a medium-sized, deer-hunting canine that can persist in a wide range of habitats, including suburbs and the edges of cities.

Although hybridization between wolves and eastern coyotes has long been suspected, the evidence became compelling only after a series of studies of DNA from various populations of canines in North America. In 2000, Paul Wilson and his colleagues argued that wolves from southeastern Canada are not closely related to the gray wolves of western North America, but are more closely related to red wolves of the southeastern United States.[64] In fact, they found that the genetic similarity between eastern Canadian wolves and red wolves is so great that they concluded that these represent different populations of the same species, a small, deer-eating wolf (the "eastern wolf," *Canis lycaon*) that is more closely related to coyotes than to the larger gray wolves of the West. The genetic analysis indicated that western coyotes and small eastern wolves split from a common ancestor only 150,000 to 300,000 years ago, but that they have evolved independently of the gray wolf for 1 to 2 million years. This would explain why eastern wolves hybridize with coyotes. Gray wolves of western North America are not known to mate with coyotes, and in fact they usually kill any coyotes they can catch. In Ontario gray wolves sometimes hybridize with eastern wolves, however, so there is some genetic exchange among all three types of canines.

This basic picture of wolf and coyote evolution was subsequently supported by other molecular genetics studies.[65] The hypothesis that there was a distinctive species of eastern wolf has profound implications for understanding how ecosystems functioned in North America before European settlement. The implication is that one of the signature species of the eastern deciduous forests was a relatively small wolf that specialized on white-tailed deer. North and west of the deciduous forests, in the boreal forests of Canada and the Great Plains, this eastern wolf was replaced by the more massive gray wolf that specialized on larger prey such as moose, elk, and bison. In the Great Plains and farther west, large gray wolves lived alongside smaller coyotes, which specialized on rodents, rabbits, and deer fawns. After European settlement, widespread predator eradication and massive changes in landscapes south of the boreal forests disrupted this pattern, leading to hybridization among gray wolves, eastern wolves, and coyotes.

In 2011, however, the interpretation of the genetics of eastern wolves changed again when Bridgett vonHoldt and her colleagues published a comparison of tens of thousands of nuclear DNA loci in many wild populations of wolves, coyotes, and other canines.[66] In contrast, earlier studies had been based on relatively short sequences of DNA in cell nuclei or mitochondria. The new, more extensive genetic analysis was possible because the genome of domestic dogs (which are domesticated forms of the gray wolf) had been sequenced, providing a basis for comparison with different populations of wild canines. The analysis did not support the hypothesis that red wolves and eastern Canadian wolves belong to a separate "eastern wolf" species. Both of these wolf populations have coyote genes that apparently result from relatively recent hybridization. In the case of eastern Canadian wolves, hybridization may have occurred before European settlement, but still within the past thousand years.

If red wolves are products of recent hybridization between gray wolves and coyotes, then efforts to preserve the genetic "purity" of this endangered species do not make sense. Similarly, if the small size of eastern Canadian wolves results from coyote genes, then this population would not warrant special protection as a separate species. Nevertheless, coyote-like wolves and wolf-like coyotes that resulted from hybridization may play a critically important ecological role in eastern forests as an effective predator of deer.

In much of the northeastern deciduous forest (Atlantic Canada, New England, and New York), the large wild canines are wolf-like coyotes. They have been called "coywolves" because of the evidence that they are hybrids between coyotes and wolves, but they are basically coyotes with some wolf genetic sequences.[67] Only about one-third of the genetic composition of eastern coyotes is wolf, but this genetic contribution may be ecologically important. These hybrids have spread eastward much faster than pure coyotes spread into the deciduous forests farther south from Ohio or from the Southwest, and they have been recorded as far south as northern Virginia.[68] White-tailed deer are a major component of their diet, and they may be better adapted than pure coyotes to successfully hunt deer. The hybrids have larger skulls and more massive attachments for jaw muscles than western coyotes, and so are better equipped to kill large prey.[69]

One key question is whether coyotes have adapted to eastern forests or are primarily restricted to open habitats created by people. An intensive survey in Adirondack Park in New York using genetic fingerprinting of coyote scat (droppings) showed that coyotes were concentrated deep within the forest, away from roads and dense human settlements, but that they reached their peak abundance in forests with open canopies and along the banks of lakes and streams.[70] Thus, they could potentially become an important deer predator in deciduous forests. They depend more on small mammals and berries than do wolves, however, so they may not completely replace the missing wolves of these forests. Reintroduction of wolves in more remote areas may therefore help to control deer populations, but hybridization with the coyotes that already occupy these forests might complicate the effort.

As these hybrid coyotes become better established in the Northeast and perhaps spread southward, white-tailed deer may again have an effective year-round predator. Coyotes tend to be nuisance animals, however, killing pets, chickens, and sheep, so their potential to fill the ecological role of the extirpated wolf will partly depend on whether they become targets of eradication campaigns. Wolves have become an icon of wilderness, and are now a popular priority for protection and conservation for many North Americans. Coyotes, which live in suburbs and even city centers, do not share this status. Also, hybridization of coyotes with red wolves and eastern Canadian wolves is considered a threat for restoring genetically pure populations of wolves. Hybridization may have a positive impact, however, if it produces a new type of canine that can become the top predator in the modern eastern forests, even in regions with extensive habitat fragmentation and disturbance. As Kyle and his colleagues argue, the important goal should be to preserve the genetic diversity of wild canines in eastern North America.[71] This will permit the evolution of an effective top predator in the highly modified forests, a predator that may combine some features of coyotes and wolves. If an animal with the physique and social behavior of the wolf is still optimal in modern forests, then natural selection will favor the evolution of a similar wild canine. On the other hand, if a smaller, coyote-like predator thrives in the modern forest landscape, then natural selection will favor this type of canine, and it would be counterproductive—and probably futile—to try to oppose this evolutionary trend.

Deer Problems in European Forests

As described in Chapter 5, in the early 1900s dense populations of ungulates (red deer and European bison) in Poland's Białowieża Forest led to changes in the composition of the forest that can still be seen a hundred years later. Heavy browsing favored some tree species and reduced the density of others. High ungulate densities resulted from intensive trapping and shooting of predators (wolves and Eurasian lynx) and artificial feeding of ungulates during winter. After ungulate populations were decimated by hunting and then slowly recovered, the composition of trees in the forest showed a steady change. Clearly large grazing and browsing animals can have a major impact on the diversity and structure of European forests.

After the last glacial period, wolves and people were the major predators of European deer. Wolves were eventually eradicated in much of Europe, but deer hunting by humans was intense. By the 1300s deer in some parts of Europe were managed in "deer parks," where herds were protected from predators and poachers so that they could be hunted by royalty and the aristocracy. In England these parks were surrounded by deer-proof palisades with "deer-leaps" that permitted deer to enter but not leave the woodland.[72]

Wolves were systematically exterminated in most parts of Europe because they were a threat to game animals and to the critically important livestock used in European farming.[73] Wolves disappeared from England in the 1500s but were able to survive in Scotland much longer because they could hide in extensive forests. The passion to exterminate wolves was so great, however, that the forests were burned to deprive them of a refuge, and the last wolves disappeared by 1684. Wolves became extinct in Ireland in 1770, but persisted in France until the early twentieth century. Wolves were also systematically exterminated in Scandinavia. The last wolves were killed in 1966 in Sweden and 1973 in Norway. Ironically, legal protection soon followed extinction in both of these countries.

Although wolves were completely extirpated from most of the central and northern continent and the British Isles, they survived around the edges of Europe.[74] In Spain and Italy wolves were hunted in farming areas to protect livestock, but there was no organized effort to eradicate them in wild areas with no farming. Wolves therefore survived in small numbers in remote mountainous areas. They also survived in less settled parts of eastern Europe.

After legal protection during the twentieth century, wolf populations expanded.[75] Wolves have spread from Italy into France and from Poland into eastern Germany. Sweden and Norway have been recolonized from Finland and Russia. Wolves have benefitted not only from restrictions or bans on wolf hunting, but also from the recovery of prey populations in areas where farming and herding have declined. In some regions of southern Europe wolves live in surprisingly close proximity to human settlements, suggesting that with protection North American wolves (or coyote-wolf hybrids) could live in rural areas with farmland and towns. Wolves are even found in sheep-raising regions in rural Italy.[76] To minimize attacks by wolves, sheep are kept in permanent pastures in tight flocks protected by shepherds and massive, well-trained guard dogs. The wolves prey on roe deer, wild boar, and other wild animals, so they do not depend on livestock. Italian wolves are smaller than the wolves of western North America and northern Europe, and they live in small groups rather than large packs, so they may be ecologically similar to coyote-wolf hybrids of eastern North America.

Most European deciduous forests have no wolves or other large predators, however. As the intensity of human hunting has declined in many rural areas, deer populations have increased, leading to the same sorts of ecological problems as in Japan and eastern North America. The problems may be particularly acute in Great Britain, where deciduous woodlands are browsed and grazed by six different species of deer.[77] Two species, the roe deer and red deer, are native to England. The fallow deer was introduced following the Norman Conquest, but this can be considered a reintroduction because fallow deer were found in England during the last interglacial period and disappeared during the last glacial period. Three other species, the Chinese water deer, Reeves's muntjac, and sika deer, were introduced from Asia. All of these species have been expanding their ranges in Britain, with the introduced sika deer and muntjac showing particularly rapid rates of expansion, with increases of 5 percent and 8 percent per year, respectively, between 1972 and 2002. Given what we know about the impact of sika deer on Japanese forests, the spread of this species in Scotland and England should be a cause of concern. During the 1800s red and roe deer were largely restricted to the Scottish highlands, but they have subsequently spread through much of Britain.[78] Today as many as four species of deer may live in the same woods, and some woods have deer densities exceeding 40 per square kilometer.

Browsing by deer results in some of the same negative effects on European woodlands as those seen in Japanese and North American forests. Numerous studies comparing fenced and control plots show that grazing and browsing by deer created a more open, sunny forest floor, with lower densities of shrubs, vines, and tree seedlings and saplings. Experiments with deer exclosures in Great Britain and continental Europe show that oaks, hornbeam, and willows consistently decline with deer browsing, but beech and birches don't always decline and may increase.[79] Numerous studies also show that deer modify the understory of deciduous forests in Britain, reducing the density of common blackberry and shrubs while favoring bracken fern, grasses, and sedges.[80] In Wytham Woods in Oxfordshire, for example, the coverage of common blackberry (bramble) in the understory declined from 40 percent to only 5 percent between 1974 and 1999, while the coverage of grasses increased.[81] A comparison of 13 English woods with different deer densities showed that the density of the ground and shrub layers decreased with increasing deer density, and that the larger deer (sika and red deer) had a greater impact than smaller deer (roe deer and muntjac).[82] Some woodland herb species may be lost with heavy grazing, as suggested by a community of "tall herb" species that are largely missing from grazed woodlands but are found growing on cliff ledges that deer and livestock cannot reach.[83] These may be similar to the "rock refugia gardens" that support a diversity of wildflowers in Pennsylvania.

Although browsing and grazing by deer may have negative effects on a wide variety of animal and plant species in England, the effect on songbirds has been particularly well studied.[84] Bird populations have been monitored in Wytham Woods since 1947.[85] During the 1970s the populations of fallow deer and muntjac increased greatly, leading to loss of much of the shrub layer. The impact of the deer on woodland vegetation may have intensified after 1989, when a deer-proof fence was completed around the woods to keep the deer out of adjacent farm fields. The fences included deer leaps that permitted deer to enter but not leave the woods, in the style of a traditional deer park. Although the deer were cut off from winter cereal crops that had been their major source of food in winter, their populations remained high in the woods.

Several species of woodland birds that nest in the shrub layer showed severe declines after 1970, while most species that nest in tree cavities or on branches high in the tree canopy did not decline.[86] Species that nest in tangled bramble (common blackberry) or low, dense shrubs, such as dunnock, song thrush, garden warbler, blackcap, willow warbler, and chiffchaff, have decreased. Although it seems clear that the consistent loss of shrub-layer birds resulted from the thinning of the understory, we cannot be sure that browsing by deer was the main reason. The slow closing of the tree canopy with forest aging may have also contributed to this change.

Closed tree canopies are not an issue in recently cut coppiced woodlands that are characterized by low, woody vegetation, however. In England several species of early successional birds have declined as fewer and fewer coppices were maintained. The remaining coppices, which are often managed for conservation of early successional plant and animal species, are now threatened by heavy browsing by deer. Even though the coppice in Bradfield Woods in Suffolk has been maintained by periodic harvesting of trees, the densities of common nightingales, dunnocks, and several species of warblers dropped as the intensity of deer browsing increased.[87] Of course, these species may have declined for reasons unrelated to deer. Many of these species migrate to other regions in winter, for example, and conditions in the wintering areas may have changed.

To determine whether deer browsing was driving these changes, bird populations were compared in fenced plots and adjacent unfenced plots after three summers of growth following tree harvesting.[88] Shrubs and common blackberry dominated the vegetation inside the fenced plots, while the unfenced plots had more grass. Not surprisingly, the fenced plots had a higher density of birds that specialize in shrub habitat, such as common nightingales, dunnocks, garden warblers, and long-tailed tits.[89] The density of nightingales, which require an open habitat with low dense shrub cover and some open ground for foraging, was 15 times greater in the fenced

plots than in the unfenced plots. Also, nightingales equipped with miniature radio transmitters spent 69 percent of their time in the 6 percent of the
coppice that had been protected from deer.[90] Many of the species that had
lower densities in unfenced plots have shown widespread population declines in England, suggesting that the rapid increase in deer populations
may be having a substantial negative impact on their populations.

Lessons for Conservation

The removal of top predators from most temperate deciduous forests has
resulted in high densities of deer, leading to a cascading series of ecological changes. Some species are lost from forest ecosystems while others
are favored, but the overall trend is a reduction in the number of species
associated with mature, closed-canopy forests. In cases where browsing
and grazing by deer is particularly severe, forests are slowly converted into
open parkland or savanna, and may eventually become almost treeless areas
dominated by grasses, sedges, or ferns. Deer populations are ultimately
limited by the food supply, but this limit may be reached only after a complex forest ecosystem is replaced by a simpler savanna or even a "grazing
lawn" that can support an extremely high density of deer. In some cases
browsing by deer may lead to the resurgence of ancient habitats that have
almost disappeared from the landscape, such as heavily grazed, parklike
woodland in Europe and the oak savannas of the midwestern United States,
and this could help sustain regional biological diversity. If forests across an
entire region are subjected to heavy browsing by deer, however, then many
of the specialized species that require closed-canopy forest and forest openings will be lost. The best solution is reintroducing predators or allowing
them to spread naturally into larger forests. When restoring predators isn't
feasible, the only solution is management of deer populations by people.
Currently the only effective method of doing this is through regulated hunting. Deer are an important part of the forest ecosystem, however, and too few
deer can also lead to a loss in biological diversity. Deer populations should
be managed to sustain their populations at relatively low densities in order
to protect a diverse forest ecosystem.

CHAPTER 8

The Global Threat of Rapid Climate Change

T he deciduous forests of the northern temperate zone are among the best-studied ecosystems in the world. We are beginning to under-stand what we need to do to protect the remaining deciduous for-ests, and even how we can reassemble degraded forests or recreate forests in areas that were completely cleared. We can protect remaining old-growth forests and slowly nurture old-growth conditions in younger forests. We can reintroduce natural disturbances that sustain biological diversity or, barring that, learn how to mimic these disturbances to create habitat diver-sity. We can give a high priority to protecting unbroken expanses of forest and to reconnecting isolated forest fragments to create more viable forest ecosystems. We can reintroduce predation where predators have been exter-minated, and reintroduce large herbivorous mammals (except, of course, those that were driven to extinction). But just as ecologists have begun to marshal and perfect these conservation strategies, the entire enterprise is being undermined by pervasive global threats that cannot be easily pre-vented or managed at a local or even regional scale. Rapid climate change and the introduction of new species of plants, animals, and pathogens from other parts of the world now threaten the natural diversity of deciduous forests.

Evidence for Rapid Climate Change

During the past century the average temperature of the earth has climbed by 0.76 degree Celsius.[1] Especially in the temperate zone, where we have detailed records of how the advance of spring affects plants and animals, we know that warmer conditions have already led to detectable biological changes. The timing of springtime events—from the blooming time of flowers to the arrival of migratory birds—has advanced by several days for a large number of species.[2] An analysis of the results of studies of 677 spe-

cies in Europe and North America demonstrated that the majority had
shifted to an earlier schedule in the spring. Frog breeding, bird nesting,
flowering of woodland herbs, and leaf-out of trees have all shifted to earlier
in the spring.[3] For 172 species for which we have especially detailed seasonal records, biological responses to spring advanced by an average of 2.3
days per decade. A more comprehensive analysis of more than 100,000 time
series between 1971 and 2000 from 542 species of plants and 19 species
of animals confirmed that spring had advanced by an average of 2.5 days
per decade across Europe.[4] This was reflected in earlier flowering, budburst, and leaf unfolding for numerous species of plants. This change
was highly correlated with higher average temperatures for early spring.
Thus, there can be little doubt that plants and animals have responded
to warmer temperatures in the northern temperate zone during the past
few decades.

A remarkably long record of flowering times is available for Japanese
mountain cherries in Kyoto, Japan.[5] The blooming of this species in spring
triggers an annual cherry blossom festival that has been celebrated for more
than a thousand years. The date of the festival is set by assessing flower
buds and compensating for prevailing temperature, so it varies from year to
year and from community to community during the same year depending
on the peak flowering times of the local cherry trees. The dates of the festival for the city of Kyoto were recorded in imperial court records and diaries,
so the time that cherry blossoms opened can be reconstructed back to the
ninth century for 60 percent of the past 1,200 years. The dates of the festival
varied from late March to early May, with considerable year-to-year variation
as well as longer periods with particularly early or late blooming times (rep-
resenting warm and cold periods, respectively). The earliest dates for cherry
blossoms have occurred during the past 30 years. This pattern is partially
due to greater urbanization in the center of Kyoto, which resulted in warmer
temperatures as a result of the replacement of vegetation by pavement and
buildings. This "urban heat island" effect can be measured by comparing
the blooming time in the city with blooming times in nearby rural areas.
During the 1950s, cherry festivals were held at similar times in the center of
Kyoto and the surrounding countryside, but during the past 50 years the
dates have diverged, with festivals occurring 4 to 5 days earlier in the city
center as a result of urban warming. The time of cherry blooming has shifted
to earlier in the season in both urban and rural areas, however, so there is a
more general warming trend superimposed on the urban warming. Only

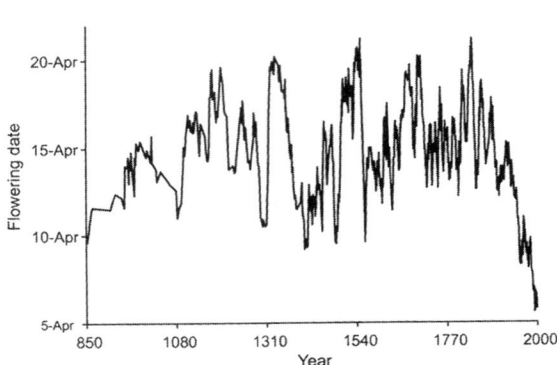

Figure 22. Changes in the average flowering times of cherry trees in Kyoto during the past 1,200 years. (Reprinted from Primack et al., 2009, with permission from Elsevier)

about one-third of the historical change in blooming times is due to urban warming.

These changes in timing in the growth, physiology, and behavior of organisms indicate that species are flexible enough to respond to changes in the climate. Given the history of alternating glacial and warm interglacial periods during the past 2.6 million years, this flexibility is not surprising. Ecologists are concerned, however, that exceptionally rapid climate change could exceed the capacity of many species to adapt, and that different paces of adaptation by different species (for example, for the timing of insect emergence and nesting by insect-eating birds) could cause a breakdown in synchrony, leading to population declines and the disruption of ecosystems. The effect of climate change on deciduous forests will depend on the response of the numerous species that make up these forests. Are these organisms already flexible enough to make physiological and behavioral adjustments to the climate? If not, can they evolve adaptations quickly enough to adjust to environmental changes? And finally, if genetic changes cannot occur quickly enough so that organisms can adapt to environmental changes in a particular place, can they shift their distributions so that they can find suitable habitats (moving northward to cooler regions, for example)? During the major climate changes at the end of the last glacial period, we know that the ranges of numerous deciduous forest species shifted northward, often disappearing from their southern glacial refuges while their overall ranges expanded.

Can Individual Organisms Adjust Quickly Enough to Survive Changes in Climate in Their Local Environment?

Although there is good evidence that organisms have adjusted their seasonal timing in response to warmer springs, it is still uncertain whether these adjustments will be sufficient. A key question is whether their seasonal timing will still be in synchrony with other seasonal events that are important to their survival. This question has probably been best studied for insect-eating birds that depend on a boom in insect abundance in the spring to feed their nestlings. Their schedule for establishing territories, building nests, and laying eggs is timed so that the peak period for feeding nestlings coincides with the peak period of insect abundance.[6] If they begin breeding earlier in response to warmer springs, then they may be out of synchrony with this crucial food supply and fewer of their nestlings would survive.

One of the best-studied populations of songbirds, the great tits of Wytham Woods near Oxford, England, demonstrates that this breakdown in synchrony does not necessarily happen. During the 47 years that this population was monitored, the average date of laying eggs has advanced by 14 days.[7] Most of this change occurred after the 1970s, and it appears to be due to a predictable response of their nesting schedule to warmer temperatures. Interestingly, the abundance of winter moth caterpillars (one of the key food sources for tit nestlings) responded to warmer temperatures in a similar way, so the peak abundance of these caterpillars still coincides with the period when tits feed nestlings even though this happens earlier in the spring. The close relationship between spring temperatures and egg laying each year as well as the lack of variability among females in egg laying dates during a particular year suggest that the response of these birds is due to their flexible response to environmental conditions, not to intense natural selection that has resulted in an evolutionary change. These birds appear to be genetically programmed to respond appropriately to warmer springs.

Another intensive study of great tits shows that their adjustment to warmer springs may not always be sufficient, however. Great tits in Hoge Veluwe, a woodland in the Netherlands, primarily feed their nestlings on caterpillars that eat oak leaves, so they raise more young when the peak period for feeding nestlings occurs after trees leaf out in the spring and before caterpillars leave the trees to pupate in the soil.[8] Although both great tit nesting and the appearance of caterpillars on leaves advanced to earlier in

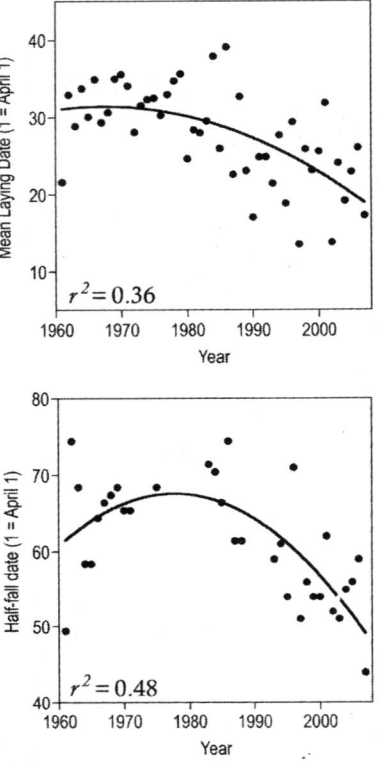

Figure 23. Changes in the average date of egg laying by great tits and the half-fall date for their main prey, winter caterpillars, in Wytham Woods near Oxford, England, between 1961 and 2007. The half-fall date is the date of peak weight of caterpillars. Both average egg-laying date and half-fall date were significantly correlated with spring temperatures, so feeding of great tit nestlings and availability of their main prey have stayed in rough synchrony in this woods as average spring conditions became warmer. (From Charmantier et al., 2008; reprinted with permission from AAAS)

the spring after 1955, the peak of caterpillar abundance now occurs before the period when the nestlings are 11 to 12 days old and need the greatest amount of food. This mismatch occurs because the caterpillars continue to respond to higher temperatures in late spring, while the breeding schedule of the birds does not change once the eggs are laid. The result is that great tits had lower rates of nest success in recent years than they did during the 1950s.

Great tits are not migratory, so they are able to respond to warmer temperatures early in the spring. This adjustment will be more difficult for migratory birds, which spend the winter far to the south, often in the tropics, where they presumably do not have any cues about an earlier spring in their northern breeding area.[9] This appears to be the case with pied flycatchers that nest in deciduous woodlands in the Netherlands and spend the winter in tropical dry forests in West Africa.[10] Migration northward from Africa is triggered by an innate annual rhythm (an internal clock located in the brain) and seasonal changes in day length. As a result, spring arrival dates in the Netherlands have not changed substantially since 1980 despite a steady increase in spring temperatures. The mean date for laying eggs has advanced by ten days, however, which means that these flycatchers are now nesting much more quickly after their return from Africa. Many females now lay eggs five days after arriving back from Africa, which is close to the minimum period needed for egg production, but not sufficiently early to ensure that the young are produced when food is abundant.[11] Consequently, these flycatchers now produce fewer young than they did twenty years ago, and populations in deciduous forests have plummeted. Populations in coniferous woods, where caterpillar populations peak more than two weeks later, did much better during the same period.

Will Organisms Evolve Quickly Enough to Keep Up with Climate Change?

The rigid migratory schedules of pied flycatchers appear to doom them to poor reproductive success as the climate becomes warmer and spring arrives earlier in northern Europe. Pied flycatchers survived numerous glacial and interglacial periods, however, so they must be able to adapt to climate change. One possibility is that they will adjust through a progressive genetic change in the timing of their departure from tropical Africa. If genetic differences cause some individual birds to leave Africa particularly early so that they arrive in Europe at the best time for nesting, then these individuals will generally produce more young. If their young inherit the migration schedule of their parents, then earlier migration times will become increasingly common until the yearly schedule is again in synchrony with the northern spring. This type of natural selection—in which populations are genetically molded by a changing environment—has now been documented in many populations of animals and plants in both laboratory experiments and wild populations.[12] It can only occur, however, for characteristics that are under genetic control. There must also be genetic variability among individuals; otherwise, natural selection cannot favor some genetic types while weeding out other genetic types.

The way in which migration is controlled has been worked out in a remarkable series of breeding experiments in several species of European warblers.[13] These experiments required raising birds in aviaries and then breeding them in flight cages for two or more generations. During spring and fall migration these birds became restless at night, when they would normally be engaged in their migratory flights. When placed in circular test chambers, they hopped in the migratory direction and whirred their wings, effectively pantomiming a nightlong migratory flight. This occurred even when the birds were hand-raised by people and were not able to learn about the migratory route from their parents or other experienced adult warblers.

These experiments reveal that the major features of migratory behavior are determined genetically. The young birds have a built-in genetic program that tells them when and where to migrate. In blackcaps (one of the best studied species of warblers), genes strongly influence whether or not particular individuals migrate, the time of year when migratory behavior begins, the duration of migratory flights, and the migratory direction.[14] When birds from populations with different migratory patterns breed together, the offspring show intermediate behavior. For example, hybrid offspring of blackcaps from Germany and the Canary Islands display neither the long duration of migratory restlessness of German birds (which must travel hundreds of miles south to Africa) nor the short period of migratory restlessness of Canary Islands birds (which are much closer to their winter areas), but an intermediate duration.

These breeding experiments also demonstrate that there is a large amount of variation in migratory patterns in blackcaps from the same population, indicating that natural selection could cause relatively rapid adjustments to a changing climate. Breeding experiments with birds from a population that includes both migratory birds and birds that are year-round residents confirmed that selective breeding (breeding only individuals that were either migratory or nonmigratory) could create a population that was completely migratory or completely sedentary within a few generations.[15] In another breeding experiment with captive blackcaps, individuals were selected for breeding if they initiated migratory activity later in the autumn.[16] After only two generations of breeding, migratory activity had shifted two weeks later. Thus, evolutionary change is one way that these birds could eventually adjust to a warmer climate.

Perhaps the most dramatic example of evolutionary changes in migratory patterns is the development of a new migratory pathway for blackcaps during the past few decades.[17] Since the 1950s, blackcaps from continental Europe began spending the winter in England in large numbers for the first time. The wintering birds had numbered bands (rings) from Germany and Austria, indicating that they were not birds that nested in Britain during the summer. Peter Berthold and his colleagues captured some of these winter-resident blackcaps and bred them in aviaries in Germany. When their young were tested in the autumn to determine their migratory direction, they consistently hopped to the west-northwest, toward England. In contrast, most of the young of local German blackcaps hopped toward the southwest, in the direction of the more typical migratory route through southern Spain. Unlike their parents, the young of birds from the English population had never made the trip to England and so did not have an opportunity to learn the correct compass direction. This suggests that they had built-in directions to the English wintering area, and that this new migratory pathway has evolved in a very short time in response to a milder winter climate in England or other environmental changes, such as an increase in popularity of bird feeders (feeding stations).

A key question, however, is whether environmental changes, including climate change, are now occurring too quickly for populations to adjust by evolving.[18] These evolutionary changes are driven by natural selection, so ultimately they result from the dominant genetic types being weeded out as a result of low survivorship or lack of breeding success. Rapid evolutionary change therefore requires high mortality or depression of reproduction, and this may eliminate the population before substantial genetic change can occur.[19] Thus, although evolutionary adjustments may occur in some species, other species may not be able to change quickly enough to survive in a changing environment.

Can Organisms Shift Their Distributions to Survive Changes in the Climate?

If a species cannot adjust its breeding behavior quickly enough in response to climate change, and if it cannot evolve quickly enough to keep up, then it may still survive by shifting its distribution into regions that have a favorable climate. As the climate warms, species could shift their ranges northward to find an appropriate climate that matches the environment where they now live. This is how many species survived the extreme changes in climate as glaciers expanded and then retreated during the Pleistocene.

As discussed in detail in Chapter 3, we know from the pollen record in lake and bog sediments that forest ecosystems changed continuously for many thousands of years following the last glacial retreat. Different species of trees spread northward into recently deglaciated areas at different rates. This process has occurred in each interglacial period, with forest ecosystems slowly reassembling. Each new incarnation of the deciduous forest had a somewhat different set of species, however. It's therefore not surprising that temperate-zone forests have few of the closely dependent relationships between two particular species that are frequent in the tropics. Instead of obligatory relationships between a particular species of flower and a particular species of pollinating bee, for example, the temperate zone is characterized by a set of pollinator species that pollinate a set of flower species. The assemblage of species in forest ecosystems has been reshuffled too often for highly dependent relationships between particular pairs of species to evolve or persist as frequently as they have in the tropical lowlands.

Ecologists have found that the distributions of some well-studied groups of organisms have already shifted northward in recent decades. Some of the best evidence for range shifts comes from bird atlases in North America and Europe. These atlas projects are completed by large numbers of volunteers who go into the field to search for birds during the breeding season to determine whether they are found in particular regions. A grid is superimposed on a map of a country, province, or state, and observers record the birds within each cell of the grid. New York state, for example, was covered with a grid consisting of more than 5,000 blocks, each of which was a 5-kilometer by 5-kilometer square. Volunteers were assigned particular survey blocks that they searched for breeding birds, recording the occurrence and evidence for breeding of each species. Atlases were completed in New York in 1980–1985 and twenty years later (2000–2005).[20] The results of these two atlases were compared to determine whether species with either a southern boundary or a northern boundary within the state shifted these boundaries northward as would be predicted as the climate steadily became warmer. Twenty-two species with southern range boundaries in New York state showed northward contraction of their ranges. These included species associated with deciduous forest, coniferous forest, and scrubby habitats, so the birds with shifting ranges were not associated with one particular habitat. The average northward shift was 5.7 kilometers per decade. The ranges of species with a northern range boundary in the state tended to expand northward during the 20 years between the atlases, but the evidence for this pattern was less clear. In general, however, New York birds appear to be responding to warmer temperatures by shifting their distributions in predictable ways.

Changes in the distribution of species in Europe usually involve southern species expanding their ranges northward. In an analysis of 99 species (9 alpine herbs in Switzerland, 59 birds in the United Kingdom, and 31 butterflies in Sweden), range boundaries moved an average of 6.1 kilometers per decade northward, or 6.1 meters per decade upward in elevation in the mountains.[21] These more southerly species were expanding their distributions farther north or farther up mountain slopes as the climate warmed. A similar pattern was detected for a wide range of different types of organisms in Great Britain, including spiders, woodlice, millipedes, fish, and many different types of insects.[22] The average northward range expansion for these species ranged from 14 to 25 kilometers per decade.

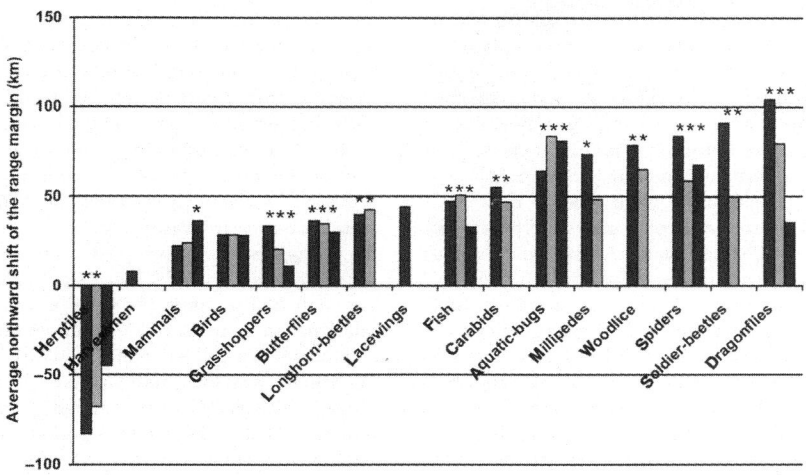

Figure 24. Average latitudinal shifts in the northern range margins over the past 40 years for 16 groups of animals in Great Britain. The three bars for each group represent the intensity of sampling, from "recorded" on the left to "well-recorded" in the middle and "heavily recorded" on the right. Asterisks indicate statistically significant changes. Only one group (herptiles, which comprises reptiles and amphibians) showed an average southward shift. (Hickling et al., 2006; reprinted with permission from John Wiley and Sons)

Impact of a Warmer Climate on Tree Distributions

Birds and many other animals have already shifted their ranges in response to warmer conditions in the temperate zone within the past few decades. Ultimately, however, many of these species require deciduous forests, not boreal coniferous forests, so expansion of their ranges northward will be limited by whether deciduous forest plants (particularly trees) spread northward. The pollen record from Europe demonstrates that the distributions of deciduous forest tree species have shifted in response to climate change during the geologically recent past.[23] Nine thousand years ago, deciduous forests with a mix of tree species were found farther south than they are today, and birch woods dominated northern Europe. As the climate became warmer deciduous tree species shifted northward, reaching their peak northward extent by about 6,000 years ago. Subsequently the climate cooled and the northern limits of their distributions shifted southward. The pattern was similar in eastern North America, where many tree species reached their maximum northern limit 6,000 years ago.[24] After that time, spruce and fir expanded their range southward as the deciduous forest retreated and the boreal forest expanded. In both Europe and North America, these changes occurred slowly, however, over thousands of years.

The warming of the earth due to increases in carbon dioxide in the atmosphere is predicted to be much more rapid than these earlier changes in climate, which usually were driven by gradual changes in the angle at which the earth tilts in relation to the sun, and the shape of the earth's orbital path around the sun.[25] If these predictions of rapid climate change are correct, then distributions of forest species will need to shift considerably during the next hundred years to stay within acceptable climate zones. Can trees, which have long generation times and limited dispersal capabilities, keep up with this rate of change?

The potential effect of climate change on particular species of trees has been studied using mathematical models called climate envelope models. These models incorporate information about the climate requirements of a particular species, with an emphasis on its tolerance of high and low temperatures and arid conditions, to determine where the species would be able to live in a future world with different climate zones. A model of this sort was developed for dominant species of trees of European deciduous forests.[26] The tolerance of each species to heat, cold, and desiccation was determined by assessing its current distribution and response to experimental conditions, such as growth in soils with different capacities for holding water. Some of the key variables in the model were soil water requirements, mean coldest-month temperature (a good measure of the overall severity of winters), and number of days in the growing season following spring leaf-out needed for growth and reproduction. For deciduous trees a mild winter without a chilling period will lead to a delay in spring budburst, reducing the growing season and hampering growth and production of seeds. Winters that are too long or too cold can also inhibit growth and reproduction, as will soils that are too dry or too saturated with water. Models based on this small set of variables roughly predict current distributions of particular species of trees in Europe. Assuming climate change due to a doubling of carbon dioxide concentrations in the atmosphere, the future distributions of three dominant tree species were predicted with the same model. With the predicted climate change, Norway spruce would disappear from most of the continent, with isolated pockets of spruce remaining on high mountains and larger blocks of spruce in northern Finland, Scandinavia, and farther east in Siberia. European beech would spread northward almost to the Arctic Ocean due to mild winters, but would contract away from the western edge of Europe because winter would be too mild close to the ocean, with no chilling period to trigger spring budburst. Pedunculate oak (English oak) would also extend its distribution far northward of its current range, but would not show a large decline in the west.

   Climate envelope models predict similar large-scale shifts in the distribution of North American trees.[27] These models were based on an analysis of locations of trees of 130 species in the United States and Canada. There were also two climate scenarios: increasing production of greenhouse gases such as carbon dioxide, and decreasing production of greenhouse gases due to energy conservation and conversion to renewable energy sources. Future tree distributions were predicted based on models that assumed either perfect dispersal to favorable locations or no dispersal away from current locations. Under all of these models, the average distribution of species shrank and their centers of distribution shifted northward. These changes are particularly severe under the no-dispersal scenario, which predicts that average geographical ranges of species will shrink by 58 percent while their centers of distribution will shift northward by 333 kilometers. When effective dispersal is assumed, then the northward dispersal is 700 kilometers and the decline in the average size of the range is only 12 percent. The authors suggest that the actual values will lie between these two extremes. Trees are likely to spread northward, but more slowly than would be ideal.

Will "Assisted Dispersal" of Species Be Necessary?

Although trees and other organisms have shifted their ranges over hundreds of kilometers as a result of climate change in the past, a major concern is that the climate is now changing so rapidly that many species will not be able to disperse into favorable areas quickly enough.[28] If this is the case, then it may be necessary to artificially move organisms into nearby regions where the climate has become favorable. This proposal is controversial, however, because it would require conservationists to introduce species into regions and habitats where they are not currently found. In the past, moving organisms to new regions has resulted in disruption of natural ecosystems. The effects of these introduced species range from very mild to catastrophic, and ecologists are understandably wary of creating catastrophic disruptions of the sort described in the next chapter.[29] Thus, if organisms are moved northward or upward in elevation as the climate changes, it will be important to move them gradually to areas just beyond their current geographical range, speeding up their dispersal into adjacent regions with newly favorable climate.[30] This would mimic the kind of dispersal and reshuffling of species in temperate-zone forests that occurred at the beginning of each interglacial period as trees spread northward into areas that were previously covered with arctic steppe or ice.

Limits to Resilience in the Face of Climate Change

Although the organisms associated with northern deciduous forests are resilient and capable of adjusting to major changes in climate, there is a limit to their resilience. The high extinction rates of plants and animals in Europe during the Pleistocene glacial advances confirm that these species can be pushed too far by climate change. Europe lost most of its tree species during this period. The great fear among ecologists is that the current period of climate change, driven by rapid industrialization, is proceeding much more swiftly and to much higher average global temperatures than has occurred during the history of living species. This could lead to extinctions even among species that have survived repeated natural glacial cycles. We know from the pollen record that trees and other plants shifted their ranges slowly over thousands of years after continental glaciers melted away, so the process appears gradual compared with the increase in global temperatures documented in recent decades. Some natural climate changes may also be rapid, however. Analysis of the annual layers of accumulated snow in glaciers in Greenland indicate that shifts from cold to warm periods can be surprisingly fast, with substantial warming occurring over a period of decades or even years at the end of the last glacial period and at the end of the temporary cold period (the Younger Dryas) that occurred near the beginning of our current warm interglacial period. Environmental models predict global warming due to greenhouse gases that is unprecedented during the period that most modern species have existed, however, and ultimately the magnitude of warming might be a greater problem than the rate of warming. Global temperatures are already approaching the highest recorded since the last glacial period, and they are projected to continue to rise steadily at a rate that will depend on the rate at which fossil fuels are burned.

Numerous studies indicate that average temperatures at northern latitudes were considerably warmer 5,000 to 6,000 years ago than they are today; glaciers were smaller, areas now covered with sea ice were open water, and the boundary between tundra and boreal forest in Canada was 300 kilometers farther north.[31] The overall cooling of the earth since that time, which was driven by changes in the earth's tilt and orbit, has been reversed during the past 125 years.[32] Most climate scientists agree that the reversal in the cooling trend is primarily due to the release of carbon dioxide and other greenhouse gases by people, and that warming will continue until average global temperatures are higher than they were 6,000 years ago.[33] Numerous climate models based on a wide range of assumptions about rates of fossil fuel use, economic growth, and human population growth predict that increasing greenhouse gas concentrations will cause a substantial increase in average global temperatures, which will create special challenges for organisms that evolved in the Pleistocene, a period dominated by long, cold glacial periods and short, warm periods.[34] Although most species can shift their ranges northward or to higher elevations, this will not be the case for species that are already restricted to mountaintops or the fringes of the Arctic Ocean. These species will have no place to move. Another vulnerable group will be species that are restricted to low-lying coastal areas such as tidal marshes that will be flooded as the sea level rises because of melting glaciers in the Arctic and in Antarctica. In the past tidal marshes and other coastal habitats gradually shifted inland in response to the advancing sea, but in many regions of the world this process will be blocked to save cities, towns, and valuable farmland adjacent to the coast. Deciduous forests should fare better than these acutely vulnerable fringe habitats, but even deciduous forest species may have difficulty adjusting.

Implications for Conservation

During the past few decades, conservationists have learned that it is not sufficient to protect a piece of forest from development and human disturbance. It is also critically important to manage the region around the forest to sustain biological diversity. The goal may be to prevent regional forest fragmentation; establish natural corridors or "greenways" between forests; or to coordinate the creation of forest disturbances in different forests to sustain a range of successional stages, from forest openings to ancient forest. We are now beginning to realize, however, that this type of regional planning, even at a large geographical scale, is insufficient. Forests are threatened by global changes that can ultimately be addressed only by international cooperation. At the local level, conservationists can only hope to ameliorate the effects of changes that occur on a continental or global scale. The causes of these changes are largely beyond their control.

Global climate change represents one of these serious global challenges, particularly for vulnerable ecosystems such as tidal marshes, alpine meadows, and sea ice communities. In temperate deciduous forests climate change will probably cause a major reshuffling of species, but we know from the pollen record in lake and bog sediments that similar changes have happened before. In fact, the tree species composition of northern forests changed continuously for thousands of years after the most recent retreat of continental glaciers. The forests of previous interglacial periods often had a different mix of species than current forests at the same sites, and during the glacial maximum most deciduous forest trees in Europe and North America survived only in relatively small southern refugia near the Mediterranean Sea or the Gulf Coast. Thus, trees and other organisms in temperate deciduous forests have survived major climate changes in the past, and they are likely to be resilient to future climate change. There are two caveats, however. Current climate change may be occurring more rapidly than during transitions from glacial to interglacial conditions, and the climate is projected to become warmer than at any time during the current interglacial period. It therefore may be necessary to incrementally assist the spread of less mobile species to more favorable climate zones. Although this will involve the introduction of species into regions where they don't currently live, it is much less risky than moving species between forest ecosystems on different continents. Also, it would simulate the natural dispersal of organisms north and south during glacial cycles. Assisted dispersal may be particularly important where people have created artificial barriers such as metropolitan corridors or extensive expanses of farmland.[35]

Of course a more general and effective response to the problems of climate change would be to reduce the release of greenhouse gases that cause global warming. A substantial lowering of greenhouse gas production will require worldwide changes in energy production and transportation systems. A shift from fossil fuels to renewable energy would not only curtail greenhouse gas emissions, but would also alleviate other problems such as particulate pollution, acid deposition, and local air pollution that affect both human health and the health of forests and other ecosystems.

CHAPTER 9

Another Global Threat: Transport of Species Between Continents

Although climate change may be a serious long-term threat, the most serious immediate threat to northern deciduous forests is the spread of pathogens and insects that can decimate or even eliminate particular species of trees from a region within a few years. Temperate deciduous forests are resilient in many ways, but they may be unusually vulnerable to this particular problem. Trees in the three isolated blocks of deciduous forest in Europe, East Asia, and eastern North America are closely related. Distinctive species of maples, oaks, beeches, hemlocks, elms, pines, and chestnuts are found on the different continents. On each continent these species have evolved defenses (usually chemical toxins) to inhibit the growth of the parasitic fungi and herbivorous insects that prey upon them. Many of these parasites and plant-eaters are specialized, attacking only elms or chestnuts or oaks. When they are introduced to a new continent, they may find close relatives of their original hosts, such as American chestnuts or sweet chestnuts rather than Chinese chestnuts. These may be suitable hosts that lack effective defenses to the new parasite or insect. The result is similar to an epidemic in a human population with no resistance to a virulent disease; a high proportion of the population may be killed. This is a problem in all three of the deciduous forest regions, but the effects have been particularly acute in Europe and North America. East Asia has a much greater diversity of trees, so it is a richer source of pathogens and insects that can devastate particular types of trees on the other continents.

Pathogens and Insects Threaten Key Tree Species

The profound effects of introduced pathogens are exemplified by the spread of chestnut blight, a parasitic fungus of chestnuts that evolved in East Asia. Chestnut fungus reached North America by 1904, probably by traveling on live chestnut trees imported from Japan or China. It rapidly spread outward from New York City to the forests of New England and upstate New York. Dispersing at a rate of 40 kilometers per year, chestnut blight reached nearly every forest occupied by wild American chestnut within 40 years.[1] Many of these forests were "oak-chestnut forests" dominated by American chestnut and several species of oaks. They rapidly lost all of their tall chestnuts, leaving oaks and hickories as the dominant trees.[2] Billions of trees were killed as the fungus spread northward to Maine and southward to Georgia, killing virtually every mature chestnut across vast forests in which 40–50 percent of the canopy trees were chestnuts.[3] Fungus spores were spread both by wood-boring beetles and by the wind, so the blight spread quickly and jumped barriers where chestnuts had been removed in an attempt to stop the spread of the disease. The economic and ecological losses were considerable because chestnuts produced an abundant and dependable crop of nuts that were eaten by people, their livestock, and a large number of wild animals. Also, chestnut wood was especially highly valued, and fast-growing chestnuts were often planted near houses. Other species replaced the missing chestnuts in the forest canopy, and chestnuts still hold on as small sprouts that grow for a few years before they are knocked back by the fungus. Most of these sprouts grow from the roots of what were chestnut seedlings when the blight removed the tall chestnuts of the forest canopy.[4] Only a small percentage of the sprouts are associated with the stumps of large chestnut trees. Chestnut sprouts form an important component of the shrub layer in some forests, so wild American chestnuts have not disappeared entirely. A key species has been removed from the forest canopy in much of the eastern deciduous forest, however. In some regions the loss of chestnut may have a more long-lasting ecological impact on forests than logging or even agricultural clearing, which were often temporary interruptions in the normal regeneration of mature forest.

The chestnut blight also spread to southern Europe, where it attacked the commercially important sweet chestnut.[5] Sweet chestnuts have largely recovered, however, because the fungus that causes the disease was infected with a virus that reduced its virulence. Sweet chestnuts are still infected with the fungus, but they typically survive and grow despite the infection. The virus was introduced to eastern North America in hopes that American chestnuts would also benefit, but unfortunately it has not reduced the virulence of the fungus enough to allow American chestnuts to recover.

Another pathogenic fungus, Dutch elm disease, had a major impact on the forests of both Europe and eastern North America. The origin of this

Figure 25. Photograph from about 1910 of ancient American chestnuts in the Great Smoky Mountains of North Carolina. (Courtesy of the Forest History Society, Durham, North Carolina, image ID# FHS3245)

disease is uncertain, but it is thought to be from Asia where a closely related species of elm fungus has been discovered.[6] Two species of fungus that cause Dutch elm disease have been identified, *Ophiostoma ulmi* and *Ophiostoma novo-ulmi*. Both are primarily spread from tree to tree by elm bark beetles. These two types of fungus caused two successive epidemics in North America and Europe. *O. ulmi* spread across Europe in the 1910s and across North America in the 1930s. It had a greater impact on European elms, killing 10–40 percent of the elms in most countries, but began to decline in virulence in the 1940s due to the spread of a virus that affected the parasitic fungus. As with chestnut blight, however, Dutch elm disease displayed no reduction in virulence in North America. In the 1940s, *O. novo-ulmi*, which is even more destructive than *O. ulmi*, began to spread independently from the Great Lakes area in North America and from eastern Europe. The Great Lakes fungi spread across North America and crossed the Atlantic to Great Britain, while the eastern European fungi spread across most of the rest of Europe. The new elm disease (*novo-ulmi*) replaced the old elm disease (*ulmi*) as it spread across both continents. Millions of elms were killed in Great Britain and other parts of Europe, and the American elms that dominated streamsides in natural areas and that shaded the residential streets of many North American cities and towns were decimated. American elms were one of the dominant trees of floodplain woodlands, so their loss had a major ecological impact along streams and rivers. In the United Kingdom the new elm disease killed more than 25 million of an estimated 30 million elms.[7] The once common English elm was especially susceptible, and, like the American chestnut, it now survives primarily in the form of root sprouts that are knocked back by the disease before they grow very tall. English elm was an important hedgerow and ornamental tree that had been introduced by the Romans. Genetic analyses indicate that the millions of English elms in the United Kingdom derive from a single genetic clone that was brought to England about 2,000 years ago.[8] It has been propagated vegetatively since that time. This clone was used in Roman vineyards and was probably planted in English vineyards by the Romans. Pruned elms were used to support grape vines. The lack of genetic diversity in the descendants of these elms due to vegetative propagation of a single clone probably increased their susceptibility to Dutch elm disease. The wych elm, which is native to England, is more resistant to the disease so more of these elms have survived.[9]

Although most of the serious tree diseases appear to have spread from East Asian forests to the forests of North America and Europe, Asian forests have also been damaged by imported pathogens. Pine wilt disease is a particularly destructive epidemic disease that has decimated pines in the forests of China and Japan. It is caused by the pine wood nematode (a tiny roundworm) that was introduced from North America in the 1930s.[10] In North America this nematode is primarily associated with dead or dying conifers, and is not known to cause pine wilt disease.[11] The nematode may have reached Japan in the early 1900s and subsequently became abundant in pine forests, particularly in the southern part of the archipelago. It has caused extensive mortality of Japanese red pines, which are a key component of secondary forests in Japan. Japanese red pines often grow with maples, oaks, and other deciduous trees in natural woodlands that are recovering from centuries of intensive cutting and harvesting of leaf litter for fertilizer.[12] These pines are often the dominant trees on dry ridge tops and steep slopes with poor soil, so the loss of large numbers of pines has major ecological consequences. Japanese red pines also play an important role in the early stages of forest regeneration, and their loss is probably ecologically disruptive. And like the loss of the tall American elms that shaded residential streets in eastern North America, or the rows of English elms along hedgerows and rural roads in England, the decline of Japanese red pines represents a severe loss in terms of the traditional cultural landscape and a sense of place. Gnarled pines are a common theme in Japanese art and carefully trained pines with elegant shapes are a standard feature of traditional gardens, so both wild and planted Japanese red pines are greatly valued.

Introduced insects can also cause major declines in forest trees. In North America the gypsy moth is particularly well known because its caterpillars periodically reach extremely high densities and defoliate large areas of forest, creating a winterlike forest canopy of bare trunks and branches.[13] Although healthy trees can grow new leaves, repeated defoliation will eventually weaken and kill trees, particularly if they are subject to other stresses such as drought. Gypsy moth caterpillars show a particular affinity to oaks, so oaks have declined in the canopy of some forests as a result of repeated caterpillar outbreaks. The composition of forests is usually not substantially affected by gypsy moths, however, and the long-term negative effects are not comparable to chestnut blight and Dutch elm disease.[14]

Another insect, hemlock woolly adelgid, has the potential of causing a dramatic change in eastern deciduous forests because it causes massive mortality in eastern hemlocks. Hemlocks are an important canopy species that grows in almost pure stands along streams and creeks, or mixed with deciduous hardwoods in mature forests, providing a coniferous component to the eastern forest that is important for many species of animals and other organisms. Woolly adelgids are tiny insects that suck the sap of hemlocks.[15] In North America they occur at such high densities on hemlock branches that they can kill the host trees within a few years. In many cases, virtually every hemlock in a forest is killed, and groves of ancient hemlocks with a dense canopy and deeply shaded forest floor are quickly converted to more open forest consisting of young black birch and red maple, species that are already ubiquitous in the surrounding region.[16] The amount of forest cover is not reduced, but the forest becomes more homogeneous. The result is an overall loss in biological diversity as species associated with hemlock, such as black-throated green warbler and Acadian flycatcher, decline.[17] The ecology of woodland streams may also change with the loss of deep shade and cool temperatures typical of streamside hemlock forests.[18]

Hemlock woolly adelgids were introduced to eastern North America before 1951, when they were first recorded in Maymont Park in Richmond, Virginia.[19] The park was once the estate of Sallie Dooley, an amateur horticulturist who collected plants from around the world. In 1911 she created a traditional Japanese garden with the help of a master gardener from Japan. It is likely that adelgids arrived on hemlocks ordered from Japan for this garden. If so, then adelgids spread slowly for several decades, a pattern that is often seen in introduced insects. After the 1950s, however, adelgids spread much more rapidly northward to New England and southward along the Appalachian Mountains.[20] The adelgid range has expanded at a rate of about 30 kilometers per year. Adelgids can reproduce asexually, so their populations grow quickly. They are transported on animals but are also small and light enough to be carried by the wind, so they spread rapidly.

I've described only a few of the potential threats to temperate deciduous forests from introduced pathogens and insects. In eastern North America, butternuts (a type of walnut) are threatened by butternut canker, ashes are threatened by emerald ash borer, white pines are threatened by white pine blister rust, and maples, birches, and other trees are threatened by Asian long-horned beetles.[21]

Strategies for Controlling the Spread of Forest Pathogens and Insects

Because of the volume of trade among the nations with deciduous forests in Asia, Europe, and North America, it is a challenge to stop fungi and insects that pose a threat to trees from spreading between continents. Given the magnitude of the economic and ecological threats posed by these introduced species, however, it is surprising how weakly most countries have regulated the importation of raw timber (which can spread fungi and wood-boring insects) and live plants (which can spread both plant diseases and insects). Part of the problem is that the enormous economic cost of invasive species is not widely appreciated, making it difficult to impose expensive inspection policies and trade restrictions. A 2011 study of the economic impact of introduced forest insects in the United States illustrated how serious this problem has become.[22] Of the 455 species of forest insects introduced into the United States from other parts of the world, 62 species have documented economic costs. Wood-boring insects had the greatest economic cost because they often kill a high proportion of infested trees. The estimated annual cost of control programs and tree removal due to invasive wood borers was $92 million for the federal government, $1.7 billion for towns and cities, $760 million for private homeowners, and $130 million for timber owners. In addition, the estimated loss in residential property value due to loss of trees was $830 million. The analysis did not take into account ecological costs such as loss of biological diversity and reduction of water quality, because these costs are difficult to estimate in dollars. Most of the costs are spread among numerous municipalities and private property owners whose interests are not well represented when trade policies and import regulations are established at the federal level. Foliage-feeding insects such as gypsy moth and sap-feeding insects such as hemlock woolly adelgid also cause substantial damage, but these were substantially less than for wood borers.

Once a tree-destroying pathogen or insect is well established, it becomes extremely difficult to control its effects. The best policy is to prevent the importation of potential threats into new continents. Wood borers are primarily imported on timber and unprocessed wood products, so more careful regulation of trade in these products could reduce their rate of spread between continents. Although more invasive forest insects have spread from East Asia to Europe and North America rather than in the other direction, and North

America has gained more forest pests from Europe than vice versa, all of these forest regions are at risk of enormous economic and ecological losses from invasive species of insects and fungi that attack trees.[23] Trade restrictions on transport of unprocessed timber and wood products (particularly packing crates) would benefit all countries with deciduous forests. Restrictions on importation of live plants or rigorous inspection, quarantine, and fumigation procedures for live plants would prevent a large set of other potentially invasive organisms from spreading, and would ultimately benefit all nations in these regions.

In the absence of regulations covering importation of all wood products and live plants, another alternative is to identify insects and pathogens that are a clear threat and then develop regulations to prevent their introduction to a new continent. One of the most damaging tree pathogens, white pine blister rust, could have been kept out of North America with a policy of this sort. The disease originated in East Asia, where pines show a high degree of resistance. After it was introduced to Europe in the 1800s, it spread rapidly. This fungus has a complex life cycle that requires the presence of two hosts, gooseberry or currant shrubs and trees in the white pine group. Gooseberries are native to Europe, and eastern white pines from North America were planted extensively in Europe, so the fungus was able to invade much of northern and western Europe. Plant pathologists warned about the danger of the fungus spreading to North America, but there were no laws restricting the importation of live plants. By the early 1900s white pine blister rust had arrived in the eastern United States, where it began to infect eastern white pines. Later it spread to forests of western North America, where it had a devastating impact on western white pines and sugar pines, killing or damaging nearly all stands of these species. The fungus was almost certainly spread from Europe to North America on nursery stock. Specific importation restrictions for white pine, gooseberry, and currant probably could have prevented this.

The European Economic Community (EEC) has begun to institute restrictions on plant products to prevent the importation of known threats to trees and forests. One serious threat is pine wilt disease, which originated in North America and has decimated pine forests in Japan and China. In 1993, the EEC required that all raw pine wood entering Europe must be kiln-dried or kiln-heated to kill the nematode that causes this disease. Another danger to European trees is oak wilt, which is caused by a fungus that is a serious threat to oaks in some regions of eastern North America.[24] Oak wilt may be native to North American forests, but it is more likely that it was introduced from high-altitude oak forests in the mountains of Mexico, Central America, and South America.[25] This disease is not known in Europe, but it represents a potential threat to the oak-dominated woodlands and forests there. After tests confirmed that fumigation with methyl bromide kills the oak wilt fungus, the EEC permitted importation of oak logs only if they had been fumigated. Some European countries also specified that all bark must be removed from oak logs before they were imported. In addition, plantations of European oaks were planted in West Virginia and South Carolina, where they could be inoculated with oak wilt fungus.[26] These oaks experienced a high mortality rate, indicating that the threat to European forests is real. These European Community (now European Union) programs have proven effective in preventing the spread of known threats. The main weakness, however, is that they are ineffective at preventing unanticipated threats represented by species that do not cause problems in their native ranges. The pine wilt nematode was recognized as a threat only because it reached East Asia before it reached Europe; in its native North America it was not even considered a pathogen. Broader regulations for importation of wood products and live plants would be more effective.[27]

Regulations in the United States now require debarking and heat-treating of most wood products to kill any pathogenic organisms or insects.[28] Wood packing materials were initially exempted from this requirement, but they are now covered.[29] Importation of live plants is not as well regulated, creating a continuing threat to both urban trees and rural forests. Leaf-feeding and sap-feeding insects and plant pathogens arrive in North America on live shrubs or trees imported from Europe or Asia, and the volume of live nursery stock imported into the United States has increased dramatically.[30] Although most imports are from Canada and therefore are less likely to carry new insect pests or pathogens, the number of plants imported from Asia has increased substantially. Inspectors with the Animal and Plant Health Inspection Service examine imported plants, but the volume of plants results in many pests getting past the inspection stations and reaching nurseries and private gardens. Australia has a more effective system that permits importation only of plants that have passed a pest risk assessment that demonstrates that they pose little risk. Even plants on this list are subject to quarantine. This "guilty until proven innocent" approach to screening plants should be considered by other nations to prevent the sort of catastrophic outbreaks that have resulted from unregulated or poorly regulated importation of plants and plant products.

Control of Forest Insects and Pathogens After They Arrive on a New Continent

Once a potentially invasive species arrives on a new continent, the best hope for eradicating it is during the first few months or years, before it is well established and widespread. The importance of a rapid response to new threats to trees has long been recognized, and in fact was deployed after gypsy moths were accidentally released in 1869 by an amateur naturalist in Medford, Massachusetts.[31] Unfortunately, nobody responded to the initial release, but there was a vigorous response when the first outbreak of gypsy moth occurred in 1890. This entailed a sustained eradication effort that involved insecticides, manual removal of egg masses, and even burning of infested woodlands. By 1900 the eradication effort appeared to be a success and the program was ended, but another outbreak occurred five years later. In this case the eradication program probably was too little, too late; it's important to respond as soon as a potential forest pest is detected, and then to sustain monitoring and eradication programs for a long period.

Intensive containment and eradication programs followed the introduction of Asian long-horned beetle into North America and Europe. This species is a major threat to both North American and European forests because it frequently kills healthy hardwood trees of a wide variety of species (maples, poplars, birches, willows, and horse chestnuts).[32] These beetles were spread in wood packaging materials to numerous ports in North America and Europe, and populations have become established in Chicago, New York, Massachusetts, New Jersey, France, Italy, Austria, and Germany.[33] Specific responses to Asian long-horned beetle beachheads vary from country to country, but they all combine rigorous monitoring with destruction of infested trees. Potential host trees are inspected for signs of the beetle from the ground with binoculars or from bucket trucks or by tree climbers. Intensive surveys for the beetles are conducted every year within 800 meters of any infected tree. Less intensive surveys are conducted in the surrounding region. All infected trees are cut down and chipped or burned. Uninfected trees of vulnerable species are either cut down and chipped or injected with insecticides that kill beetle larvae and adults. The estimated cost of this program in the United States between 1997 and 2008 was $373 million. Some eradication programs have been declared a success (which happens only after no beetles are detected after intensive monitoring for four or five years), and other programs are still in progress. Perhaps the most dangerous infestation in North America is in Worcester, Massachusetts, where long-horned beetles had already spread widely and colonized natural forests outside the urban area before they were detected.[34] Consequently, the monitoring and eradication programs had to cover a large area. Maples and other susceptible trees were cut down and chipped both in the city and in adjacent forests, removing the tree canopy of many neighborhoods and dramatically changing the composition of the forests. In the short term, the beetle didn't cause an increase in tree mortality in the forests, but the average growth rate of infested trees was lower in a small patch of forest. The long-term effects of the beetle could not be studied because all susceptible trees were eradicated as part of the control program, but Asian long-horned beetles could have a major impact on forests dominated by maples and other susceptible species. The outcome of the eradication campaigns is therefore critically important for the future of deciduous forests in eastern North America.

After an introduced species such as chestnut blight or hemlock woolly adelgid becomes widespread and abundant, complete eradication is probably no longer feasible. The strategy then must shift to reducing the impact of the invasive species. Eventually introduced insects and tree pathogens will probably become integrated into the forest ecosystem as they acquire their own predators and diseases, and as the host trees evolve resistance. This type of ecological response has already occurred with chestnut blight in Italy, where a virus infects the blight fungus, reducing its virulence to chestnut trees so that many live long enough to reproduce.

The Promise and Risks of Biological Control

Native predators and parasites may eventually adapt to a new host, such as hemlock woolly adelgid, but this process will take a long time and will not protect hemlock forests in the short term. In the meantime, the two species of hemlocks in eastern North America (eastern hemlock and Carolina hemlock) may be lost from many forests, along with any other organisms that are closely associated with hemlocks. The loss of these important species might be prevented if specialized predators and parasites of the invasive species can be located in its original habitat and released in the threatened habitat.

Using living organisms to control an invasive species (biological control) is potentially less environmentally disruptive than using chemical compounds that are poisonous not only to the invader, but also to a wide range of other organisms. Unfortunately biological control acquired a bad reputation because of numerous ill-conceived attempts to use generalized predators (such as small Asian mongooses) to control pests. Often the introduced predator became a threat to native species and had little impact on the targeted pest population.[35] Biological control programs are prone to cause severe ecological damage unless they are planned and tested with extreme care. The critical criterion for a potential biological control agent is that it should be highly specific to the invasive species so it does not threaten native species. This requires a detailed study of the life cycle of the biological control agent as well as careful tests in the laboratory or greenhouse to determine whether it attacks closely related or otherwise similar native species.

When hemlock woolly adelgid began killing hemlocks in southern New England, Mark McClure, an entomologist at the Connecticut Agricultural Experiment Station, responded almost immediately by arranging to travel to Japan to study this invasive species in its natural habitat. His trip was partly supported by the Steep Rock Association, a nonprofit land trust in Washington, Connecticut.[36] With the help of Japanese biologists, he visited 76 widely scattered sites where hemlocks grew in southern Honshu.[37] About half of the sites were forests and the others were ornamental plantings in gardens and parks. He found the adelgids on two species of hemlocks (northern Japanese hemlock and southern Japanese hemlock). He discovered that adelgid densities were low and the native hemlocks were healthy, as were ornamental eastern hemlocks imported from North America. He also discovered several species of insects that prey on adelgids. The best initial candidate for a biological control agent was a nonpredatory mite called *Diapterobates humeralis*, which eats the white, woolly wax of the adelgid egg sac, destroying the eggs. Unfortunately, it turned out to be difficult to raise large numbers of these mites in the laboratory because they have low reproductive rates, and the mites did not show the same affinity to the woolly filaments around adelgid eggs on eastern hemlocks that they had shown on the native Japanese hemlocks.[38]

Fortunately, a small black lady beetle called *Sasajiscymnus tsugae* turned out to have greater potential as a biological control agent.[39] Kensuke Ito and Tadahisa Urano of the Forestry and Forest Products Research Institute in

Japan sent samples to Connecticut. While working as a postdoctoral fellow with Mark McClure, Carole Cheah discovered that this lady beetle possessed the key characteristics needed for biological control. It has a high reproductive rate with two or more generations per year, so it can be reared in captivity in large numbers for release. Also, it primarily feeds on hemlock woolly adelgids of all life stages, and its annual cycle is highly synchronized with the annual cycle of the adelgids.[40] After careful testing in the lab under quarantine, these lady beetles were first released in forests in Connecticut in 1995. At many sites where beetles were released, the density of adelgids fell precipitously. Moreover, hemlocks at these sites stopped dying, and ten years after the release they had recovered their dense canopies.[41] The introduced lady beetles were recovered at these sites, indicating that they had become established.

In 1995, Michael Montgomery, an entomologist with the U.S. Forest Service, collaborated with Chinese researchers to search for hemlock woolly adelgid predators in the mountain forests of southwestern and central China where three species of hemlocks grow.[42] In China hemlocks are found primarily on steep, forested mountainsides in remote areas, so finding hemlock woolly adelgids was a major challenge. They found 50 species of predatory lady beetles at the sites where adelgids existed. Identifying these predators was also a challenge because 21 of the 50 species turned out to be new to science.[43] Three of these lady beetle species (all in the genus *Scymnus*) preyed on adelgids and had other characteristics that made them potentially effective biological control agents.[44] Although adults of these species sometimes feed on aphids, the larvae die if they cannot feed on hemlock woolly adelgid eggs. They were also tested to ensure that they do not pose a threat to native North American species of adelgids. Conservationists were particularly concerned about whether they would attack woolly alder aphid, a native species that is the main prey of the carnivorous caterpillars of harvester butterflies.[45] Fortunately, when Chinese and Japanese lady beetles were offered woolly alder aphids, they did not attack them frequently and could not survive to maturity by eating them, so they do not represent a threat to this native adelgid.

The two most promising species of *Scymnus* lady beetles have been reared in captivity and released in field trials in eastern North America to determine whether they are effective at controlling adelgids. In the initial field trials, lady beetles were released onto hemlock branches infested with adelgids. The branches were then bagged to contain the beetles. Other bagged branches without lady beetles serve as controls. The results were promising, indicating that the beetles can sharply reduce the population growth rate of adelgids, particularly for adelgid populations that are in the initial stages of infesting a tree. Unfortunately, however, these lady beetles apparently did not become established after they were released in adelgid-infested hemlock stands. The predatory beetles could not be found a year after they were released in woods infested with adelgids.

Still another small beetle, *Laricobius nigrinus*, was discovered feeding on hemlock woolly adelgids on western hemlocks in British Columbia. Adelgids have not caused widespread mortality in western hemlocks because this species is naturally resistant and there are effective adelgid predators in western forests. DNA analyses show that the hemlock woolly adelgids of western North America are genetically distinct from those of Asia, which means they are native to western North America.[46] In contrast, the adelgids of eastern North America are genetically similar to those in southern Japan, confirming that they were introduced from Japan.

Based on laboratory experiments and field trials in Virginia, the *Laricobius* beetles from British Columbia may be effective for controlling hemlock woolly adelgids in eastern North America. They have already been released in large numbers to protect hemlocks at numerous sites in eastern North America, including stands of old-growth hemlocks in Great Smoky Mountains National Park. They have become well enough established at some sites that they can be harvested, providing a source of beetles for release at other sites with infestations. It's still unclear whether they will consistently prevent adelgids from reaching densities that severely damage hemlocks.[47] *Laricobius* beetles and other predators appear to be more effective in New England than in southern forests because periodically cold winters in New England knock adelgid populations down to a low level, and predators then inhibit their recovery by eating a high proportion of the survivors. When *Laricobius* beetles were released at one southern site (Hemlock Hill in North Carolina), however, they preyed on adelgid egg sacs at a high rate and hemlock trees displayed sustained recovery.[48]

Herbivorous insect populations may also be controlled by pathogens, such as fungi. In an effort to find effective fungal pathogens for hemlock woolly adelgid, numerous insect-killing fungi were collected from hemlock woolly adelgids in both the eastern United States and southern China.[49] Each of these was cultured and then tested in the laboratory on hemlock woolly adelgids and other insects. Field trials in Massachusetts demonstrated promising results; when fungal spores were sprayed on hemlocks, adelgid populations declined.

This multi-pronged effort to find effective predators and pathogens is probably the most effective short-term response once a highly destructive pathogen has become established. The hemlock effort required research in the native range of the adelgid in Japan and China, close collaboration with researchers in those countries, and carefully controlled experiments on the prospective control agents while they were still in strict quarantine in the laboratory to ensure that they specialize on adelgids, do not threaten native species of insects, and are effective at reducing population growth in their prey. The final step was field trials, first with bagged branches or trees, and then with the release of large numbers of predatory insects in stands of hemlock infested with hemlock woolly adelgid. Even successful predators and pathogenic fungi probably will not eradicate adelgids, but they may reduce their populations so they do not reach high enough densities to kill their hosts. They would then become more like native herbivorous insects, which are normally kept in check by a combination of plant defenses, predators, and pathogens.

In Connecticut, where some of the earliest mass releases of the *Sasajiscymnus* lady beetles took place, biological control may have already proven effective. After watching virtually all of the hemlocks die in a study area where I monitor bird populations in the Connecticut College Arboretum, I was surprised to see that hemlocks were not only surviving but recovering along three hiking trails in the same region. The hemlock crowns were filling in with needles; the branch tips had bright green new growth; and there was little sign of adelgids. It turned out that all of these trails are close to release sites for *Sasajiscymnus* lady beetles from Japan. Quantitative analyses by Cheah confirmed my casual impressions.[50] These sites had minimal tree mortality after 2006 following releases of large numbers of lady beetles in 1995–2001. Measurement of hemlock foliage transparency (which indicates the openness of tree crowns) showed that the hemlock foliage at lady beetle release sites was as dense as at sites that had never been infested with adelgids, and much denser than at infested sites where no lady beetles had been released.

Sometimes there is a time lag between the release of a predator or pathogen and a noticeable effect on the population of the target species. The gypsy moth may provide an extreme example. In 1989, parts of New England were in the early stages of a gypsy moth outbreak that was expected

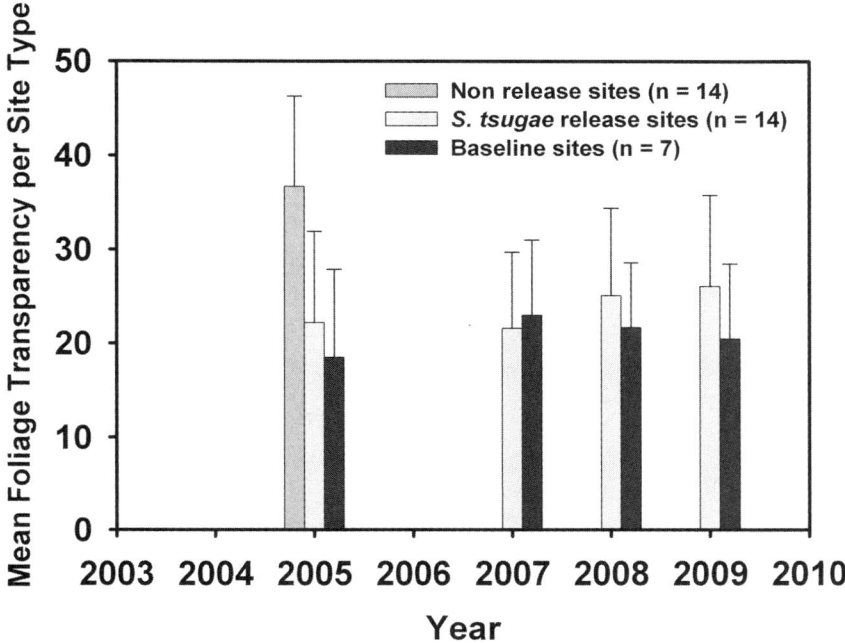

Figure 26. Average foliage transparency of American hemlocks at sites infested with hemlock woolly adelgid where predatory lady beetles had been released (S. *tsugae* release site), sites with woolly adelgids but no released predators (non-release sites), and sites not affected by woolly adelgids (baseline sites). Low transparency values indicate healthy, full hemlock crowns with dense needles. (Cheah, 2010; reprinted with permission of C. Cheah, the Connecticut Agricultural Experiment Station)

to cause extensive defoliation in oak-dominated forests. Suddenly late-stage caterpillars began dying. Branches of trees were festooned with hanging, dead caterpillars, and defoliation of the forest came to an abrupt halt. The dead caterpillars were filled with spores of the pathogenic fungus *Entomophaga maimaiga*. This fungus was purposively introduced to North America from Japan in 1909 in an effort to control gypsy moths, but the program was considered a failure because the fungus could not be located after it was released into the wild. It began to have a major effect on gypsy moth populations only in 1989, after an amazing 80-year delay.[51] It is unclear whether the fungus was spreading slowly during this period or if it was accidentally introduced a second time. The rapid spread of this fungus in 1989 was apparently triggered by a very wet spring that coincided with the incipient gypsy moth outbreak.[52] Since 1989 this fungus has spread to most parts of eastern North America where gypsy moths are found, and it may have prevented gypsy moth outbreaks or greatly reduced their impact. Outbreaks of gypsy moths occasionally cause defoliation in Japan, however, where both the moth and the parasitic fungus are native, so it is likely that gypsy moths will continue to be a problem in North America.[53]

The success of biological control efforts depends on rapid but careful research to respond to a newly invasive species. Introduction of biological control agents is already highly regulated in most countries, and field trials are permitted only after careful screening. Ironically, biological control agents would not be needed as frequently if importation of live plants required the same level of care and testing.[54]

Breeding Resistant Trees

In East Asia hemlocks and chestnuts thrive in regions with hemlock woolly adelgids and chestnut blight partly because the Asian trees are naturally resistant to attack by these species. When adelgids and chestnut blight were introduced to new continents, however, they colonized different species of hemlocks and chestnuts that lack sufficient defenses, which often take the form of chemical defenses that make plant tissues toxic to predators or parasites. The new hosts may evolve natural resistance, but this could require hundreds of years. It will never happen, of course, if the host is driven to extinction. Plant breeders have attempted to speed up this process by locating and breeding resistant trees, or by transferring genetic information from resistant tree species to vulnerable tree species either through traditional breeding methods or, more recently, through genetic engineering.

One important priority is to retain individual trees that display strong resistance to an invasive insect or pathogen so that they can produce progeny with this trait. Western white pines that are resistant to white pine blister rust have been developed by not harvesting "plus trees" that appear to be resistant.[55] Unfortunately, however, a common response to insects and diseases that cause extremely high mortality in trees is to assume that all individuals of the vulnerable species are doomed, and to preemptively harvest them. This approach was used when chestnut blight swept across eastern North America, so many naturally resistant individuals may have been cut down before they had a chance to survive the blight. More recently, the same approach has been used in some localities in response to hemlock woolly adelgid.[56] The prophecy that all hemlocks in a forest are doomed becomes self-fulfilling if the response is to cut them all down.

Several major breeding programs have attempted to develop American chestnuts that are resistant to chestnut blight. The American chestnuts that survived the blight have a relatively low level of natural resistance, so only a few researchers have worked on breeding pure American chestnuts that survived the blight.[57] Instead, most breeding programs rely on hybridizing American chestnuts with Chinese chestnuts in an attempt to develop a tree that looks and grows like an American chestnut, but has the natural resistance of the Asian species. Decades of hybridization produced some trees that are resistant to the blight, but they typically have the short, bushy growth form of Chinese chestnuts, not the tall, wide-spreading profile of American chestnuts. The best hybrid (the "Clapper tree") has the height and shape of an American chestnut, but its resistance ultimately turned out not to be strong enough to survive the blight.

The original breeding programs were so unsuccessful that these efforts were abandoned in the 1960s. In 1983, Charles Burnham (whose specialty was the genetics of corn) and other scientists established the American Chestnut Foundation to initiate a new breeding program.[58] Instead of just hybridizing Chinese and American chestnuts and screening the offspring for the perfect combination of characteristics, the American Chestnut Foundation used a procedure called backcrossing. Backcrossing, which is frequently used for improving crops, started with a first generation of Chinese chestnut and American chestnut hybrids. These were screened for resistance to chestnut blight, and the most resistant hybrids were retained for breeding. These hybrids were then bred (backcrossed) with pure American chestnuts, and the second-generation offspring were screened for resistance. The ultimate goal is to produce a tree that is almost identical genetically to a pure American chestnut except that it retains the blight resistance of the Chinese chestnut ancestor. Like the first-generation hybrids, the resistant second-generation hybrids were crossed with pure American chestnuts. This process was repeated three times, resulting in trees that were genetically 94 percent American chestnut, so they have the tall growth form and wide-spreading branches of this species. The small amount of Chinese chestnut genetic information in these trees includes the crucial genetic coding that results in blight resistance. Additional breeding could increase the proportion of American chestnut genes even more, but the current genetic lines are considered close enough to pure American chestnuts to permit field

Figure 27. A large backcrossed chestnut tree at Meadowview Research Farms in Virginia, where American chestnuts that are resistant to chestnut blight have been bred. Flowers are bagged to control pollination. (Courtesy of the American Chestnut Foundation)

trials. In the final step before field trials, different genetic lines resulting from backcrossing were bred together ("intercrossed") to increase resistance to chestnut blight. Beginning in 2010, hybrid chestnuts from this program were planted at sites in Tennessee, North Carolina, Virginia, and Pennsylvania. These early plantings were protected by deer-proof fences, an indication of how the forests have changed since tall chestnuts disappeared a century ago.

These resistant chestnuts were produced by traditional plant breeding. Eventually this lengthy, labor-intensive process will probably be superseded by the use of molecular genetics techniques ("molecular engineering") to create American chestnuts with only a few specific genes for disease resistance that were transferred from Chinese chestnuts.[59] The foundation for achieving this goal is already being built with a major project to sequence the entire genome of Chinese chestnut in order to identify the genes responsible for blight resistance.[60] Once these genes are identified, then hybrid chestnut seedlings could be screened for resistance more efficiently, which would speed up traditional breeding. Ultimately, these resistance genes could be inserted into American chestnut genomes to produce a resistant genetic line.

If chestnuts are successfully reintroduced across eastern North America, the result will be profound ecological changes that may challenge our assumptions about "normal" ecological processes in North American deciduous forests. The large white blossoms produced by chestnuts in early summer will provide a copious new source of nectar for pollinating insects deep within the forest.[61] Because chestnuts produce nuts every year, nut-eating species such as eastern gray squirrels, eastern chipmunks, and white-footed mice may sustain relatively stable, high population densities rather than boom-and-bust cycles that result from their response to occasional massive crops of acorns or beechnuts.[62] Also, young chestnuts grow well in fairly shady conditions, so they will compete well with shade-tolerant trees such as red maple for space in the canopy, eventually changing the composition of the forest canopy. Like other tree species, American chestnuts supported numerous species of specialized insects that do not use other hosts.[63] Some of these specialists (such as many of the leaf-mining moths) feed on other species of chestnuts from Asia and Europe, so they have survived on planted ornamental chestnuts and presumably would quickly recolonize American chestnuts. In contrast, seven species of moths (including the American chestnut moth) specialized narrowly on American chestnuts, and these species may already be extinct. If so, then the restored American chestnuts will be missing some of the specialized herbivorous insects that evolved with them over a long period of time. Paul Opler suggested that this might make them more susceptible to attack by more generalized leaf-eating insects.[64] All of these hypothesized effects of chestnut reintroduction are speculative, but if chestnuts regain their place as a major component of the canopy of the "oak-chestnut" forests of eastern North America, we can expect major ecological changes.

A Longer View of the Loss of Tree Species

The catastrophic decline of particular tree species due to insects or pathogens is primarily due to long-distance trade in wood products and live plants. Even before transoceanic trade began, however, tree species suffered infrequent catastrophic declines. The pollen record in lake and bog sediments indicates that hemlocks declined suddenly throughout North America about 5,400 years ago.[65] They almost disappeared from many sites, resulting in an increase in the frequency of birches, oaks, maples, American beech, and other types of trees. At most sites hemlock populations recovered after 1,000 to 2,000 years, showing that the tree species can eventually bounce back after being decimated on a continental scale.

The pollen record reveals a similar pattern for elms in Europe. Elm pollen declined in sediments that were deposited about 5,000 years ago in northern Europe, from northern Germany to Britain and Ireland.[66] Analysis of annual sediment layers in a lake in Norfolk show that the decline occurred abruptly, with a 73 percent reduction in elm pollen during a period of only six years. The record from numerous other sites cannot be so finely calibrated, but they are consistent with an abrupt decline. A sudden population crash is consistent with the hypothesis that a pathogen or insect spread across northern Europe and the British Isles, causing widespread mortality of elms. Remains of the elm bark beetle that carries Dutch elm disease have been discovered in sediments from the mid-Holocene when the decline occurred, so an early outbreak of Dutch elm disease is a possible cause of the decline. There is no direct evidence for this, however. The period around 5,000 years ago was marked by changing conditions in northern Europe that may have contributed to the elm decline. The climate became more seasonal, with warmer summers and colder winters, and there was a major cultural shift from hunting and gathering to planting cereals and raising livestock. Changes in the climate and creation of clearings in the forest may have contributed to the elm decline, but the strong evidence for a sudden decline that was synchronous across large areas of northern Europe suggests that the direct cause was a pathogen or insect outbreak.

As with hemlocks in North America, it required elms more than a thousand years to recover their former abundance at some sites, and they never recovered in some regions of England. This rate of recovery does not endanger the forest if pathogenic outbreaks occur only very rarely (once in several thousand years) and only affect a single species of tree, but it becomes a problem when new pathogens sweep across the forest every few decades.

Other Invasive Species

I have focused on introduced insects and fungi that kill trees because these cause particularly obvious and severe changes in forest ecosystems. The diversity and stability of deciduous forests are also threatened by other types of introduced species, however, including animals, plants, and animal pathogens. These are typically species that people have transported from one isolated deciduous forest region (in East Asia, Europe, or North

America) to another. The majority of introduced species appear to be innocuous either because they are largely confined to artificial habitats created by people or because they behave much like similar native species. A few introduced species become invasive, displaying rapid, exponential growth resulting in substantial declines in native species that are their prey, hosts, or competitors. Some cause structural changes in forest ecosystems that indirectly affect numerous species. Like the blights and sap-sucking insects that knock out particular species of trees, the small number of species that become invasive can threaten the diversity and stability of forest ecosystems.

An example is white-nose syndrome, an epidemic disease caused by a pathogenic fungus that is causing extremely high mortality in bats in eastern North America. The fungus is also found in Europe (where it does not noticeably affect bat populations), and may have been spread by people exploring caves on both continents.[67] The disease was first discovered in North America in 2006, and it has already killed millions of cave-dwelling bats. It affects hibernating bats by causing them to awaken from torpor. Each arousal requires a large amount of energy from their fat deposits, and consequently the bats run out of energy before the end of the winter. One sign of the disease is bats flying around on winter days, apparently hunting for food. Also, the dead bats in caves are emaciated, suggesting that they died from lack of food. This disease causes such high mortality among hibernating bats that it is feared that it will cause the extinction of several species that were already considered threatened for other reasons. The disease may even threaten the previously abundant little brown myotis.[68] Bats such as the little brown myotis are important aerial insect-eaters in deciduous forest ecosystems, foraging over the forest canopies, along streams, and in forest openings.[69] The ecological effects of severe reductions in their abundance on forest ecosystems are difficult to predict.

Invasive plants do not cause the abrupt, catastrophic population crashes associated with epidemic diseases and herbivorous insects, but they may gradually erode the diversity of ecosystems. Some invasive plants spread through the forest understory, displacing a variety of native species. A more frequent pattern is that invasive species from another continent will thrive in forest openings, delaying or even preventing forest succession so that the forest is slowly converted to low shrub or vine-dominated vegetation.[70] Invasive plants have spread into deciduous forests on all three continents, and are often a major concern for forest managers. Japanese barberry, for example, can spread quickly through the understory of hardwood forests in eastern North America, displacing native species of shrubs and herbs and inhibiting the growth of tree seedlings. Ironically, in the 1920s Japanese barberry was used to replace common barberry, a related European species that had been introduced as an ornamental shrub soon after English settlement of North America.[71] Common barberry not only turned out to be an invasive shrub, but it also carried black stem grain rust, which infects wheat crops. After a national campaign to mobilize the general public to destroy both planted and naturalized common barberry was largely successful, the U.S Department of Agriculture recommended Japanese barberry (which does not carry the rust) as a safe substitute. Japanese barberry subsequently became an invasive species, however, forming dense thickets on the forest floor that reduced the density of native shrubs such as blueberry and huckleberry.[72] It also changes soil characteristics, increasing soil pH and the availability of nitrogen.[73] These changes favor the growth of Japanese barberry, and may be one factor that makes this species a successful invader of undisturbed mature forests.

White-tailed deer avoid browsing on barberry foliage, so this shrub spreads particularly quickly in forests with high densities of deer.[74] Anne Eschtruth and John Battles showed that white-tailed deer accelerate the spread of this species and two other invasive plants (Japanese stiltgrass and garlic mustard) in eastern North America. They set up 400 small enclosures (each with a nearby control plot) in ten hemlock forests near the Delaware River in Pennsylvania and New Jersey.[75] All three invasive plants increased more rapidly in control plots (where native plants declined due to heavy browsing by deer) than in enclosures that were protected from deer. The expansion of invasive species was greatest at sites where the canopy had been thinned by defoliation from hemlock woolly adelgid. Thus, both high deer densities and an invasive defoliating insect contributed to the rapid spread of these invasive species in the forest understory. Presumably deer are feeding on more palatable native herbs and tree seedlings, removing competitors of the less palatable invasive species.

Invasive plant species have also caused environmental problems in Europe and East Asia. In England, for example, introduced pontic rhododendrons have spread through oak woodlands. The impact on the understory is similar to that of Japanese barberry in American woodlands.[76] In Japan introduced species are a particularly severe problem in streamside habitats, where they threaten the diversity of native plant communities.[77] Introduced species occupy approximately 15 percent of streamside habitats. Most of the invasive plants along streamsides (37 percent of all invasive species, and 64 percent of the widely distributed invasive species) are native to North America. Some of the more important North American invasives are late goldenrod, black locust, bur cucumber, and giant ragweed. Introduced plants are also common in forested nature reserves in Jiangsu Province in eastern China.[78] Again, most of these introduced species (56 percent) are from North America. In these reserves invasive plants are found primarily in habitats disturbed by people, but two species of *Veronica* from western Eurasia (Persian speedwell and ivyleaf speedwell) have spread into the understory of mixed coniferous-deciduous forest.

Introduced animals are also an important problem in some deciduous forest ecosystems. Eastern gray squirrels from eastern North America have displaced the native Eurasian red squirrel from most of Britain.[79] They have also damaged trees such as beech and sycamore by stripping off the bark, and they threaten common hazel by harvesting hazelnuts while they are still green. Also, the introduction of North American raccoons to Japan is likely to have a major impact on the populations of a variety of small animals because raccoons are highly efficient, generalized predators.[80]

Implications for Conservation

The two most important global changes that threaten temperate deciduous forests are climate change and the worldwide transport of living organisms that result from world trade. Although climate change has received much more attention in recent years, northern forests are more immediately and severely threatened by the spread of organisms from other continents. In fact, introduced invasive species are eroding diversity in a wide variety of ecosystems throughout the world, and the severity of their impact is not sufficiently appreciated.[81] The most severe threats to northern deciduous forests are from specialized pathogens and plant-eating insects that are spread from one deciduous forest region to another. Because of their long isolation, the deciduous forest regions of Europe, Asia, and North America have closely related species of organisms that may be susceptible to the same diseases and insect pests, but do not necessarily share the same defenses against these threats. This sets the stage for catastrophic population declines.

These declines do not always result in extinction, but they may effectively remove a dominant and ecologically important species from a forest ecosystem. It only took a few years for American chestnuts, English elms, and Japanese red pines to disappear after the arrival of new pathogens. By the time these introduced pathogens were well established, there was little that local conservationists could do other than document the rate of decline and assess the repercussions for the rest of the ecosystem. A sudden plague of this sort can undo the hard work of hundreds of local conservation efforts. Decades of protecting old-growth stands of eastern hemlock, for example, are now threatened by the rapid spread of hemlock woolly adelgid. Also, conservation successes resulting from years of public education about the ecological importance of insectivorous bats and the key role of caves for protecting hibernating bats have been undermined almost overnight by the spread of white-nose syndrome.

Like global climate change, these problems seem intractable because they require a major economic commitment based on international agreements. But also like global climate change, all of the potential parties to an agreement are threatened by the problem and have an incentive to participate. Invasive introduced organisms are a particular threat to the industrialized nations of the three major continental regions with deciduous forests, all of which are vulnerable to the introduction of species from the other continents. The solution is much stricter limitations on movement of live organisms. More effective international agreements have already been established for raw wood products that harbor bark insects and pathogenic fungi, but stricter controls are needed for live plants. Current regulations often ban the importation of species that are already well-established invasive species, but new species that are not yet known to be invasive pose an even greater potential threat. Even live plants that are unlikely to become invasive may carry pathogenic fungi, and stricter inspection and quarantine regulations could prevent the spread of new blights. Another long-term approach that would help prevent the spread of invasive species is a greater emphasis on local wood production, and use of native species and established cultivars in ornamental planting.[82] The transport of large numbers of live trees and shrubs from one continental deciduous forest region to another is comparable to playing a game of Russian roulette with our forests.

CHAPTER 10

Blending Conservation Strategies from Three Continents

T
he deciduous forests of East Asia, Europe, and eastern North America are remnants of a far more extensive band of deciduous forest that once extended around the Northern Hemisphere. Millions of years of separation resulted in the evolution of different sets of plant and animal species on the three continents, but the basic structure and ecological workings of deciduous forests remained remarkably similar. These forest landscapes began to change profoundly during the past 5,000 years, however, as people converted much of the forest to farmland, pasture, and human settlements. Although the great extent of agricultural clearing was similar on the three continents, the emerging patterns of land use were distinctly different. Eventually there were efforts to protect and restore forests and biological diversity on each continent, but the motivations and methods were heavily influenced by cultural differences. Although there is now frequent communication among conservationists and researchers on the three continents, the conservation priorities and strategies on each continent are still heavily influenced by basic differences in perceptions of natural beauty and proper land management.

Protecting and Preserving Wild Nature in North America

In 1967, the environmental historian Roderick Nash lucidly explained the central role of the concept of wilderness in American culture in *Wilderness and the American Mind*.[1] He described how Americans came to value wilderness in the nineteenth century just as the last frontiers were being settled and developed. An influential minority of educated people began to work to preserve the last herds of bison, wild mountains, and uncut forests. Wild natural landscapes were valued as a substitute for the culturally rich landscapes of Europe. The first national parks were centered on monumental geological formations such as high waterfalls, deep canyons, immense rock formations, and spouting geysers. In place of the Roman ruins, castles, and cathedrals of Europe, Americans had impressive natural wonders that became important symbols of national identity.[2] In the early twentieth century, after the mass production of automobiles and extensive construction of paved roads, the national parks became important pilgrimage destinations for American families, who planned road trips around visiting a series of western parks.

The perception that the grandeur of natural landscapes made the United States special began to emerge in the early nineteenth century, and was reinforced by the dramatic paintings of wilderness scenes by Thomas Cole, Frederic Church, Asher Durand, Albert Bierstadt, Thomas Moran, and other artists of the Hudson River School.[3] But natural landscapes were rapidly being converted to farms, which in many regions were then replaced by houses, mills, and factories. As a result, a movement to preserve special natural landscapes as an important component of the national heritage began to emerge. In its early form, however, this movement did little to protect forests except those that were monumental in themselves (groves of coastal redwoods or giant sequoias in California) and those that happened to grow on the rim of the Grand Canyon or the floor at Yosemite Valley.[4]

Other perspectives on the importance of natural areas were slowly gaining traction in the nineteenth century, however, and increasingly wild forests were valued even if they did not contain deep canyons or gigantic waterfalls. The Transcendentalist writings of Ralph Waldo Emerson and Henry David Thoreau argued that wild areas are important as a place of spiritual calm and renewal, and Thoreau wrote that some wilderness areas should be protected as public parks. These works had little influence beyond intellectual circles, but they were popularized by the books and articles of John Muir.[5] Muir argued for the complete preservation of wilderness areas to permit spiritual renewal for an increasingly urban population. Although he helped protect the monumental geological formations of Yosemite Valley, he also argued for the importance of protecting forested wilderness.

Muir's arguments gained popularity in the late nineteenth century at the same time that Gifford Pinchot and others were arguing for the adoption of scientific, sustainable forestry to ensure a future supply of timber products in the United States. These two revolutionary conservation movements immediately came into conflict because of their incompatible goals. Muir and the Sierra Club that he founded emphasized the importance of preserving undisturbed wilderness accessible only by foot, horse, or canoe.

In contrast, the scientifically trained foresters focused on intensive and continuous management of forests to provide a sustainable source of timber, game, and water. These two visions could not be realized in the same forest (although they could be achieved in different forests, as was eventually demonstrated).

It's surprising that the ideal of preserving pure, undisturbed nature became so widely popular in the United States, a country that had developed with a romantic view of replacing wilderness with farms, towns, and cities. In the early nineteenth century European travelers commented on how little most American settlers valued the natural beauty of forests or trees, and how thoroughly they replaced tall forests with raw fields filled with girdled, dead trees, tree stumps, and split rail fences, leaving virtually no trees standing.[6] By the late 1800s, however, the rapidity and scale of the forest destruction and the shock of the extermination or near extermination of seemingly unlimited numbers of passenger pigeons and American bison made Americans more aware of the loss of natural beauty and diversity. This coincided with growing concern about the cultural implications of the closing of the last open frontiers. The availability of unsettled land and wild places was thought to be central to the self-reliance, independence, and innovation that characterized American culture.[7] Wilderness reserves would provide a place where Americans could continue to develop and test these national characteristics.

In the 1940s efforts to protect wilderness were also provided with a scientific justification when Aldo Leopold argued that the best way to understand and manage disturbed ecosystems is to study ecosystems that have not been disturbed.[8] He wrote that effective land management requires "a base datum of normality, a picture of how healthy land maintains itself as an organism."[9] Only small remnants of many once extensive natural ecosystems had escaped development, and these remaining wild areas were invaluable for scientific research. Moreover, it was important to have representative samples of each type of ecosystem in each region of the world. An extensive wilderness area in the Rocky Mountains will not substitute for a wilderness area in the eastern deciduous forest because interactions among species, patterns of forest succession, adaptations to natural disturbances, nutrient cycling, and other basic ecological processes are likely to differ.

To some extent, the emphasis on preserving large stretches of undisturbed wilderness and "virgin" forests did relatively little to protect eastern deciduous forests, most of which were second-growth forests that grew back after timber harvesting or farm abandonment. Even the most remote

forests in eastern North America had an obvious history of human disturbance and thus did not appear to qualify as wilderness. One notable exception
to this attitude resulted in the preservation of the forests of the Adirondack
Mountains in upstate New York. Although much of this forest had been cut
or disturbed in various ways, there were still extensive tracts of old-growth
forest. Also, the forests and lakes of the Adirondacks had become a major
recreation area for New Yorkers by the second half of the nineteenth century. It was a wild, mountainous landscape where people could hike, canoe,
hunt, and fish. Remarkably, 12,146 square kilometers (3 million acres) were
preserved in Adirondack Park in 1892.[10] Although the New York legislature
was persuaded to establish the park primarily because of the importance of
the Adirondack forests in providing a stable source of water for the Erie Canal
and the Hudson River, the movement to save the park originated among
people who visited the region to hunt, fish, or admire the beauty of the landscape. The key importance of wilderness preservation was demonstrated in
1894, when a new state constitution included a provision that the unsettled
parts of Adirondack Park would be "forever kept as wild forest land," and that
harvesting trees would not be allowed.[11]

Although Yellowstone National Park was established in 1871 and numerous other national parks followed during the next 50 years, nearly all of
these were in western North America. None of the national parks protected
an extensive area of the floristically diverse eastern deciduous forest. Acadia
National Park in Maine was established in 1916, but much of the habitat in
this park is boreal coniferous forest and rocky shoreline.[12] The first national
parks with extensive deciduous forest in the United States, Great Smoky
Mountains National Park and Shenandoah National Park, were approved in
1926 but were not established until the 1930s.[13] One reason for the delay
was that the proposed eastern parks were on private rather than government land, so they had to be purchased from private landowners and timber
companies.[14] The parks were approved with the understanding that pur-
chasing the land would be the responsibility of the states and private individuals, not the federal government. The funds to purchase land for these
parks were contributed by the states (North Carolina, Tennessee, and Virginia), the Rockefeller family, and residents in the three states.

Great Smoky Mountains National Park is an especially important deciduous forest reserve because it is so large (2,115 square kilometers) and
supports more than 1,500 species of plants, including 125 species of native
trees (more species than are found in Europe).[15] The park protects some of

Figure 28. *A Twilight in the Adirondacks*, painted by Sanford Robinson Gifford in 1864, depicts the heavily forested Adirondack Mountains of New York. A small campsite lit by a fire is the only sign of human presence in the expanse of wilderness. (Courtesy of the Adirondack Museum)

the most impressive stands of old-growth deciduous forest in eastern North America. Unlike most western parks, however, large areas within Great Smoky Mountains and Shenandoah national parks had been lumbered or farmed. Hence these parks are mainly covered with mature second-growth forests with, as in most extensive eastern forests, clear signs of past agricultural land use.[16]

In 1891, a year before Adirondack Park was established, the U.S. Congress passed a timber bill that gave the president the authority to set aside public land in forest reserves.[17] Within ten years 47 million acres of forest reserves had been established in the western United States. Unlike Adirondack Park and the national parks, these forest reserves (which later became the national forests) were managed for sustainable production of timber and other natural resources. Timber harvesting, hunting, livestock grazing, and dams were permitted in the national forests.[18] The "wise use" of natural resources for long-term economic return was the guiding principle for managing the national forests, and—in contrast to the national parks—recreation and preservation of natural beauty were not originally considered important priorities. Preservation of old-growth forest, in particular, was considered antithetical to "scientific forestry" because old trees produce fewer board-feet of timber per year than do young trees. For foresters trained at Yale University and other universities with schools of forestry, the ideal was a vigorous stand of young, fast-growing trees that were harvested sustainably on a short rotation. Many foresters perceived this type of highly managed productive forest as the only way to avert an impending timber shortage that would hobble the economy and diminish the standard of living of Americans. Foresters also were trained to perceive homogeneous stands of straight trees as more esthetically attractive than natural forests filled with snags, diseased trees, curved trunks, and tangled vegetation, resulting in an antipathy toward natural forests.

Given the traditional utilitarian approach of the U.S. Forest Service, it's ironic that many national forests include wilderness areas where human disturbance is restricted even more than in most national parks. While he was working for the Forest Service in the early 1900s, Aldo Leopold, who later became widely known for his writing about the importance of preserving and restoring intact natural ecosystems, successfully argued for the establishment of wilderness areas in two national forests in New Mexico.[19] This set the stage for a successful campaign by Robert Marshall, who became head of the Forest Service's Division of Recreation and Lands, to protect large wilderness areas in the national forests.[20] By 1964 the U.S. Forest Service had designated 14 million acres of roadless wilderness areas.[21] Logging was prohibited in these areas. In contrast to the national parks, which traditionally have made easy public access to the main features of a park a major priority, no roads or buildings were allowed in wilderness areas. The central goal of the wilderness areas was not only to preserve the wilderness, but also to preserve the opportunity for a wilderness experience. Robert Marshall's criterion for a wilderness was the absence of buildings, roads, and railroads in an area "sufficiently spacious that a person may spend at least a week or two of travel in them without crossing his own tracks."[22] This meant that visitors must travel by foot, horse, or canoe, experiencing a taste of the independence and challenge of the original frontier. The wilderness would be an escape from the pressures of modern urban society that would help people stay physically fit and develop self-reliance and a deep appreciation for nature.

The Forest Service wilderness areas were more firmly protected after 1964, when the Wilderness Act was passed.[23] This act enabled Congress to designate wilderness areas on public lands. About 40 percent of these wilderness areas are in national forests (which had a total of 55 wilderness areas in 2005), while the rest are in national parks, Bureau of Land Management lands, and other federal lands. Although wilderness areas in national forests include large tracts that are above timberline (the so-called "rock-and-ice" wilderness areas), they also include extensive forests. This is especially true in eastern national forests, where wilderness areas protect large expanses of deciduous forest.

When I worked in a wilderness area as an undergraduate student, I discovered that the Forest Service took the wilderness ethic seriously. I had a summer position on a trail crew in the Selway-Bitterroot Wilderness Area in Montana. We were not permitted to carry chain saws or other mechanical equipment during weeklong hikes into the wilderness area. A mule train carried our equipment, and we cleared trails using axes and shovels. Except for the creation of narrow trails for hikers and horses, the goal was to not disrupt the wildness and quietness of the landscape. At that time the only major exceptions were for forest fires, when planes, helicopters, and even temporary roads might be used to fight a fire.

The basic ideal of preserving pristine wild nature was also a guiding principle of land preservation for many private, nonprofit conservation groups such as the National Audubon Society, the Nature Conservancy, and local land trusts. These organizations established nature reserves that were smaller than national parks or forests, but large enough to encompass an interesting natural area. One cannot lose oneself for two weeks in these reserves, but it is possible to explore wild nature for a morning or a day. These smaller reserves are especially important in the eastern deciduous forest, where there are fewer national parks and wilderness areas than in western North America.

The American conservation movement, with its distinctive emphasis on wilderness, was largely an indigenous effort. Although it was triggered by Romanticism in art and literature that originated in Europe, it developed a distinctive philosophy that drew on the frontier culture of the United States.[24] The works of Emerson, Thoreau, Muir, Leopold, and the popular nature writers they inspired resonated with many Americans who were shocked by the rapid destruction of forests, wildlife, and wild places.[25] This loss was especially poignant because of the growing perception that natural landscapes as exemplified in the national parks were a distinctive component of national identity, comparable in importance to the historic buildings, monuments, and ruins for Europeans. All of these concerns converged to provide sufficient support for preserving some of the last undeveloped regions of the continent in a natural state. The definition of a "natural state" has shifted over time, particularly with changing attitudes about large predators and wildfires, but the ideal of preserving pristine wilderness has remained central to the preservation of natural areas in North America.

Wilderness Preservation in the Light of Modern Ecological Research

Until Aldo Leopold emphasized the importance of preserving intact ecosystems so that scientists can figure out how natural ecosystems work, the wilderness preservation movement was not based on a scientific understanding of ecosystems or their value for future research. The central concerns were esthetic and spiritual. It is therefore interesting to consider how well this approach to conservation works to protect natural ecosystems and biological diversity in light of the scientific insights presented in this book.

Protection of extensive national parks and wilderness areas clearly helps address one of the key problems with sustaining functioning forest ecosystems, which is fragmentation of natural habitats by roads, residential developments, and industrial facilities. Officially designated wilderness areas in national forests and national parks do not have roads or buildings, and are large enough to support top predators and to sustain large-scale natural disturbances such as wildfires and blowdowns caused by windstorms. They are also large enough to sustain beaver ponds and populations of forest bird species that need unbroken expanses of forest. Scientific research confirmed the importance of preserving large natural areas that cover entire valleys or mountain ranges. Smaller nature reserves that protect patches of old-growth forest or second-growth woodland have not fared as well for several reasons: they are too small to support some key species such as top predators; they can be destroyed by a single disturbance such as a tornado or hurricane; and they are easily penetrated by species that thrive at the forest edge, including small predators, parasitic cowbirds, and invasive plant species.

Wilderness areas are often large enough to support large predators that control deer populations, and reintroduction of predators can be considered if these have been extirpated. This is a critically important advantage when managing deciduous forests. Also, the interiors of wilderness areas typically have a low density of invasive plant species because they are far away from the disturbed areas that are the source of these species. Unfortunately, however, the extensive size and remoteness of wilderness areas does not protect them from tree-killing pathogens and insects such as chestnut blight and hemlock woolly adelgid.

Ironically, modern ecological and archaeological research undermined the notion that wilderness areas are "pristine" natural places that have never been disturbed by people. Humans have molded natural landscapes since the end of the last glacial period by eliminating some species and favoring others, and by changing the frequency of fires. This makes it difficult to define what is truly "natural," presenting a problem for restoration programs that set a goal of restoring natural conditions rather than simply protecting threatened species or sustaining biological diversity. As a result, American conservationists often struggle fruitlessly with the question of what type of "natural state" should be the goal for protection and restoration.

The wilderness concept has led to special problems when small nature reserves are managed as if they were miniature wilderness areas. Minimal management that consists primarily of trail maintenance may work well in the 5,427-square-kilometer Selway-Bitterroot Wilderness Area in Montana, but can be a disastrous failure in a 100-acre natural area plagued by high deer densities, no large predators, invasive plant species, high densities of

medium-sized predators, and an insufficient frequency of natural disturbances to maintain oaks and other shade-intolerant species. In this case the
hands-off approach used in a wilderness area can result in a collapse in the
diversity of forest species and ultimately to the replacement of deciduous
forest by another type of ecosystem dominated by grasses or shrubs.

Another problem is that an emphasis on large natural areas that fulfill a
preconceived notion of wilderness or wild nature may not encompass restricted habitats that support specialized species that may be missing from
even very large protected areas. In the northeastern United States these localized habitats include limestone outcrops, serpentine bedrock, and sandy
soils.[26] A focus on threatened species will protect the species associated
with each of these habitats better than a focus on protecting wild, roadless
areas.

On balance, however, the emphasis on wilderness has been an effective
approach for preserving functioning natural ecosystems and a broad range
of species. It resulted in protected areas that are large enough to support sustainable populations and to absorb the natural disturbances that eventually
hit all natural areas. The criticism that "wilderness" is an artificial concept—
that virtually all ecosystems have been shaped by human activities—is irrelevant to this conclusion. Large expanses of natural habitat are much more
likely to be sustainable and resilient than small nature reserves embedded in
farmland or residential areas. Also, as Aldo Leopold argued, they are more
likely to show us how the natural diversity of a region is perpetuated over the
long term, providing lessons that can be applied to small reserves and more
developed landscapes.

Protecting Humanized Landscapes in Europe

In the 1970s, René Dubos, a well-known scientist and writer, argued that
sensibly cultivated landscapes were a more desirable environment for people than wild nature, and that sound conservation does not depend on
preservation of wilderness areas.[27] This perspective derived from his child-
hood in the countryside of Île de France in northern France, surrounded by
a landscape of carefully managed fields, pastures, and woods.[28] The soil of
this region remains fertile after 2,000 years of intensive agriculture because of careful stewardship of cropland and pastures. Not only is this
landscape productive in terms of providing for human needs, but it is also
beautiful and biologically diverse. Dubos considered it a much better habi-

tat for people than the primeval forest that covered the region thousands of years ago.

In the 1940s Dubos moved to the United States, where he became a highly respected microbiologist at Harvard University and later at Rockefeller University. He was well known for a series of popular books on epidemic diseases, microbiology, and environmental issues. It is doubtful that Dubos would have ever needed to confront the wilderness issue if he had remained in France rather than becoming a naturalized citizen of the United States. If he had stayed in France and became interested in protecting nature, his efforts would have been directed at the "humanized" nature of his youth. In the United States, however, he was forced to confront an environmental movement that viewed only wild nature as worthy of protection.

In an essay he published in *Science* in 1976, he downplayed the importance of preserving wilderness areas.[29] In a brief paragraph he mentions that some wilderness should be protected because wilderness is productive (in terms of energy captured by photosynthesis) and supports high biological diversity, and because occasional contact with primeval nature is important for people. This paragraph is anomalous, however, because the rest of the essay emphasizes the desirability and superiority of properly managed artificial landscapes. He argues that we can fashion a gardenlike "humanized" world that would be more attractive and conducive to human life than undisturbed wilderness.

Much of the ecological preservation and restoration in western Europe is consistent with Dubos' vision of a properly managed semi-natural landscape. Traditionally, European conservationists have not been overly concerned about restoring wild nature to its original shape. Instead they frequently manage highly artificial habitats such as farmland, wood pasture, and coppice to protect rare and threatened species, and to maintain an attractive and evocative cultural landscape. In contrast, until recently conservationists in the United States displayed little interest in protecting biological diversity in farms, ranches, tree plantations, or other heavily managed habitats.

In Europe even woodlands are typically artificial, managed habitats. Although deciduous woodlands in Europe and North America are ecologically similar, the woodlands of Europe—particularly northwestern Europe—are much more heavily managed than most deciduous forests in eastern North America. Forests are periodically thinned by selective removal of trees. Trees that are stunted, diseased, or dead, or that have curved or bent trunks, are removed, leaving an open understory under a canopy of straight, tall trees.

This means the dead and decayed wood that supports much of the forest's biological diversity is rare. One of the surprising results is that German foresters install numerous nest boxes to maintain populations of birds that originally nested in woodpecker cavities and other natural holes in dead or diseased trees. These natural nest sites are not available in woods that are efficiently managed for wood production, so the nest boxes are needed to support populations of birds that feed on leaf-eating insects that would reduce the growth rate of trees.[30] Also, in many regions of northwestern Europe deciduous forests have been replaced with plantations consisting of orderly rows of conifers, all of the same age and species. Not surprisingly, given their prior experience with pruned and manicured forests, many Europeans perceive North American forests, with their dense understory, decaying logs, barkless snags, and dead branches, as chaotic and untidy.

England has some well-known ancient forests (defined as woodlands more than 400 years old) that consist of a mixture of native deciduous trees. Historically, these were managed in different ways and with different levels of intensity. Some were "wood pastures," which are open areas with scattered old trees that resulted from heavy grazing by livestock and deer. During the nineteenth century, Romanticism in art and literature emphasized the beauty and spiritual significance of large, old trees and ancient groves, and many particularly impressive individual trees were given names and became tourist attractions. This did not save some ancient forests, such as parts of Sherwood Forest, from being converted to plantations of exotic conifers, however. Between 1950 and 1975 British foresters focused on maximizing wood production, and ancient trees were viewed as unproductive.[31] Ancient woods were destroyed despite an active and influential conservation movement in England, suggesting that wild nature had less standing than other priorities (such as protecting threatened species of birds).

The remaining ancient forests are now carefully protected, but the goal is not necessarily to restore them as natural, old-growth forests. The main focus is on saving the species associated with old trees and preserving particular types of semi-agricultural landscapes such as wood pastures, pollard groves, and coppices. Wood pastures traditionally consisted of open grazing areas with either scattered timber trees (which are tall and wide spreading) or pollards. Pollards are trees that are periodically pruned of all major branches above two to four meters to provide forage for livestock without killing the tree.[32] The sprouts that grow on the chopped-off stub of the tree trunk are too high for browsing cattle and deer to reach. After centuries of pruning, the trees are converted to massive, squat trunks with vigorous young branches sprouting out of their tops, so a grove of pollards probably does not resemble any natural landscape that occurred before human management. Nevertheless, wood pastures with pollards are important for conserving biological diversity. Many species are associated with the open parkland habitat in wood pastures, and the ancient trees support a diversity of lichens and invertebrates that require the old bark and decayed wood that are found on very old trees.[33]

Maintenance of a wood pasture with pollards requires both pruning and grazing.[34] Without periodic pruning, groups of pollards become odd-looking woods consisting of tall trees with massive bases. The tree canopy expands and the grasses and herbs are shaded out. If grazing is stopped, shrubs and young trees replace the open ground cover of grass and herbs. Although the wood pasture in Epping Forest was protected in 1878, it took on the character of a young second-growth forest after pruning of pollards and livestock grazing were prohibited. In some wood pastures, however, the increasing abundance of deer has reversed some of the negative effects of livestock removal.

Perhaps the best example of managing a traditional but artificial type of woodland to sustain biological diversity is the effort to save coppice.[35] In coppice, trees are cut at or near ground level and the stump is permitted to resprout with numerous small branches. This is a traditional source of firewood, poles, and other small pieces of wood. The resulting landscape looks more like a shrubland than a forest, and it mainly supports species of wildflowers and animals that depend on open, sunny conditions with low woody vegetation.[36] This is true even when scattered tall trees called standards remain standing in the coppice. As management of coppice declined because of lack of commercial demand for small branches for firewood and traditional woodcraft, many species that require low woody vegetation also declined.[37] Eurasian woodcock, common nightingale, and marsh tit, for example, are now less abundant because of the loss of traditional coppice and other shrubby habitats.[38]

Forest rides are another highly artificial feature of English forests that have a high priority for conservation. These are wide strips of herb-rich grassland that cross woodlands. Rides are maintained by periodic mowing to provide easy access to the woods for hunting and riding. They are a special focus for conservation because they support rare species of butterflies and plants that need sunny woodland clearings.[39] Although rides may be ecological counterparts of natural glades in a forest that has not been disturbed by humans, the focus of conservation management is on preserving these broad corridors, not to recreate natural glades such as beaver meadows or forest openings created by long-extinct elephants.

The maintenance of rides, coppices, pollards, and wood pastures for conservation restores a cultural landscape rather than a natural landscape. In 2000, Frans Vera published his revolutionary hypothesis that much of Europe was once covered with open woods similar to wood pasture because of browsing and grazing by large mammals that are now extirpated, but wood pastures were maintained for conservation long before forest managers realized that they might be simulating a natural system.[40] The main goal is to sustain a traditional landscape and the species of fungi, plants, and animals that it supported. Oliver Rackham, who has written numerous well-respected books on ecology and conservation of British woodlands, argued that we don't know enough about the original forests and other natural habitats of Great Britain to try to restore them. The efforts to restore "wildwood" (unmanaged forests) that began to gain popularity in England in the 1990s are therefore misguided.[41] Instead, he recommended responding to "practical observation" of how particular management practices affect the key species and features that make the woodland worth protecting. "Conservationists do no service to woodland if they try to remake it in the image of what they imagine wildwood was like," he argued.[42]

Rackham's pragmatic approach to preserving ecological diversity is commonplace in Europe but unusual in North America, where ecological restoration implies restoring, to the extent possible, the original natural ecosystem. Consequently British ecologists—along with most ecologists in Europe—do not waste a lot of time debating what the natural ecosystem was like before the forests were cleared for agriculture, or which particular period in the ecological history of the site is the best target for restoration. This difference in restoration goals arose partly because North American ecologists have more accurate information about the nature of forests before agricultural clearing (if not before humans had major direct and indirect impacts on forests), but it also reflects the higher priority for conserving wild nature as opposed to "humanized" nature in North America.

The European approach of protecting and nurturing biological diversity in cultural landscapes is particularly well reflected in the efforts to protect species that are declining because of the intensification of agriculture. During the twentieth century fallow fields, hedgerows, and field edges with a rich diversity of native plants and animals disappeared in many parts of Europe because fields were expanded to accommodate large equipment used in plowing, planting, and harvesting. This change and other agricultural changes such as use of pesticides and synthetic fertilizers led to steep declines in populations of birds, pollinating insects, and other organisms that were once common in rural landscapes.[43] Instead of creating nature reserves with meadows to protect these species, however, the main conservation effort in the United Kingdom and other European Union countries has focused on paying farmers to change their farming practices to accommodate declining species, often by reviving traditional farming methods.[44] Biological diversity is essentially treated as a "farm product," so these conservation measures help to support farmers and protect valued rural landscapes. English farmers, for example, are paid to delay mowing to avoid destroying nests of grassland birds. They are also paid to reduce the intensity of grazing on pastures and to retain stubble in grain fields during the winter to provide food for seed-eating birds.[45] Some of these programs focus on management of farmland for a single high-priority species, and the Netherlands tried paying farmers for the number of young produced by threatened species of meadow birds.[46] Programs of this sort have met with mixed success in enhancing the diversity of plants, insects, and birds, and additional research on their effectiveness is needed.[47]

Practicing nature conservation and food production "on the same ground" was proposed in England as early as 1969, but this approach did not become a fully developed program with substantial funding until the 1990s.[48] Farm subsidies were designed to simultaneously reduce overproduction of crops, support farmers and the rural economy, and improve the natural environment. Thus a major component of conservation in the United Kingdom and other European Union countries is protecting biological diversity on working farms. Often the goal is not only to save populations of particular species of birds, butterflies, bees, and plants, but also to preserve traditional rural landscapes that are rich with cultural and historical meaning and confer a "sense of place" to particular regions of Europe.

Conservation of Humanized Landscapes in the Light of Modern Ecological Research

In judging the wilderness approach to conservation, my standard was how well it works to protect biological diversity and well-functioning natural ecosystems. These are the general goals of conservation biologists, who try to understand and develop ecologically sound methods to sustain natural ecosystems. For the first goal (protecting biological diversity), the European approach of managing all habitats, however artificial, appears to be effective. All species, including inconspicuous fungi on the bark of ancient trees or small herbs growing on the edge of wheat fields, are considered important and are potential targets of habitat management to sustain their populations. This approach appears to fail at achieving the second goal, protecting well-functioning natural ecosystems, however. The humanized ecosystems may be diverse, but they are not "well-functioning" in terms of being self-sustaining. In fact they can only be sustained with continual human engagement, such as pruning trees every few years or mowing hay meadows each year after the nesting season for birds. Conservation begins to resemble agriculture, and in fact is frequently done in conjunction with food and wood production. Superficially, this seems very different from reintroducing wolves and allowing wildfires to burn in order to reestablish a functioning natural ecosystem in Yellowstone National Park. Even the North American approach of restoring natural ecosystems has gradually moved to actively managing populations and manipulating habitats, however. Creating a habitat similar to a presettlement forest may require controlling white-tailed deer populations, cutting down trees to create canopy gaps to provide light for shade-intolerant plants, and mowing a floodplain meadow to simulate a beaver meadow. In the temperate deciduous forest, there are few wilderness areas or national parks that are large enough to be potentially self-sustaining. Top predators are missing from smaller forests, and natural disturbances such as wildfires or storms are too infrequent to provide sufficient habitat for early successional species.

One major advantage of the European approach is that large populations of organisms can be protected over extensive areas at relatively little cost because many of the conservation efforts take place on "working land." Maintenance of farmland and pasture are primarily supported by production of food and fiber, and agricultural subsidies provide incentives to modify farming methods to prevent erosion of biological diversity. Instead of paying farmers to take land out of production, agricultural subsidies are used constructively to prevent the loss of species by shifting to less intensive and productive farming methods. And often the conservation measures can directly contribute to agricultural efficiency and cost-effectiveness (by increasing the number of pollinating insects or insect-eating birds, for example). This same general approach can be used in working forests that are managed for wood production. This pragmatic focus on figuring out how to save and restore threatened species wherever they are found (regardless of whether the setting is natural or artificial) is potentially effective, but the resulting landscapes may not look like anything found in pre-agricultural Europe. Europeans argue, however, that these cultural landscapes—small villages set in a mosaic of pastures, fields, hedgerows, and woodlands—are worthy of preservation.

The scarcity of large, unmanaged natural areas in Europe is a limitation, however, because it is frequently difficult to understand how to manage forest species without knowing the context of their evolution. The extensive boreal forests of Finland and Scandinavia, for example, have been converted to remarkably homogeneous stands of young pine or spruce that are managed by planting and intensive thinning before a final clearcut harvest.[49] As a result, many species that depend on dead wood, old trees, or particular stages of forest succession are in trouble. Forest managers are now attempting to restore natural features to these plantations. To find a natural counterpart to highly managed coniferous forests, however, Swedish and Finnish biologists had to travel to remote spruce forests in Siberia.[50] They found that Russian forests have 33 times as many snags, 46 times as many fallen dead trees, and 8 times as many large living trees as managed forests in Sweden. Upland areas away from rivers are covered with spruce forests, large swaths of which are periodically destroyed by fires and windstorms. Along free-flowing rivers, however, the floodplain is covered with birches and poplars that seldom burn. All of this information was useful for figuring out what ecological processes were missing in managed forests, causing a loss of biological diversity. Unfortunately, however, there are frequently no clear wild versions of natural habitats in Europe to provide this type of perspective. This is particularly true for lowland deciduous forests, which grow in areas favored for farming. Białowieża Forest in Poland has been the subject of numerous ecological studies because it is one of the few extensive areas of lowland temperate forest where natural ecological processes have played out for more than a hundred years.

Nature in Miniature: Conservation in Japan

The elegance of Kyoto's temples and palaces is enhanced for naturalists by vivid images of wild animals and plants. Although some images are stylized, most are accurate depictions of recognizable species. A small room in Zuihō-in, a temple in the Diatoku-ji monastery complex, is centered on a screen with a remarkably precise and evocative painting of a male Baikal teal, one of the most beautiful Asian ducks. Other temples in Japan feature detailed and accurate imprints of ferns on the sliding paper screens that cover windows. At Nishi Honganji, a seventeenth-century temple in central Kyoto, paintings on large sliding doors depict three species of cranes foraging in rice fields. To see red-crowned, white-naped, and hooded cranes today, one must travel to Hokkaido or southern Kyushu, but in the 1600s they must have been found close to Kyoto. Another room is devoted to 68 exquisitely painted Eurasian tree sparrows flying in flocks and feeding on the ground in various realistic postures.

Perhaps the most evocative images of wild plants and animals in Japanese art are Zen Buddhist ink paintings. With a few black brush strokes, these accurately capture the courtship flight of displaying white wagtails or the tense hunting pose of a perched common kingfisher. Images of plants, mushrooms, and wild animals are also common on everyday objects. Although iconic and highly symbolic autumn maple leaves, spring cherry blossoms, and monogamous pairs of cranes are common motifs, traditional ceramic bowls and paper fans may also have surprisingly accurate and delicate depictions of particular types of insects, marine invertebrates, or wildflowers. The traditional artists of Japan show a sensitivity to the distinctiveness and behavior of different wild species that is difficult to find in Western art until the nineteenth century, when naturalist-artists such as John James Audubon began painting. Also, Japanese art often focuses on the "small, intimate and gentle aspects of nature" (such as dragonflies, mushrooms, and ferns) rather than massive or grandiose scenes.[51] The emphasis on sensitive images of nature in Japanese art, along with the simulated natural landscapes in Japanese gardens and traditional activities such as cherry blossom viewing, accounts for the perception among both Japanese and Westerners that Japanese culture is exceptionally attuned to nature.

Many visitors to Japan take the high-speed train from Tokyo to Kyoto or Osaka, passing through a densely populated and heavily industrialized coastal plain. For first-time visitors who are expecting to see scenes like those in a traditional Japanese landscape painting, this trip is a shock. At times it seems that the entire landscape is encased in concrete and asphalt. Rivers have been converted to straight canals with concrete walls bordered by concrete walkways. The seacoast is fortified with concrete barriers, often in the

Figure 29. *Two Wagtails*, an ink painting by Kano Sanraku (late sixteenth or early seventeenth century, Japan). The postures and movement of displaying wagtails are depicted with vivid accuracy. (Owned by Sanso LLC, used by permission; digital image provided by University of Michigan, Department of the History of Art, Visual Resources Collections)

shape of giant interlocking children's jacks. Rice fields are islands of green surrounded by concrete canals, blocks of apartment buildings, and paved roads. Only steep hills have small patches of bamboo or second-growth pines and oaks, and even here the slopes above roads are encased in concrete. How does all of this fit with the traditional love of nature? Has this tradition been destroyed by economic progress and materialism?

The view from a high-speed train is deceptive on two counts. First, the visitor is only seeing the coastal plain, a relatively narrow strip of flat land where the great majority of Japanese live and where nearly all economic activity takes place. A short trip inland will bring one to heavily forested hills and mountains and small traditional towns and villages. Still another perspective is gained by disembarking at a station and walking through the streets in a typical residential neighborhood. Behind walls that front on the street, one glimpses elegant gardens. The garden styles display remarkable variation, with some emphasizing perennial flowers while others feature shrubs and small trees or even collections of bonsai trees. Many of these gardens emulate ancient temple gardens by displaying miniature simulations of forest and mountain landscapes.

Traditional Japanese gardens reflect an emphasis on viewing nature from a particular vantage point rather than on actively exploring it.[52] A traditional temple or palace garden is viewed from a veranda where one can sit and sip green tea from an elegant ceramic cup, protected by a roof from gentle rain. Cloudy skies and falling rain enhance the many shades of green of different types of shrubs and trees, so a sunny day is not required for contemplative enjoyment of the beauty of the garden. Even larger "stroll gardens" emphasize a series of garden views that are hidden and suddenly revealed as one moves along a path.[53] These scenes are often miniature replicas of well-known natural scenic views in Japan or China.

Passively viewing a beautiful natural scene from a fixed point in this manner is reflected in many Japanese traditions. An ancient aristocratic tradition was to float in a boat in the middle of a lake or river to view the full moon and its reflection. Even today people in Japanese cities and towns pause from their usual busy schedule each year during several days when spring cherry blossoms reach their peak. They set out food and drinks on blankets in local parks and socialize under branches covered with cherry blossoms, surrounded by fallen pink petals. Companies in Osaka send young employees to stake out places under cherry trees in local parks until the trees are in full bloom, when all employees are released from work to

Figure 30. View of the garden at Ninna-ji Temple in Kyoto, Japan. (Photograph by Karen Askins)

join them to eat, drink sake, and enjoy the blossoms.[54] Television weather forecasts follow the front of blossoming cherries as it advances northward so that cherry-viewing parties can be planned. Earlier in the spring many people also make a special effort to view blossoming Japanese "plum" (Japanese apricot) trees, which have delicate blossoms in a wide range of colors.

The tradition of viewing nature from a fixed site carries over to visits to wilder landscapes. People make pilgrimages to view waterfalls or particular rock formations in a lake or in the sea. One of the standard sites to visit in Kyoto is the Moon Crossing Bridge in Arashiyama. Looking upriver, visitors see the smooth-flowing Katsura River moving through steep, wooded hillsides that are pink with cherry blossoms in the spring and scarlet with maple leaves in the autumn. A large expanse of deciduous forest was protected to preserve this view, and new cherry trees are planted on the slopes to enhance spring cherry blossom viewing. The forested slopes are expansive enough to support troops of Japanese macaques and a variety of forest birds.

Many of the hillsides and mountain slopes around Kyoto (a total of 60 square kilometers) have been protected since 1934 to retain the surround-

Figure 31. View from the Moon Crossing Bridge in Arashiyama, Kyoto, in spring, when cherry trees are blooming on the hillsides above the river.

ings of ancient Buddhist temples, Shinto shrines, and palaces that ring the city. Many of these hills and mountains are important as backdrops to historically important buildings or as "borrowed scenery" for ancient gardens. Traditional Japanese gardens are often designed to frame a view of a mountain, a range of hills, or a body of water.[55] Entire forested hills have been preserved to serve as a backdrop for imperial palace gardens on the edge of the city.

Sometimes famous traditional viewpoints incorporate a sufficiently large landscape to effectively result in wilderness preservation. For example, the view from Kappa Bridge, a wooden suspension bridge over the clear, sinuous Azusa River, incorporates heavily forested slopes and a range of rocky peaks that form part of the Northern Alps. The importance of the standard view upriver is emphasized by turning and looking downriver, to a cluster of hotels, restaurants, and concession stands. Farther downriver the road bringing visitors to Kappa Bridge was built with a remarkable amount of concrete, with elaborate tunnels, roofs, and walls to protect the road from landslides, but above the bridge the river flows freely through a natural forest.

The mountains framed by the view from Kappa Bridge are protected inside Chūbu-Sangaku National Park, and are a favorite site for backpacking.

Views do not necessarily or usually incorporate expansive landscapes, however. Appreciation of nature in Japan appears to be heavily influenced by traditional gardens, which create intricate landscapes in small spaces.[56] A traditional garden presents a miniaturized simulation of a wild landscape, with rocks arranged to represent mountains, waterfalls, or islands, and shrubs and trees pruned to create the illusion of forests.[57] The principles of creating the illusion of expansive landscapes in a small space (using small rocks behind much larger rocks to create the appearance of great distance, for example) date back more than a thousand years to the Heian Period, when Kyoto was the cultural and political center of Japan. Even the simple Zen gardens constructed of rocks and sand like the rock garden at Ryōanji in Kyoto display a miniaturized landscape, such as rocky islands in wave-swept sea.[58] Animals are an important component in many gardens. In addition to koi (colorful domesticated carp), turtles, frogs, dragonflies, kingfishers, and other wild animals are valued components of the garden view. Finally, gardens are carefully designed to be attractive and evocative in each of the four seasons, with special attention to blossoming trees and shrubs in spring, and the various shades of scarlet, orange, and yellow of autumn foliage.[59]

Appreciating small and delicate natural objects seems to carry over into natural history. Japanese appreciate and admire insects in a way that is unusual in Europe and North America. A common pastime for children even in urban areas is to capture and identify insects, and many children keep insects in captivity.[60] Large department stores sell equipment for catching and keeping insects. Some of the most popular insects, such as singing crickets, singing grasshoppers, and rhinoceros beetles, are bred in captivity and sold in department stores and shops. Insectariums (insect zoos) are popular destinations for many children. Japanese retain some of their interest in insects as adults; many residents of Tokyo know how to identify the sounds of the six species of cicadas that call from urban trees and parks.[61]

Appreciation of nature in confined spaces is reflected in nature reserves, which are often remarkably small by European (much less North American) standards. Katano Kamoike, a reserve on the Japan Sea coast that is important for winter-resident waterfowl, covers only 10 hectares (25 acres). It is internationally recognized as an important wetland because of the large numbers of Baikal teal and other waterfowl, as evidenced by its designation as a Ramsar Site, and it has an excellent interpretation center. Tokyo Port Wild

Bird Park is another famous natural area that is relatively small. It incorporates tidal flats, freshwater ponds, cattail marsh, planted woods, and a miniature rice field, all in the space of 27 hectares (67 acres). My colleague Reiko Kurosawa compared this reserve to a bonsai garden tray, with a miniaturized array of Japanese habitats. The focus is on inclusiveness and immaculate beauty, not spaciousness. Like Katano Kamoike and many other nature reserves in Japan, Tokyo Port park has a visitor's center with large windows overlooking a body of water, so nature is primarily viewed from a fixed point. A particularly attractive visitor's center at Kiritappu Wetland in Hokkaido has a coffee shop on the upper floor so one can sit and sip coffee or tea and look out over an expanse of marshes where pairs of red-crowned cranes forage. This is reminiscent of the experience of sitting on a veranda overlooking a traditional garden while drinking tea. In contrast to the Tokyo Port Bird Park, however, the Kiritappu Wetland is a large reserve, like many of the nature reserves and national parks in Hokkaido. This suggests that the small size of nature reserves in more heavily developed areas of Japan reflects the scarcity of undeveloped land on the coastal plain rather than a preference for small reserves. The small reserves are carefully protected and highly valued, however, because they bring many urban residents into contact with wild nature. These reserves may be particularly effective because the Japanese are educated to appreciate the beauty of nature in small spaces.

Political Constraints on Japanese Conservation Efforts

National conservation organizations in Japan are neither as politically influential nor as well financed as those in the United States or the United Kingdom, but this does not reflect a lack of interest in conservation. Thousands of local conservation groups work to protect nature in their neighborhoods and towns. These groups are energetic and well organized, and they can be very effective at protecting small natural areas. They also restore natural habitats and monitor populations of threatened species. When I visited Hotokenuma Marsh in 2001, I was surprised to learn that management of the marsh was the responsibility of a local bird club and the town government of Misawa. This marsh is an internationally important Ramsar Wetland Site that supports threatened species such as the Japanese marsh warbler and Japanese reed bunting. Volunteers, many of whom are experienced naturalists, work on marsh restoration and population monitoring. They are doing an effective job of managing the site, but in Europe or North

America a site of this importance would almost certainly be managed by a government agency or a national conservation organization with a paid, professional staff.

Another example of a local community mobilizing to restore natural habitats is the dragonfly pond program in the heavily industrialized city of Yokohama. The town government works with local community groups to build or restore small ponds to create habitat for dragonflies. Dragonflies are frequently featured in traditional Japanese art and poetry, and they are one of the most popular groups of insects.[62] People have become concerned about sharp declines in dragonfly abundance and diversity, however, as urbanization, wetland destruction, and more intensive management of rice fields have destroyed the habitats that many species require. Consequently, dragonflies are now used as an important indicator of environmental quality in Japan. Forty-eight of the 180 species of dragonflies in Japan have been classified as threatened or near threatened, so there is a need for habitat restoration.[63] Local groups in Yokohama carefully landscape and maintain artificial ponds to create a wide variety of habitats needed by different species of dragonflies. In the Honmoku Citizens' Park, for example, a sterile concrete-lined pond was renovated by adding earth to the banks, creating a stream, and planting a wide variety of native aquatic plants.[64] By paying particular attention to specific microhabitats such as sunny banks needed by particular species, the pond managers were able to boost the number of dragonfly species from 3 to 27. This pond became a center for cultural and educational activities. Similar ponds were built at numerous schools in Yokohama, where they have become the focus for science education centered on biological diversity and water quality, and this idea has spread to other parts of Japan. Because hundreds of dragonfly ponds have been created, the pond restoration program has a substantial impact on dragonfly conservation.

There are thousands of neighborhood groups, small regional conservation societies, and town governments that work on local conservation projects. In the late 1990s there were more than 4,500 nongovernmental environmental organizations, most of which focused on local nature conservation, recycling, and organic food projects.[65] But Robert Mason argues that most environmentalists in these groups are inclined to ask, "What can we do about local environmental problems?" rather than questioning national policies and priorities that cause widespread environmental damage. Consequently, the numerous people who belong to these small groups have little power to stop large-scale construction projects that eliminate or fragment natural habitats. There are no well-funded national organizations that can effectively contest regulations and legislation at the national level.

Although lack of influence of national conservation organizations is partly due to the local focus of most conservation groups, it also reflects legal constraints on the activities and finances of nonprofit organizations, few of which have tax-exempt status.[66] Until recently nongovernmental organizations had to achieve capital assets and membership numbers that greatly exceeded those of most environmental groups in order to qualify for tax-exempt status. If a nonprofit organization achieved tax-exempt status, then it was required to become affiliated with an appropriate government ministry that could influence the activities and board membership of the organization. Given these constraints, it isn't surprising that politically powerful independent conservation organizations did not develop in Japan.

The influence of conservation organizations was also limited because of the enormous political power of the construction industry, which depends on highly influential political support from rural prefectures for building new highways, bridges, dams, and museums.[67] These construction projects are one of the main sources of employment in rural areas, and the result is that few parts of the country's interior have not been penetrated by large, over-engineered roads, many with little traffic. Large construction projects (such as massive hotel complexes) are even built in national parks. The influence of the "construction state" (the powerful alliance of rural politicians, Transportation Ministry bureaucrats, and construction companies) results in Japan having a much higher rate of concrete use per capita than other highly industrialized countries.

In the early 1970s environmental groups had broader support and more political influence because rapid industrial development had resulted in intolerable levels of air pollution and water pollution.[68] After several major disasters in which hundreds of people died and thousands became ill from air pollution (industrial asthma) and mercury poisoning from contaminated fish, the political pressure became so great that the national legislature (the Diet) quickly passed stringent controls on air and water pollution.[69] In 1970 the "Pollution Diet" passed 14 major environmental laws, and the Environment Agency was established the following year.[70] By 1972 Japan had "among the most stringent environmental regulations in the world."[71] This environmental revolution primarily focused on human health, however, and not on preserving natural ecosystems such as forests. Also, the anti-pollution movement was organized by more than 3,000 environmental organizations that displayed little coordination.[72] They did not coalesce into a permanent national group that could counter environmental threats throughout the country.

The effectiveness of nongovernmental environmental groups in Japan has improved since the late 1990s. In 1998 the Non-profit Organization Law was passed, making it easier to form nonprofit organizations.[73] This was partly in response to the effectiveness of nonprofit groups in rescue and relief operations following the Kobe earthquake. Also, international environmental norms now require participation by nongovernmental organizations in environmental decision making, so the Japanese government has been required by international agreements to engage nonprofit environmental groups in some projects. Government budget cuts have also made collaboration with environmental groups more attractive. A measure of the growing influence of nonprofit groups is the success of a movement to end logging in national parks. In 1996 environmental groups were able to stop most of the planned harvesting of old-growth forest in Shiretoko National Park in Hokkaido.[74] The opposition to harvesting ancient forests in Shiretoko and other national parks led to a change in policy about timber harvesting in national parks. Japanese environmental groups are slowly becoming more effective at protecting the remaining natural areas in the country.

Although these national political considerations are undoubtedly paramount in explaining why so few large natural forests have been protected in Japan, the traditional emphasis on nature in miniature may be an underlying cause. If visits to a garden, local pond, or small patch of mudflat provide a sufficiently satisfying natural experience, then there may be little support for stopping a major dam to save an extensive forest. On the basis of a survey of Japanese citizens on their attitudes and knowledge of conservation issues and in-depth interviews with Japanese scientists, officials, conservationists, journalists, and foresters, Stephen Kellert concluded that "Japanese appreciation of nature was found to be idealized and constrained," focusing on particular species or experiences with nature. Many respondents favored natural beauty in highly controlled settings such as gardens over experience with wild nature.[75] Also, love of nature was frequently based on enjoyment of art rather than direct experience with wild nature, and the love of natural beauty was not based on ecological considerations and did not lead to an attitude of ethical responsibility for protecting nature. Thus, appreciation of natural beauty among the Japanese did not always translate into concern about the future of forests and other natural environments.

Conservation of "Nature in Miniature" in the
Light of Modern Ecological Research

An emphasis on preserving particular landscape views or microhabitats that support favored species of small organisms has clear limitations in terms of preserving intact, functioning ecosystems. Species that require large expanses of unbroken habitat, such as large predators and some forest birds, will not be well protected by this approach. Also, this approach does not permit preservation of a large enough area to permit natural disturbances to produce a mosaic of different patches of forest at different stages of recovery from disturbance. Even for small organisms it may result in isolated populations with limited genetic diversity that are prone to entirely disappearing because they are not connected to other populations. A large network of small populations can persist if migration can occur among them, but this will not happen if the populations are isolated by large expanses of development.

On the other hand, the Japanese emphasis on small organisms and specific microhabitats can result in effective conservation of insects and other invertebrates that receive little attention in North America. Public awareness of cicadas, butterflies, dragonflies, and other insects—and concern about declines in their populations—is much greater in Japan than in most other countries. The Japanese approach also permits preservation of pockets of nature in highly developed urban environments that are critically important for public education about biological diversity. These small islands of woodland, marsh, tidal flat, and streamside are also critically important for migratory birds in transit between nesting areas in Siberia and northern Japan and wintering areas in Taiwan and Southeast Asia.

Blending the Conservation Approaches of the Three Continents

The different approaches to conservation of deciduous forests in Japan, North America, and western Europe reflect deep differences in culture and traditional land use rather than intrinsic differences in forest ecosystems. The basic conservation problems such as forest fragmentation, suppression of natural disturbances, and loss of top predators are remarkably similar on the three continents. The various approaches to solving conservation problems all have strengths and weaknesses, and the most effective solution would draw on lessons from all three continents.

First, however, it is important to emphasize that the three different approaches I've described (focusing on preservation of wilderness, humanized regional landscapes, or miniaturized natural landscapes) are not truly restricted to a single continent. There has been extensive communication about ecology and conservation among conservationists on the different continents, and some approaches have arisen independently on two or three continents. Wilderness areas are valued by some people in Europe and Asia as a result of exposure to American conservation ideals or an indigenous interest in mountaineering, hiking, or (in Japan) traditional spiritual pilgrimages to wild mountains or seacoasts. This led to protection of some wilderness areas in Eurasia. Shiretoko National Park in far northern Hokkaido, for example, is a long peninsula surrounded by the ocean that is accessible only by hiking trails. Signs at the trailhead warn hikers to beware of brown bears, which can be dangerous when startled. Experiencing the interior of this park is certainly a wilderness experience. The circular hiking trail in Bosque de Muniellos, a 5,488-hectare nature reserve in the Cantabrian Mountains of northern Spain, also provides a wilderness experience. Like Shiretoko National Park, Bosque de Muniellos supports a population of brown bears. To protect ancient oak forests and alpine meadows, only 20 people are permitted to hike this trail each day, so reservations are required.

Also, just as wilderness preservation is not restricted to North America, the effort to save "humanized nature" in the form of traditional rural landscapes is not completely restricted to Europe. A similar approach is emerging in Japan with the movement to preserve "satoyama."[76] Satoyama is a highly diverse landscape consisting of terraced rice fields, coppice, tall woodlands, and villages with small gardens. Numerous aquatic, early successional, and woodland species thrive in the highly managed fields and woods in traditional satoyama landscapes, and many of these species are now in trouble. Pesticide use, replacement of traditional channels with concrete irrigation systems, and conversion of a mosaic of different types of land use into broad fields suitable to mechanized farming have greatly reduced both the biological diversity and the beauty of these agricultural landscapes. As a result, in many parts of Japan volunteers with local conservation groups are working to restore traditional satoyama landscapes. There is also an emerging interest in preserving biological diversity in rural agricultural landscapes in some parts of North America, such as rural New England, but this is still in a very early stage of development.[77] The Nature Conservancy has taken a lead in working with farmers and ranchers to arrange conservation easements so that working land is protected from development, and to manage the land for both agricultural production and biological diversity. Conservation of small microhabitats and insects takes place not only in Japan, but also on the other two continents, particularly in the United Kingdom and other European countries. Despite these areas of similarity and overlap, however, the best-developed and most heavily emphasized approaches to protecting biological diversity differ substantially in Japan, North America, and northwestern Europe. Conservationists in these three regions have much to learn from one another.

It is important to try to protect large, continuous deciduous forest reserves where natural ecological processes can take place. In many regions this will mean reconnecting forest patches and managing for old-growth conditions. It may also require reintroduction of missing species such as predators, woodland wildflowers, or old-growth fungi. If natural disturbances such as river flooding cannot be restored on a large enough scale, then habitat management to create meadows and young forest may be needed. The North American experience can serve as a model for restoring wild forest (if not wilderness) that will sustain the complex interactions among species that characterized forests before agricultural clearing.

The applicability of the wild forest approach is limited in many parts of the world, however, including much of the eastern United States and Canada. Many of the rural landscapes that replaced the forest are esthetically appealing and biologically diverse, and they deserve to be protected and managed for conservation. The European experience of managing rural areas to preserve both a traditional rural culture and a rich diversity of species in both woodlands and open habitats can serve as a model for eastern North America and East Asia. This approach is just as appropriate for terraced rice fields in China or dairy farms and hay meadows in Vermont as it is for traditional European farming areas.

The Japanese tradition of appreciating nature where you find it, in a garden, a minute pond, or a small park with cherry trees in bloom, is critically important for educating people in urban environments about the importance of natural beauty and biological diversity. Children who tend a dragonfly pond and monitor different species of colorful dragonflies are more likely to understand the importance of preserving natural ecosystems. They also contribute directly to protecting small organisms such as insects that receive relatively little attention in many other countries. In contrast, the American approach of emphasizing wilderness areas and large reserves has proved particularly ill-suited to preserving small invertebrates and plants with specialized habitat requirements, as the lack of public response to catastrophic declines in some bumblebee species and other pollinating insects demonstrates.[78] The Xerxes Society is one of the few groups in North America that focuses on conservation of insects and other invertebrates, many of which are critically important in the functioning of natural ecosystems. Interest in watching and identifying dragonflies and butterflies is growing in North America, and endangered species of butterflies and tiger beetles are the focus of conservation efforts, but even invertebrates that are obviously critically important to ecosystem functioning such as pollinators still receive relatively little attention from conservationists.

Despite the monumental threats from habitat fragmentation, introduced forest pathogens and insects, climate change, and heavy browsing by deer, the forests of the northern temperate zone can be sustained and restored as complex and diverse ecosystems. These forests have proven remarkably resilient through both the deep time of paleontologists and archaeologists and the historical time of historians and ecologists. They survived a succession of glacial advances in small southern refuges, and they have rebounded in many regions after farming was abandoned. Most of the remaining forest ecosystems are heavily damaged and many are on a trajectory toward conversion to a scrubbier, less diverse habitat, however, so their preservation will require intervention by conservationists. Effective intervention will depend on protection of threatened species and on gaining a better understanding of how forest ecosystems work.

Conservationists generally manage ecosystems by informed "tinkering" (trial and error or "adaptive management"). To paraphrase Aldo Leopold, the first rule of intelligent tinkering is to "keep every cog and wheel."[79] Thus, the first priority is to keep all of the components of the original deciduous forest ecosystems by preventing any species, however small and apparently insignificant, from becoming extinct. The next priority, which is well under way, is to understand the natural processes that maintain a functioning forest with a full set of species. These processes include occasional catastrophic disturbances such as canopy blowdowns as well as the steady working of energy flow, nutrient cycling, predation, and plant succession. Whenever feasible, we should permit these natural processes to function without management. In many cases, however, key species and natural processes have disappeared from a forest ecosystem, and managers must simulate these in some way, playing the role once played by wolves or beavers or spring floods. With careful management of this sort, we can restore summer-green forests over large areas of East Asia, western Europe, and eastern North America, ensuring a sustainable supply of forest products while protecting some of the most beautiful and diverse ecosystems on earth.

Appendix of Scientific Names

PLANTS

Alder	*Alnus spp.*
Allegheny blackberry	*Rubus allegheniensis*
American beech	*Fagus grandifolia*
American chestnut	*Castanea dentata*
American elm	*Ulmus americana*
Anemone	*Anemone spp.*
Ash	*Fraxinus spp.*
Aspen	*Populus spp.*
Bald cypress	*Taxodium distichum*
Barley	*Hordeum vulgare*
Beech	*Fagus spp.*
Birch	*Betula spp.*
Black birch (sweet birch)	*Betula lenta*
Black cherry	*Prunus serotina*
Black locust	*Robinia pseudoacacia*
Black oak (eastern black oak)	*Quercus velutina*
Black tupelo (black gum)	*Nyssa sylvatica*
Black walnut (eastern black walnut)	*Juglans nigra*
Blackberry	*Rubus spp.*
Blueberry	*Vaccinium spp.*
Bracken fern (bracken)	*Pteridium aquilinum*
Breadfruit	*Artocarpus altilis*
Bur cucumber	*Sicyos angulatus*
Burnweed	*Erechtites hieracifolius*
Butternut	*Juglans cinerea*
Canada mayflower	*Maianthemum canadense*
Carolina hemlock	*Tsuga caroliniana*
Cherry	*Prunus spp.*

Chestnut	*Castanea spp.*
Chinese chestnut	*Castanea mollissima*
Coast redwood	*Sequoia sempervirens*
Common ash	*Fraxinus excelsior*
Common barberry	*Berberis vulgaris*
Common blackberry	*Rubus fruticosus*
Common hazel	*Corylus avellana*
Common millet	*Panicum mileaceum*
Common periwinkle	*Vinca minor*
Common wood fern	*Dryopteris intermedia*
Corn (maize)	*Zea mays*
Cottonwood	*Populus spp.*
Cryptomeria	*Cryptomeria japonica*
Currant	*Ribes spp.*
Dawn redwood	*Metasequoia glyptostroboides*
Dwarf bamboo	*Sasa spp.*
Eastern hemlock	*Tsuga canadensis*
Eastern white pine	*Pinus strobus*
Elm	*Ulmus spp.*
English elm	*Ulmus minor var. vulgaris*
	(U. procera)
European beech	*Fagus sylvatica*
European hornbeam	*Carpinus betulus*
European larch	*Larix decidua*
European mountain ash (rowan)	*Sorbus aucuparia*
Fig	*Ficus spp.*
Fir	*Abies spp.*
Flowering dogwood	*Cornus florida*
Foxtail millet	*Setaria italica*
Garlic mustard	*Alliaria petiolata*
Giant ragweed	*Ambrosia trifida*
Ginkgo	*Ginkgo biloba*
Golden chestnut	*Castanopsis spp.*
Gooseberry	*Ribes spp.*
Goosefoot	*Chenopodium berlandieri*
Hackberry	*Celtis spp.*
Hay-scented fern	*Dennstaedtia punctilobula*
Hazel	*Corylus avellan*

Hemlock	*Tsuga spp.*
Hickory	*Carya spp.*
Hinoki cypress	*Chamaecyparis obtusa*
Holly	*Ilex spp.*
Honey locust	*Gleditsia triacanthos*
Hornbeam (European hornbeam)	*Carpinus betulus*
Horse chestnut	*Aesculus spp.*
Huckleberry	*Gaylussacia spp.*
Indian pipe	*Monotropa spp.*
Ivyleaf speedwell	*Veronica hederifolia*
Jack pine	*Pinus banksiana*
Japanese apricot	*Prunus mume*
Japanese barberry	*Berberis thunbergii*
Japanese beech	*Fagus crenata*
Japanese cherry birch	*Betula grossa*
Japanese lawngrass	*Zoysia japonica*
Japanese mountain cherry	*Prunus jamasakura*
Japanese red pine	*Pinus densiflora*
Japanese stiltgrass	*Microstegium vimineum*
Larch	*Larix spp.*
Late goldenrod	*Solidago altissima*
Laurel	*Cinnamomum spp.*
Lime (basswood)	*Tilia spp.*
Longleaf pine	*Pinus palustris*
Maple	*Acer spp.*
Marsh elder	*Iva annua*
Northern Japanese hemlock	*Tsuga diversifolia*
Norway maple	*Acer platanoides*
Norway spruce	*Picea abies*
Oak	*Quercus spp.*
Pedunculate oak (English oak)	*Quercus robur*
Persian speedwell	*Veronica persica*
Pin cherry	*Prunus pensylvanica*
Pine	*Pinus spp.*
Pontic rhododendron	*Rhododendron ponticum*
Poplar	*Populus spp.*
Ragweed	*Ambrosia spp.*
Red maple	*Acer rubrum*

Red pine	*Pinus resinosa*
Red trillium	*Trillium erectum*
Rice (domestic rice)	*Oryza sativa*
Rice (wild wetland rice)	*Oryza rufipogon*
Sassafras	*Sassafras albidum*
Scots pine	*Pinus sylvestris*
Shagbark hickory	*Carya ovata*
Silver grass (Miscanthus)	*Miscanthus sinensis*
Small-leaved lime	*Tilia cordata*
Snowberry	*Symphoricarpos spp.*
Solomon's seal	*Polygonatum biflorum*
Southern beech	*Nothofagus cunninghamii*
Southern Japanese hemlock	*Tsuga sieboldii*
Spruce	*Picea spp.*
Squash	*Cucurbita spp.*
Stinging nettle	*Urtica dioica*
Striped maple	*Acer pensylvanicum*
Sugar maple	*Acer saccharum*
Sugar pine	*Pinus lambertiana*
Sugi	*Cryptomeria japonica*
Sunflower	*Helianthus annuus*
Sweet chestnut	*Castanea sativa*
Sweetgum	*Liquidambar spp.*
Sycamore	*Platanus spp.*
Sycamore maple	*Acer pseudoplatanus*
Tamarack	*Larix laricina*
Trailing arbutus	*Epigaea repens*
Trillium	*Trillium spp.*
Tuliptree	*Liriodendron tulipifera*
Violet	*Viola spp.*
Walnut	*Juglans spp.*
Western hemlock	*Tsuga heterophylla*
Western white pine	*Pinus monticola*
Wheat	*Triticum spp.*
White ash	*Fraxinus americana*
White oak (eastern white oak)	*Quercus alba*
White pine (eastern white pine)	*Pinus strobus*

Willow	Salix spp.
Wood anemone	Anemone quinquefolia
Wych elm	Ulmus glabra
Yellow birch	Betula alleghaniensis
Zelkova	Zelkova spp.

FUNGI

Black stem grain rust	Puccinia graminis
Butternut canker	Sirococcus clavigignenti-juglandacearum
Chestnut blight	Cryphonectria parasitica
Dutch elm disease	Ophiostoma ulmi and O. novo-ulmi
Oak wilt fungus	Ceratocystis fagacearum
White-nose syndrome fungus	Pseudogymnoascus destructans
White pine blister rust	Cronartium ribicola

INSECTS

American chestnut moth	Ectoedemia castaneae
Asian long-horned beetle	Anoplophora glabripennis
Bumblebees	Bombus spp.
Elm bark beetle	Scolytus spp.
Emerald ash borer	Agrilus planipennis
Gypsy moth	Lymantria dispar
Harvester butterfly	Feniseca tarquinius
Hemlock woolly adelgid	Adelges tsugae
Leafcutter ant	Atta spp.
Rhinoceros beetles	Subfamily Dynastinae
Scallop-shell moth	Hydria prunivorata
Winter moth	Operophtera brumata
Woolly alder aphid	Paraprociphilus tessellatus

OTHER INVERTEBRATES

| Pine wood nematode | Bursaphelenchus xylophilus |

BIRDS

Acadian flycatcher	*Empidonax virescens*
American redstart	*Setophaga ruticilla*
Arctic tern	*Sterna paradisaea*
Bachman's warbler	*Vermivora bachmanii*
Baikal teal	*Anas formosa*
Barred owl	*Strix varia*
Black-and-white warbler	*Mniotilta varia*
Black-throated blue warbler	*Setophaga caerulescens*
Black-throated green warbler	*Setophaga virens*
Black woodpecker	*Dryocopus martius*
Blackburnian warbler	*Setophaga fusca*
Blackcap	*Sylvia atricapilla*
Blue-and-white flycatcher	*Cyanoptila cyanomelana*
Blue-winged warbler	*Vermivora cyanoptera*
Brown creeper	*Certhia americana*
Brown-headed cowbird	*Molothrus ater*
Brown shrike	*Lanius cristatus*
Brown thrasher	*Toxostoma rufum*
Bull-headed shrike	*Lanius bucephalus*
Capercaillie (Eurasian capercaillie)	*Tetrao urogallus*
Carrion crow	*Corvus corone*
Cerulean warbler	*Setophaga cerulea*
Chestnut-sided warbler	*Setophaga pensylvanica*
Chiffchaff (common chiffchaff)	*Phylloscopus collybita*
Coal tit	*Periparus ater*
Common kingfisher	*Alcedo atthis*
Common nightingale	*Luscinia megarhynchos*
Common quail	*Coturnix coturnix*
Common whitethroat	*Sylvia communis*
Common yellowthroat	*Geothlypis trichas*
Crested tit	*Lophophanes cristatus*
Crow	*Corvus spp.*
Downy woodpecker	*Picoides pubescens*
Dunnock	*Prunella modularis*
Eastern crowned leaf warbler	*Phylloscopus coronatus*
Eastern towhee	*Pipilo erythrophthalmus*

Eurasian eagle-owl	*Bubo bubo*
Eurasian golden oriole	*Oriolus oriolus*
Eurasian jay	*Garrulus glandarius*
Eurasian nuthatch	*Sitta europaea*
Eurasian tree sparrow	*Passer montanus*
Eurasian woodcock	*Scolopax rusticola*
Eurasian wren	*Troglodytes troglodytes*
European stonechat	*Saxicola rubicola*
Field sparrow	*Spizella pusilla*
Garden warbler	*Sylvia borin*
Golden-winged warbler	*Vermivora chrysoptera*
Gray catbird	*Dumetella carolinensis*
Gray's grasshopper-warbler	*Locustella fasciolata*
Great gray shrike (northern shrike)	*Lanius excubitor*
Great spotted woodpecker	*Dendrocopos major*
Great tit	*Parus major*
Hooded crane	*Grus monacha*
Hooded crow	*Corvus cornix*
Hooded warbler	*Setophaga citrina*
Ivory-billed woodpecker	*Campephilus principalis*
Jackdaw (Eurasian jackdaw)	*Corvus monedula*
Japanese bush-warbler	*Cettia diphone*
Japanese green pigeon	*Sphenurus sieboldii*
Japanese green woodpecker	*Picus awokera*
Japanese marsh warbler (marsh grassbird)	*Megalurus pryeri*
Japanese reed bunting (ochre-rumped bunting)	*Emberiza yessoensis*
Jungle crow (large-billed crow)	*Corvus macrorhynchos*
Kentucky warbler	*Geothlypis formosa*
Lesser whitethroat	*Sylvia curruca*
Linnet	*Carduelis cannabina*
Little cuckoo (lesser cuckoo)	*Cuculus poliocephalus*
Long-tailed tit	*Aegithalos caudatus*
Magnolia warbler	*Setophaga magnolia*
Marsh tit	*Poecile palustris*
Masked grosbeak (Japanese grosbeak)	*Eophona personata*
Meadow bunting	*Emberiza cioides*
Meadow pipit	*Anthus pratensis*
Melodious warbler	*Hippolais polyglotta*

Narcissus flycatcher	*Ficedula narcissina*
Northern goshawk	*Accipiter gentilis*
Northern hawk-cuckoo	*Hierococcyx hyperythrus*
Northern waterthrush	*Parkesia noveboracensis*
Oriental cuckoo	*Cuculus optatus*
Ovenbird	*Seiurus aurocapilla*
Pacific wren	*Troglodytes pacificus*
Passenger pigeon	*Ectopistes migratorius*
Pied flycatcher (European pied flycatcher)	*Ficedula hypoleuca*
Pileated woodpecker	*Dryocopus pileatus*
Prairie warbler	*Setophaga discolor*
Pygmy woodpecker	*Dendrocopos kizuki*
Red-backed shrike	*Lanius collurio*
Red crossbill	*Loxia curvirostra*
Red-crowned crane (Japanese crane)	*Grus japonensis*
Red-eyed vireo	*Vireo olivaceous*
Red-shouldered hawk	*Buteo lineatus*
Rook	*Corvus frugilegus*
Russet sparrow	*Passer rutilans*
Short-toed treecreeper	*Certhia brachydactyla*
Siberian blue robin	*Larvivora cyane*
Skylark (Eurasian sky lark)	*Alauda arvensis*
Song thrush	*Turdus philomelos*
Swainson's thrush	*Catharus ustulatus*
Thick-billed shrike (tiger shrike)	*Lanius tigrinus*
Whinchat	*Saxicola rubetra*
White-backed woodpecker	*Dendrocopos leucotos*
White-eyed vireo	*Vireo griseus*
White-naped crane	*Grus vipio*
White wagtail	*Motacilla alba*
Willow flycatcher	*Empidonax traillii*
Willow tit	*Poecile montanus*
Willow warbler	*Phylloscopus trochilus*
Winter wren	*Troglodytes hiemalis*
Wood thrush	*Hylocichla mustelina*
Worm-eating warbler	*Helmitheros vermivorum*
Yellow-breasted chat	*Icteria virens*
Yellowhammer	*Emberiza citrinella*

MAMMALS

African elephant	*Loxodonta africana*
Asian black bear	*Ursus thibetanus*
Asian elephant	*Elephas maximus*
Aurochs	*Bos primigenius*
Beaver (American beaver)	*Castor canadensis*
Beaver (Eurasian beaver)	*Castor fiber*
Bison (American bison)	*Bison bison*
Bison (European bison)	*Bison bonasus*
Black bear (American black bear)	*Ursus americanus*
Bobcat	*Lynx rufus*
Brown bear	*Ursus arctos*
Chinese water deer	*Hydropotes inermis*
Cougar	*Puma concolor*
Coyote	*Canis latrans*
Dire wolf	*Canis dirus*
Domestic cat	*Felis catus*
Eastern chipmunk	*Tamias striatus*
Eastern cottontail	*Sylvilagus floridanus*
Eastern gray squirrel	*Sciurus carolinensis*
Eastern wolf	*Canis lycaon*
Elk (wapiti)	*Cervus elaphus*
Eurasian lynx	*Lynx lynx*
Eurasian red squirrel	*Sciurus vulgaris*
European beaver	*Castor fiber*
European bison (wisent)	*Bison bonasus*
Fallow deer	*Dama dama*
Fisher	*Martes pennanti*
Giant beaver	*Castoroides spp.*
Gray wolf	*Canis lupus*
Hippopotamus	*Hippopotamus amphibius*
Howler monkey	*Alouatta spp.*
Japanese giant flying squirrel	*Petaurista leucogenys*
Japanese macaque	*Macaca fuscata*
Japanese marten	*Martes melampus*
Japanese serow	*Capricornis crispus*
Little brown myotis	*Myotis lucifugus*

Lynx (Eurasian lynx)	*Lynx lynx*
Mammoth	*Mammuthus spp.*
Mastodon	*Mammut americanum*
Mongoose	*Herpestidae*
Moose (elk in Europe)	*Alces alces*
Mountain lion (cougar, puma)	*Puma concolor*
Murr water buffalo	*Bubalus murrensis*
Naumann's elephant	*Palaeoloxodon naumanni*
New England cottontail	*Sylvilagus transitionalis*
Prairie-dog	*Cynomys spp.*
Raccoon	*Procyon lotor*
Raccoon dog	*Nyctereutes procyonoides*
Red deer	*Cervus elaphus*
Red fox	*Vulpes vulpes*
Red wolf	*Canis lupus rufus (Canis rufus)*
Reeves's muntjac	*Muntiacus reevesi*
Roe deer	*Capreolus capreolus*
Saber-toothed cat	*Smilodon fatalis*
Short-faced bear	*Arctodus simus*
Short-horned water buffalo	*Bubalus mephistopheles*
Sika deer	*Cervus nippon*
Southern flying squirrel	*Glaucomys volans*
Straight-tusked elephant	*Palaeoloxodon antiquus*
Tarpan	*Equus ferus (Equus caballus)*
Tiger	*Panthera tigris*
White-footed mouse	*Peromyscus leucopus*
White-tailed deer	*Odocoileus virginianus*
Wild boar	*Sus scrofa*
Wildebeest (blue wildebeest)	*Connochaetes taurinus*
Yabe's giant deer	*Sinomegaceros yabei*

Notes

CHAPTER 1. Parallel Worlds

1. The deciduous forest regions of North America, Europe, and East Asia are the main focus of this book. My analysis does not encompass the distinctly different deciduous forests of the tropics, or the temperate deciduous forests of the Southern Hemisphere (Schmaltz, 1991). It also does not emphasize research on relatively small areas of deciduous forest in western North America and the mountains of Mexico and Central America (Barnes, 1991; Röhrig, 1991).

2. Latham and Ricklefs, 1993.

3. Qian and Ricklefs, 1999.

4. Graham, 1972a.

5. Li, 1952.

6. Wen, 1999.

7. Tiffney, 1985a, 1985b; Qian and Ricklefs, 1999.

8. Qian and Ricklefs, 1999.

9. The metaphor of an ecological theater and an evolutionary play comes from the book with that title by G. Evelyn Hutchinson (1965).

CHAPTER 2. Origins of the Deciduous Forest

1. Parrish et al., 1987.

2. Falcon-Lang et al., 2004.

3. Parrish et al., 1987.

4. Skelton et al., 2003: 148–156; Beerling, 2007: 200–202.

5. Falcon-Lang et al., 2004.

6. Skelton et al., 2003: 90–91.

7. David Beerling's book *The Emerald Planet* includes a detailed and engaging review of attempts to understand ancient polar forests (Beerling, 2007: 132–137).

8. Royer et al., 2005.

9. Beerling, 2007: 137.

10. Graham, 1999: 156.

11. Parrish et al., 1987; Graham, 1999: 158.

12. Graham, 1999: 157.

13. Parrish et al., 1987; Prothero, 2009: 2–5.

14. Skelton et al., 2003: 159.

15. Russell, 2009: 222–228.

16. Graham, 1999: 158, Falcon-Lang et al., 2004.

17. Skelton et al., 2003: 90.

18. Russell, 2009: 230–231, 238.

19. Russell, 2009: 230–232.

20. Parrish et al., 1987.

21. Parrish et al., 1987.

22. Parrish et al., 1987.

23. Fiorillo and Gangloff, 2000.

24. Russell, 2009: 247–249.

25. Wolfe and Upchurch, 1986.

26. Sarjeant and Currie, 2001; Sweet and Braman, 2001; Prothero, 2009.

27. Skelton et al., 2003: 312–325.

28. Skelton et al., 2003: 329–334.

29. Prothero, 2009: 127.

30. Nichols and Johnson, 2008; Skelton et al., 2003: 297–300; Schulte et al., 2010.

31. Signor and Lipps, 1982; Nichols and Johnson, 2008: 8–12.

32. Nichols and Johnson, 2008: 60–66.

33. Thornton, 1996: 57–58, 109.

34. Wing, 2004.

35. Wolfe and Upchurch, 1986.

36. Nichols and Johnson, 2008: 69–90.

37. K is the standard abbreviation for the Cretaceous, and T is the abbreviation of the Tertiary, a term that was previously used to refer to all except the final 1.8 million years of the Cenozoic. The Tertiary is now divided into the "Paleogene" and "Neogene" Periods, and the boundary is sometimes called the K-Pg (Paleogene) boundary.

38. Nichols and Johnson, 2008: 192.

39. A similar pattern, with a decline in pollen abundance and a marked increase in the abundance of fern spores, occurs at the K-T boundary in marine sediments in Hokkaido, Japan (Saito et al., 1986).

40. Nichols and Johnson, 2008: 227.

41. Labandeira et al., 2002.

42. Graham, 1999: 161.

43. Wolfe, 1987.

44. Wolfe, 1987; Wolfe and Upchurch, 1987.

45. Johnson and Ellis, 2002; Nichols and Johnson, 2008: 57.

46. Johnson and Ellis, 2002; Ellis et al., 2003.

47. Johnson and Ellis, 2002.

48. Wilf et al., 2006.

49. Wappler et al., 2009.

50. Wolfe, 1987.

51. Wing, 2004.

52. Wing et al., 2005.

53. Wolfe, 1972.

54. Graham, 1972b.

55. Wolfe, 1972; Wolfe, 1987.

56. Ruddiman and Kutzbach, 1991.

57. Barnes, 1991.

58. Graham, 1999: 250.
59. McCartan et al., 1990; Graham, 1999: 249–250.
60. Tanai, 1972.
61. Wolfe, 1979; Davis, 1983.
62. Latham and Ricklefs, 1993; Harrison et al., 2001.
63. Graham, 1999: 37–38; Skelton et al., 2003: 174–175.
64. Graham, 1999: 274.
65. Braun, 1950.
66. Braun, 1950: 39–56; Delcourt, 2002: 37–38.
67. Delcourt, 2002: 129–139.
68. Graham, 1999: 285.
69. Davis, 1983.
70. Delcourt, 2002: 148–149.
71. Davis, 1983.
72. Wolfe, 1979; Delcourt and Delcourt, 1987: 374–376.
73. Latham and Ricklefs, 1993.
74. Wolfe, 1979.
75. Delcourt, 2002: 147–149.
76. Davis, 1969.
77. Grimm and Jacobson, 2004.
78. Delcourt and Delcourt, 1987: Chapter 5, 359–361.
79. Delcourt and Delcourt, 1987: 165, 231, 356–358.
80. Delcourt and Delcourt, 1987: 374–381.
81. Graham, 1999: 323.
82. Harrison et al., 2001.
83. Ching, 1991.
84. Qian and Ricklefs, 2000.

CHAPTER 3. Deciduous Forests After the Arrival of People

1. Martin, 1967; Martin, 2005.
2. Grayson and Meltzer, 2003.
3. Gill et al., 2009.
4. Martin and Steadman, 1999.
5. Grayson and Meltzer, 2003.
6. Graham, 1990; Guthrie, 1990; Stuart, 1999.
7. Grayson, 2008.
8. Gill et al., 2009, 2012; Johnson, 2009.
9. Burney et al., 2003.
10. Stuart, 1999.
11. Owen-Smith, 1987; Martin and Steadman, 1999.
12. Stuart, 1999.
13. Norton et al., 2010.
14. Elvin, 2004: 31–32.
15. Owen-Smith, 1987.
16. Owen-Smith, 1987; Waldram et al., 2008.
17. Owen-Smith, 1987.
18. Williams, 2003: 15–19.

19. Askins, 2002; Williams, 2003: 30–32.
20. Pyne, 1982: 48; Williams, 2003: 30–31.
21. Mellars and Reinhardt, 1979.
22. Williams, 2003: 23.
23. Pyne, 1982: 66.
24. Russell, 1983.
25. Delcourt and Delcourt, 1987: 378–379; Shaffer, 1992: 62.
26. Williams, 2003: 44–45.
27. Williams, 2003: 46.
28. Williams, 2003: 48–49.
29. Delcourt and Delcourt, 2004: 36–42.
30. Delcourt and Delcourt, 2004: 79.
31. Shaffer, 1992: 28–37.
32. Williams, 2003: 71.
33. Fuller and Qin, 2009.
34. Jones and Liu, 2009.
35. Zheng et al., 2009.
36. Shu et al., 2010.
37. Lu et al., 2009.
38. Fuller and Qin, 2009.
39. Totman, 2000: 28–30.
40. Fuller and Qin, 2009.
41. Mann, 2005.
42. Totman, 2000: 28–29.
43. Thirgood, 1981.
44. Thirgood, 1981: 37–39.
45. Quotation from Critias III (Thirgood, 1981: 36).
46. Thirgood, 1981: 3.
47. Grove and Rackham, 2001: 156–161.
48. Williams, 2003: 79–84.
49. Williams, 2003: 83–84; Pinto, 2010.
50. Williams, 2003: 102.
51. Williams, 2003: 108–111.
52. Williams, 2003: 134–136.
53. Rackham, 1986: 72–88.
54. Shaffer, 1992: 38–42.
55. Shaffer, 1992: 7.
56. Shaffer, 1992: 44–46.
57. Shaffer, 1992: 56.
58. Shaffer, 1992: 51–62.
59. Delcourt and Delcourt, 2004: 122; Benson et al., 2009.
60. Benson et al., 2009.
61. Benson et al., 2009.
62. Shaffer, 1992: 87–89.
63. Shaffer, 1992: 58–60; Hudson, 1997: 289–290.
64. Hudson, 1997: 179–180.
65. Shaffer, 1992: 91.
66. Delcourt and Delcourt, 2004: 104–109.

67. Delcourt and Delcourt, 2004: 90–95.

68. Delcourt and Delcourt, 2004: 95.

69. Williams, 2003: 140–142.

70. Elvin, 2004.

71. Elvin, 2004: 9–17. This analysis of the shrinking geographical range of elephants is based on research by Wen Huanran.

72. Totman, 2000: 34–35.

73. Totman, 1989: 18.

74. Totman, 1989: 18.

75. Totman, 1989: 14–15.

76. Shidei, 1974.

77. Shidei, 1974.

78. Totman, 1989: 27–28.

79. Maekawa, 1974.

80. Totman, 1989: 39.

81. Totman, 1989: 52–56; Williams, 2003: 238.

82. Totman, 2000: 203–215.

83. Totman, 1989: 56–69.

84. Williams, 2003: 238.

85. Totman, 1989: 79.

CHAPTER 4. Decline of Natural Forests and the Invention
      of Sustainable Forestry

1. Translation from *Mémoires concernant l'histoire, les sciences, les artes, les moeurs, les usages & des Chinois, par les Missionaires de Pékin* (Paris: Nyon, 1776–1786) by Elvin (2004: 463).

2. Elvin, 2004: 462–467.

3. Elvin, 2004: 460–461.

4. Elvin, 2004.

5. Elvin, 2004: 470–471.

6. Elvin, 2004: 471.

7. Shapiro, 2001: 80–83.

8. Shapiro, 2001: 67–70.

9. Xu, 2011.

10. Williams, 2003: 173–175.

11. Williams, 2003: 203–206.

12. Williams, 2003: 273–275.

13. Williams, 2003: 206–209, 273–275.

14. Williams, 2003: 237.

15. Totman, 1989: 84–85.

16. Totman, 1989: 84–97.

17. Totman, 1989: 116–129.

18. Totman, 1989: 135.

19. Totman, 1989: 161–169.

20. Totman, 1989: 132–133.

21. Tsutsui, 2003.

22. Yamaura et al., 2009.

23. Cronon, 1983: 90.
24. Cronon, 1983: 85–89.
25. Deneven, 1992.
26. Silverberg, 1968: 50–73.
27. Shaffer, 1992.
28. Thorson, 2002: 126–127.
29. Hall et al., 2002.
30. Foster, 1992.
31. Donahue, 2004: 54–73.
32. Donahue, 2004: 64.
33. Kulikoff, 1986: 47–50.
34. Pimm and Askins, 1995.
35. Pimm and Askins, 1995; Williams, 2003: 412–413.
36. Williams, 2003: 317–324.
37. Williams, 2003: 386–389.
38. Lewis, 2005: 99.
39. Lewis, 2006: 103.
40. Nash, 2001: 106–109.
41. Nash, 2001: 61–62, 108, 116–121.
42. Rackham, 1986: 93–97.

CHAPTER 5. Giant Trees and Forest Openings

1. Faliński and Falińska, 1986: 152.
2. Angelstam et al., 2001.
3. Faliński and Falińska, 1986: 152.
4. Milecka et al., 2009.
5. Niklasson et al., 2010.
6. Bernadzki et al., 1988; Faliński and Falińska, 1986: 377–382.
7. Bernadzki et al., 1988.
8. Bernadzki et al., 1988.
9. Kuijper et al., 2010a.
10. Kuijper et al., 2010b.
11. Kuijper et al., 2010b.
12. Pigott, 1975.
13. Faliński and Falińska, 1986: 152.
14. Pigott, 1975; Thomas and Packham, 2007: 393.
15. Bobiec et al., 2011.
16. Bobiec et al., 2012.
17. Vera, 2000: 245–274; Vines, 2002.
18. Mitchell and Cole, 1998. Oak pollen is also regularly found in pollen deposits in Białowieża beginning in the middle Holocene (Milecka et al., 2009).
19. Vera, 2000: 307.
20. Vera, 2000: 301–306.
21. Vera, 2000: 246.
22. Bernadzki et al., 1988.
23. Runkle, 2000.

24. Tyrell and Crow, 1994.

25. Thomas and Packham, 2007: 391.

26. Faliński, 1988.

27. Bobiec et al., 2000.

28. Wesołowksi, 2007.

29. Runkle, 1982.

30. Chiver et al., 2011.

31. Greenberg and Lanham, 2001; Chiver et al., 2011.

32. Fuller, 2000.

33. Annand and Thompson, 1997; Robinson and Robinson, 1999; Heltzel and Leberg, 2006.

34. Askins, 2002: 93–95.

35. Loeb and O'Keefe, 2006.

36. Menzel et al., 2002.

37. Fukui et al., 2011.

38. Kricher, 2011: 215–220.

39. Fuller, 2000.

40. Seaton, 1996; Kershner and Leverett, 2004: 184–185.

41. Peterson and Pickett, 1995; Kershner and Leverett, 2004: 48–50.

42. Peterson and Pickett, 1995.

43. Peterson and Pickett, 1995.

44. Oliver, 1981.

45. Frelich, 2002: 159–162, 184–186.

46. Bonan and Shugart, 1989.

47. Foster and Motzkin, 2003; Lorimer and White, 2003.

48. Runkle, 1996.

49. Bormann and Likens, 1979; Parshall and Foster, 2002.

50. Boose et al., 2001.

51. Foster and Boose, 1992.

52. Boose et al., 2001.

53. Henry and Swan, 1974.

54. Shimatani and Kubota, 2011.

55. Yamashita et al., 2002.

56. Parshall and Foster, 2002; Lorimer and White, 2003.

57. McShea et al., 2007.

58. Lorimer and White, 2003.

59. Cogbill et al., 2002.

60. Fuller et al., 1998; Foster et al., 2002.

61. Thomas and Packham, 2007: 239–240.

62. Meier et al., 1996.

63. Bellemare et al., 2002.

64. Peterken and Game, 1984.

65. Selva, 1996.

66. Rose, 1992.

67. Selva, 1996.

68. Petranka et al., 1993.

69. Herbeck and Larsen, 1999.

70. Haney and Schaadt, 1996.
71. Leverett, 1996.
72. Stahle, 1996.
73. Tyrell and Crow, 1994.
74. Litvaitis et al., 2006.
75. Litvaitis and Litvaitis, 1996.
76. Litvaitis, 1993; DeGraaf and Yamasaki, 2001: 319–320.
77. DeGraaf and Yamasaki, 2001: 320; Litvaitis et al., 2006.
78. Litvaitis et al., 2006.
79. Litvaitis, 2003.
80. Litvaitis, 1993.
81. Kovach et al., 2003.
82. Godin, 1977: 71–72.
83. Fuller and DeStefano, 2003.
84. White, 1998.
85. Prather and Smith, 2003.
86. Askins et al., 2007a.
87. King et al., 2001.
88. Costello et al., 2000.
89. Annand and Thompson, 1997.
90. Litvaitis, 1993.
91. Dettmers, 2003.
92. Askins, 1993.
93. Askins, 1993.
94. Głowaciński and Järvinen, 1975; Ehrlich et al., 1994.
95. Fuller and Rose, 1995: 86–89.
96. Fuller and Rose, 1995: 99–101.
97. Paquet et al., 2006.
98. Hewson and Noble, 2009.
99. Fuller et al., 2007a.
100. Kurosawa, 2009.
101. Reiko Kurosawa, personal communication.
102. Imanishi, 2002; Amano and Yamaura, 2007.
103. Yamaura et al., 2009.
104. Harris, 1984.

CHAPTER 6. Forest Islands and the Decline of Forest Birds

1. Askins, 1990.
2. Niering and Goodwin, 1962.
3. Niering and Goodwin, 1962, Hemond et al., 1983.
4. Butcher et al., 1981; Askins and Philbrick, 1987.
5. Askins et al., 1990.
6. Hickey, 1937.
7. Leck et al., 1988; Askins et al., 1990; Litwin and Smith, 1992.
8. Robbins, 1979.
9. Briggs and Criswell, 1978; Lynch and Whitcomb, 1978.

10. Wilcove, 1988; Blodget et al., 2009; Holmes, 2011.

11. Faaborg et al., 1995.

12. Burke and Nol, 1998.

13. Temple and Cary, 1988; Porneluzi et al., 1993; Paton, 1994.

14. Robinson et al., 1995a.

15. Kress, 1983.

16. Reviewed in Whitaker and Warkentin, 2010.

17. Anders et al., 1998; Whitaker and Warkentin, 2010.

18. Betts et al., 2008.

19. Betts et al., 2008.

20. Askins et al., 1987.

21. Austen et al., 2001; reviewed in Askins et al., 1990.

22. Robbins et al., 1989.

23. Robinson et al., 1995b.

24. Friesen et al., 1995; Ford et al., 2001.

25. Sands et al., 2012.

26. Hagan, 1992.

27. Anderson, 2008.

28. Lynch, 1989; Robbins et al., 1992; Wunderle and Waide, 1993.

29. Faaborg et al., 2010.

30. Rappole, 1995: 147.

31. Eaton, 1995; Hamel, 2000.

32. Sauer et al., 2008.

33. Askins, 1993; Sauer et al., 2008.

34. Askins et al., 2000.

35. Higuchi et al., 1982.

36. Askins et al., 1987.

37. See Kurosawa and Askins (1999) for a description of how these categories were defined for Japanese birds.

38. Kurosawa and Askins, 2003.

39. Higuchi, 1998.

40. Kurosawa and Askins, 1999; Askins et al., 2000.

41. Kawaji, 1994; Wada, 1994; Ueta, 1998; Kurosawa and Askins, 1999, 2003.

42. Karasawa et al., 1991.

43. Ikeda, 2002.

44. Yamaura et al., 2009.

45. Amano and Yamaura, 2007; Yamaura et al., 2009.

46. Yamaura et al., 2009.

47. Higuchi and Morishita, 1999; Namba et al., 2010.

48. Bayard and Elphick, 2010.

49. Roth and Johnson, 1993; Gale et al., 1997; Friesen et al., 1999.

50. Thompson et al., 2002; Thompson, 2007.

51. Robinson et al., 1995a.

52. Thompson et al., 2002.

53. Matthysen, 1999; Bellamy et al., 2000.

54. Bellamy et al., 2000.

55. Askins et al., 1990, 2000; Kurosawa and Askins, 2003.

56. Ford, 1987.
57. Chris Elphick, personal communication.
58. Ford, 1987; Woolhouse, 1987.
59. Whitcomb et al., 1981; Askins et al., 1987.
60. Ford, 1987.
61. Yamauchi et al., 1997; Askins et al., 2000.
62. Opdam et al., 1985; van Dorp and Opdam, 1987.
63. Opdam et al., 1985; Cramp et al., 1985: 863.
64. Mörtberg, 2001.
65. This effect is called the passive sampling hypothesis (Connor and McCoy, 1979). Woolhouse (1983) and Askins et al. (1990) discuss the importance of considering this hypothesis when analyzing bird distributions in forest fragments.
66. Matthysen, 1999.
67. Matthysen and Adriaensen, 1998.
68. Mönkkönen and Welsh, 1994; Fuller et al., 2007b.
69. Mönkkönen and Welsh, 1994.
70. Tomiatojc, 2000; Yalden and Albarella, 2009: 60.
71. Yalden and Albarella, 2009: 56–60.
72. Mönkkönen and Welsh, 1994.
73. Hino, 1990; Mönkkönen and Welsh, 1994.
74. Ford, 1987.
75. Cieślak, 1985; Kurosawa and Askins, 2003.
76. Martin and Clobert, 1996, Wesołowski, 2007.
77. Brewer and MacKay, 2001.
78. Brewer and MacKay, 2001.
79. Chesser et al., 2010.
80. Brewer and MacKay, 2001.
81. Armstrong, 1955: 7.
82. Armstrong, 1955: 12–16.
83. Ford, 1987.
84. Hinsley et al., 1996; Bellamy et al., 2000.
85. Berg, 1997; Frank and Battisti, 2005.
86. Ford, 1987.
87. Wesołowski, 1983.
88. Hejl et al., 2002.
89. Wesołowski, 1983.
90. Rosenberg and Raphael, 1986; Lehmkuhl et al., 1991; McGarigal and McComb, 1995; George and Brand, 2002.
91. Brazil and Yabuuchi, 1991: 209.
92. Kurosawa and Askins, 2003.
93. Wesołowski, 2007.
94. Gregory et al., 2007.
95. Hopkins and Kirby, 2007.
96. Hewson et al., 2007.
97. Sanderson et al., 2006; Heldbjerg and Fox, 2008.
98. Hewson and Noble, 2009.
99. Hewson and Noble, 2009.
100. Wesołowski and Tomialojć, 1997.

CHAPTER 7. Missing Wolves and the Decline of Forests

1. Takatsuki and Gorai, 1994; Takatsuki and Ito, 2009.
2. Miyaki and Kaji, 2004, 2009.
3. Takatsuki and Gorai, 1994; Takatsuki and Ito, 2009.
4. Tsuji and Takatsuki, 2004.
5. Takatsuki and Gorai, 1994.
6. Takatsuki and Ito, 2009.
7. McNaughton, 1984; Askins, 2002.
8. Takatsuki, 2009; Takatsuki and Ito, 2009.
9. Ito et al., 2009.
10. Takatsuki and Ito, 2009.
11. Ito and Takatsuki, 2009.
12. Akashi and Nakashizuka, 1999.
13. Elvin, 2004; Kitchener and Yamaguchi, 2010.
14. Mech and Boitani, 2003.
15. Walker, 2005: 98–102.
16. Knight, 2003: 194–195.
17. Walker, 2005: 9.
18. Walker, 2005: 66–78.
19. Knight, 2003: 194; Walker, 2005: 6–7.
20. Walker, 2005: 137–146.
21. Walker, 2005: 148–151.
22. Walker, 2005: 151–156.
23. Knight, 2003: 199; Walker, 2005: 113–118, 211–213.
24. Knight, 2003: 202.
25. Knight, 2003: 230–231.
26. In his book *Waiting for Wolves in Japan*, John Knight (2003) describes the impact of these large wild animals on the farmers, foresters, and hunters of rural Japan.
27. Knight, 2003: 216–223.
28. House of Japan, 2011.
29. Terborgh et al., 2001.
30. Terborgh et al., 2001.
31. Boerner and Brinkman, 1996.
32. Tilghman, 1989; Horsley et al., 2003.
33. Tilghman, 1989.
34. Horsley et al., 2003.
35. Tilghman, 1989.
36. Marquis, 1981.
37. Reviewed in Russell et al., 2001.
38. Russell et al., 2001.
39. Rooney and Dress, 1997.
40. Rooney and Dress, 1997.
41. Comisky et al., 2005.
42. Comisky et al., 2005.
43. Goetsch et al., 2011.
44. Rooney and Waller, 2003.

45. Russell et al., 2001.
46. deCalesta, 1994.
47. McShea et al., 1995.
48. Healy, 1997.
49. Tremblay et al., 2005.
50. Ditchkoff and Servello, 1998.
51. Royo et al., 2010; McCabe and McCabe, 1997.
52. Caughley, 1981.
53. Caughley, 1981: 8.
54. Caughley, 1981: 14.
55. Rutberg, 1997.
56. Healy, 1997.
57. Healy, 1997.
58. Côté et al., 2004.
59. Ripple and Beschta, 2003.
60. Ripple and Beschta, 2003, 2007.
61. Ripple and Beschta, 2007.
62. Phillips et al., 2003.
63. Parker, 1995: 46–57.
64. Wilson et al., 2000.
65. Kyle et al., 2006; P. J. Wilson et al., 2009; Bozarth et al., 2011.
66. vonHoldt et al., 2011.
67. Chambers, 2010; Way et al., 2010.
68. Bozarth et al., 2011.
69. Kays et al., 2010.
70. Kays et al., 2008.
71. Kyle et al., 2006.
72. Rackham, 1980: 191–193.
73. Boitani, 2003.
74. Boitani, 2003.
75. Boitani, 2003.
76. McNamee, 1996.
77. Ward, 2005.
78. Fuller and Gill, 2001.
79. Gill and Beardall, 2001.
80. Kirby, 2001.
81. Gill and Fuller, 2007.
82. Gill and Fuller, 2007.
83. Kirby, 2001.
84. Fuller and Gill, 2001; Hopkins and Kirby, 2007. Fuller and Gill (2001) summarize a series of papers on the effect of deer browsing on different organisms in a special issue of Forestry devoted to this topic.
85. Perrins and Overall, 2001.
86. Perrins and Overall, 2001.
87. Gill and Fuller, 2007.
88. Gill and Fuller, 2007.
89. Holt et al., 2011.
90. Holt et al., 2010.

CHAPTER 8. The Global Threat of Rapid Climate Change

1. Hurrell and Trenberth, 2010; Wormworth and Şekercioğlu, 2011.
2. Root et al., 2003.
3. Parmesan and Yohe, 2003.
4. Menzel et al., 2006.
5. Primack et al., 2009.
6. Both, 2010.
7. Charmantier et al., 2008.
8. Visser et al., 2006.
9. Both, 2010.
10. Both and Visser, 2001.
11. Both, 2010.
12. Karell et al., 2011; Newman et al., 2011: 150–152.
13. Gwinner, 1996.
14. Berthold and Terrill, 1991.
15. Berthold and Terrill, 1991.
16. Pulido et al., 2001.
17. Berthold et al., 1992.
18. Gienapp et al., 2007.
19. Sheldon, 2010.
20. Zuckerberg et al., 2009.
21. Parmesan and Yohe, 2003; Brommer and Møller, 2010.
22. Hickling et al., 2006.
23. Huntley and Prentice, 1993.
24. Webb et al., 1993.
25. Sykes et al., 1996.
26. Sykes et al., 1996.
27. McKenney et al., 2007.
28. Neilson et al., 2005.
29. Fazey and Fischer, 2009; Ricciardi and Simberloff, 2009.
30. Vitt et al., 2009.
31. Ruddiman, 2008: 245–247.
32. Ruddiman, 2008: 326–341.
33. Ruddiman, 2008: 341.
34. Newman et al., 2011: 32–44.
35. Lawler et al., 2013.

CHAPTER 9. Another Global Threat: Transport of Species
            Between Continents

1. Liebhold et al., 1995.
2. Kricher and Morrison, 1988: 58.
3. Thomas and Packham, 2007: 213.
4. Paillet, 2002.
5. Rackham, 2006: 338.
6. Brasier and Buck, 2001.
7. Thomas and Packham, 2007: 209–210.

8. Gil et al., 2004.

9. Thomas and Packham, 2007: 210.

10. Rackham, 2006: 339.

11. Liebhold et al., 1995.

12. Shidei, 1974.

13. Liebhold et al., 1995.

14. Liebhold et al., 1995.

15. Thomas and Packham, 2007: 199–201.

16. Orwig et al., 2002.

17. Tingley et al., 2002.

18. Webster, 2012.

19. Havill and Montgomery, 2008.

20. Orwig et al., 2002.

21. Thomas and Packham, 2007: 196–199; Liebhold and McCullough, 2011.

22. Aukema et al., 2011.

23. Niemelä and Mattson, 1996.

24. Liebhold et al., 1995; Gibbs, 2003.

25. Juzwik et al., 2008.

26. Gibbs, 2003.

27. Gibbs, 2003.

28. DeNitto, 2003.

29. U.S. Animal and Plant Health Inspection Service: www.aphis.usda.gov /import_export/plants/plant_imports/wood_packaging_materials.shtml.

30. Liebhold et al., 2012.

31. Liebhold et al., 1995.

32. Haack et al., 2010; Liebhold and McCullough, 2011.

33. Haack et al., 2010.

34. Dodds and Orwig, 2011.

35. Simberloff and Stiling, 1996.

36. McClure, 1995; Carole Cheah, personal communication.

37. McClure, 1995.

38. McClure and Cheah, 1999.

39. Cheah, 2011.

40. McClure and Cheah, 1999; Cheah, 2011.

41. Cheah, 2011.

42. Montgomery and Keena, 2011.

43. Onken and Reardon, 2011.

44. Montgomery and Keena, 2011. The three species are *Scymnus camptodromus*, *S. sinuanodulus*, and *S. ningshanensis*.

45. Butin et al., 2004.

46. Havill et al., 2011.

47. Onken and Reardon, 2011.

48. McDonald et al., 2011.

49. Costa, 2011.

50. Cheah, 2011.

51. Liebhold et al., 1995; Hajek, 2004.

52. Hajek, 2004.

53. Hajek, 1999.

54. Liebhold et al., 1995.
55. Liebhold et al., 1995.
56. Kizlinski et al., 2002.
57. Burnham, 1988; Horton, 2010.
58. Hebard, 2001; Jacobs, 2007; Horton, 2010.
59. Horton, 2010.
60. Forest Health Initiative, 2012: www.foresthealthinitiative.org/genomics .html.
61. Horton, 2010.
62. Dalgeish and Swihart, 2012.
63. Opler, 1978.
64. Opler, 1978.
65. Fuller, 1998.
66. Parker et al., 2002.
67. Warnecke et al., 2012.
68. Fenton, 2012.
69. Krusic et al., 1996.
70. Fike and Niering, 1999.
71. Silander and Klepeis, 1999.
72. Ehrenfeld et al., 2001.
73. Ehrenfeld et al., 2001.
74. Krusic et al., 1996.
75. Eschtruth and Battles, 2008.
76. Rackham, 2006: 267.
77. Miyawaki and Washitani, 2004.
78. Dong et al., 2011.
79. Rackham, 2006: 420.
80. Ikeda, 2002.
81. Vitousek et al., 1996.
82. Liebhold et al., 1995.

CHAPTER 10. Blending Conservation Strategies from Three Continents

1. Nash, 2001.
2. Cox et al., 1985: 135–137.
3. Nash, 2001: 82–83.
4. Cox et al., 1985: 137; Williams, 1989: 406.
5. Nash, 2001: 125–140.
6. Cox et al., 1985: 53–54; Williams, 1989: 121–122.
7. Nash, 2001: 141–156.
8. Leopold, 1949: 194–198.
9. Leopold, 1949: 196.
10. Nash, 2001: 116–121.
11. Williams, 1989: 406–407.
12. Runte, 1987: 114–115.
13. Runte, 1987: 116–117.
14. Cox et al., 1985: 407.
15. Pierce, 2000: xiv.

16. Brown, 2000: 174–175.
17. Williams, 1989: 410–411.
18. Williams, 1989: 415–421.
19. Cox et al., 1985: 227.
20. Cox et al., 1985: 227–229.
21. Williams, 1989: 458–459.
22. Cox et al., 1985: 229.
23. Lewis, 2005: 145–147.
24. Nash, 2001: 44–49.
25. Nash, 2001.
26. Anderson and Ferree, 2010.
27. Nash, 2001: 241.
28. Dubos, 1972: 135–152.
29. Dubos, 1976.
30. Bruns, 1960.
31. Rackham, 2006: 60.
32. Rackham, 2006: 16.
33. Rackham, 1980: 201–202.
34. Rackham, 1980: 201–202.
35. Rackham, 2006: 16.
36. Rackham, 2006: 211–213.
37. Mason, 2007.
38. Fuller et al., 2007a; Hinsley et al., 2007; Hoodless and Hirons, 2007.
39. Peterken, 1996: 310, 398–400, 457–458.
40. Vera, 2000.
41. Rackham, 2006: 430.
42. Rackham, 2006: 86.
43. J. D. Wilson et al., 2009: 114–125, 242–276.
44. Askins et al., 2007b: 32–33.
45. J. D. Wilson et al., 2009: 286–302.
46. Askins et al., 2007b: 33.
47. Kleijn et al., 2006.
48. J. D. Wilson et al., 2009: 284–288.
49. Askins, 2002: 138–142.
50. Angelstam, 1996; Askins, 2002: 142–143.
51. Kalland and Asquith, 1997: 16.
52. Kalland and Asquith, 1997: 17–18.
53. Young and Young, 2005: 124–125, 156–157.
54. Young and Young, 2005: 48.
55. Young and Young, 2005: 20.
56. Kalland and Asquith, 1997.
57. Young and Young, 2005: 20.
58. Young and Young, 2005: 108–109.
59. Young and Young, 2005: 48–53.
60. Laurent, 2001.
61. Short, 2000: 97–98.
62. Primack et al., 2000.
63. Kadoya et al., 2009.

64. Primack et al., 2000.
65. Mason, 1999.
66. Broadbent, 1998: 292; Mason, 1999; Schreurs, 2002: 244–245.
67. Mason, 1999; Kerr, 2001: 13–50.
68. Broadbent, 1998; Forrest et al., 2008.
69. Huddle et al., 1975; George, 2001.
70. Mason, 1999; Schreurs, 2002: 44–46.
71. Schreurs, 2002: 47.
72. Mason, 1999.
73. Mason, 1999.
74. Watanabe, 2008.
75. Kellert, 1991.
76. Kobori and Primack, 2003; Takeuchi et al., 2003.
77. Foster, 2002.
78. Grixti et al., 2009.
79. Leopold, 1953: 147.

References

Akashi N., and T. Nakashizuka. 1999. Effects of bark-stripping by sika deer (*Cervus nippon*) on population dynamics of a mixed forest in Japan. *Forest Ecology and Management* 113:75–82.

Amano T., and Y. Yamaura. 2007. Ecological and life-history traits related to range contractions among breeding birds in Japan. *Biological Conservation* 137:271–282.

Anders A. D., J. Faaborg, and F. R. Thompson III. 1998. Postfledging dispersal, habitat use, and home-range size of juvenile Wood Thrushes. *Auk* 115:349–358.

Anderson M. G. 2008. Conserving forest ecosystems: guidelines for size, condition and landscape requirements. Pages 213–220 in R. A. Askins, G. D. Dreyer, G. R. Visgilio, and D. M. Whitelaw, ed. *Saving Biological Diversity: Balancing Protection of Endangered Species and Ecosystems.* New York: Springer.

Anderson M. G., and C. E. Ferree. 2010. Conserving the stage: climate change and the geophysical underpinnings of species diversity. *PLoS One* 5:doi:10.1371/journal.pone.0011554.

Angelstam P. 1996. The ghost of forest past: natural disturbance regimes as a basis for reconstruction of biologically diverse forests in Europe. Pages 287–337 in R. M. DeGraaf and R. I. Miller, ed. *Conservation of Faunal Diversity in Forested Landscapes.* London: Chapman & Hall.

Angelstam P., M. Breuss, G. Mikusinski, et al. 2001. Effects of forest structure on the presence of woodpeckers with different specialisation in a landscape history gradient in NE Poland. Pages 25–38 in D. Chamberlain and A. Wilson, ed. *Avian Landscape Ecology: Pure and Applied Issues in the Large-Scale Ecology of Birds.* International Association for Landscape Ecology, U.K.

Annand E. M., and F. R. Thompson III. 1997. Forest bird response to regeneration practices in central hardwood forests. *Journal of Wildlife Management* 61:159–171.

Armstrong E. A. 1955. *The Wren.* London: Collins.

Askins R. A. 1990. *Birds of the Connecticut College Arboretum: Population Changes over Forty Years.* Connecticut College Arboretum Bulletin 31:1–43.

Askins R. A. 1993. Population trends in grassland, shrubland, and forest birds in eastern North America. Pages 1–34 in D. Power, ed. *Current Ornithology,* vol. 11. New York: Plenum Press.

Askins R. A. 2002. *Restoring North America's Birds: Lessons from Landscape Ecology.* Second ed., New Haven: Yale University Press.

Askins R. A., F. Chávez-Ramírez, B. C. Dale, et al. 2007b. Conservation of grassland birds in North America: understanding ecological processes in different regions. *Ornithological Monographs* 64:1–46.

Askins R. A., H. Higuchi, and H. Murai. 2000. Effect of forest fragmentation on migratory songbirds in Japan. *Global Environmental Research* 4:219–229.

Askins R. A., J. F. Lynch, and R. Greenberg. 1990. Population declines in migratory birds in eastern North America. Pages 1–57 in D. M. Power, ed. *Current Ornithology*, vol. 7. New York: Plenum Press.

Askins R. A., and M. J. Philbrick. 1987. Effect of changes in regional forest abundance on the decline and recovery of a forest bird community. *Wilson Bulletin* 99:7–21.

Askins R. A., M. J. Philbrick, and D. S. Sugeno. 1987. Relationship between the regional abundance of forest and the composition of forest bird communities. *Biological Conservation* 39:129–152.

Askins R. A., B. Zuckerberg, and L. Novak. 2007a. Do the size and landscape context of forest openings influence the abundance and breeding success of shrubland songbirds in southern New England? *Forest Ecology and Management* 250:137–147.

Aukema J. E., B. Leung, K. Kovacs, et al. 2011. Economic impacts of non-native forest insects in the continental United States. *PLoS One* 6:doi:10.1371/journal.pone.0024587.

Austen M. J. W., C. M. Francis, D. M. Burke, and M. S. W. Bradstreet. 2001. Landscape context and fragmentation effects on forest birds in southern Ontario. *Condor* 103:701–714.

Barnes B. V. 1991. Deciduous forests of North America. Pages 219–344 in E. Röhrig and B. Ulrich, ed. *Ecosystems of the World 7: Temperate Deciduous Forests*. New York: Elsevier.

Bayard T. S., and C. S. Elphick. 2010. How area sensitivity in birds is studied. *Conservation Biology* 24:938–947.

Beerling D. J. 2007. *The Emerald Planet: How Plants Changed Earth's History*. New York: Oxford University Press.

Bellamy P. E., P. Rothery, S. A. Hinsley, and I. Newton. 2000. Variation in the relationship between numbers of breeding pairs and woodland area for passerines in fragmented habitat. *Ecography* 23:130–138.

Bellemare J., G. Motzkin, and D. R. Foster. 2002. Legacies of the agricultural past in the forested present: an assessment of historical land-use effects on rich mesic forests. *Journal of Biogeography* 29:1401–1420.

Benson L. V., T. R. Pauketat, and E. R. Cook. 2009. Cahokia's boom and bust in the context of climate change. *American Antiquity* 74:467–483.

Berg Å. 1997. Diversity and abundance of birds in relation to forest fragmentation, habitat quality and heterogeneity. *Bird Study* 44:355–366.

Bernadzki E., L. Bolibok, B. Brzeziecki, et al. 1998. Compositional dynamics of natural forests in the Białowieża National Park, northeastern Poland. *Journal of Vegetation Science* 9:229–238.

Berthold P., A. J. Helbig, G. Mohr, and U. Querner. 1992. Rapid microevolution of migratory behaviour in a wild bird species. *Nature* 360:668–670.

Berthold P., and S. B. Terrill. 1991. Recent advances in studies of bird migration. *Annual Review of Ecology and Systematics* 22:357–378.

Betts M. G., A. S. Hadley, N. Rodenhouse, and J. J. Nocera. 2008. Social information trumps vegetation structure in breeding-site selection by a migrant songbird. *Proceedings of the Royal Society B: Biological Sciences* 275:2257–2263.

Blodget B. G., R. Dettmers, and J. Scanlon. 2009. Status and trends in an extensive western Massachusetts forest. *Northeastern Naturalist* 16:423–442.

Bobiec A., E. Jaszcz, and K. Wojtunik. 2012. Oak (*Quercus robur* L.) regeneration as a response to natural dynamics of stands in European hemiboreal zone. *European Journal of Forest Research* doi:10.1007/s10342-012-0597-6.

Bobiec A., D. P. J. Kuijper, M. Niklasson, et al. 2011. Oak (*Quercus robur* L.) regeneration in early successional woodlands grazed by wild ungulates in the absence of livestock. *Forest Ecology and Management* 262:780–790.

Bobiec A., H. van der Burgt, K. Meijer, et al. 2000. Rich deciduous forests in Białowieża as a dynamic mosaic of developmental phases: premises for nature conservation and restoration management. *Forest Ecology and Management* 130:159–175.

Boerner R. E. J., and J. A. Brinkman. 1996. Ten years of tree seedling establishment and mortality in an Ohio deciduous forest complex. *Bulletin of the Torrey Botanical Club* 123:309–317.

Boitani L. 2003. Wolf conservation and recovery. Pages 317–344 in L. D. Mech and L. Boitani, ed. *Wolves: Behavior, Ecology, and Conservation.* Chicago: University of Chicago Press.

Bonan G. B., and H. H. Shugart. 1989. Environmental factors and ecological processes in boreal forests. *Annual Review of Ecology and Systematics* 20:1–28.

Boose E. R., K. E. Chamberlin, and D. R. Foster. 2001. Landscape and regional impacts of hurricanes in New England. *Ecological Monographs* 71:27–48.

Bormann F. H., and G. E. Likens. 1979. Catastrophic disturbance and the steady state in northern hardwood forests: a new look at the role of disturbance in the development of forest ecosystems suggests important implications for land-use policies. *American Scientist* 67:660–669.

Both C. 2010. Food availability, mistiming and climatic change. Pages 129–147 in A. P. Møller, W. Fiedler, and P. Berthold, ed. *Effects of Climate Change on Birds.* New York: Oxford University Press.

Both C., and M. E. Visser. 2001. Adjustment to climate change is constrained by arrival date in a long-distance migrant bird. *Nature* 411:296–298.

Bozarth C. A., F. Hailer, L. L. Rockwood, et al. 2011. Coyote colonization of northern Virginia and admixture with Great Lakes wolves. *Journal of Mammalogy* 92:1070–1080.

Brasier C. M., and K. W. Buck. 2001. Rapid evolutionary changes in a globally invading fungal pathogen (Dutch elm disease). *Biological Invasions* 3:223–233.

Braun E. L. 1950. *Deciduous Forests of Eastern North America.* Philadelphia: Blakiston.

Brazil M., and M. Yabuuchi. 1991. *The Birds of Japan.* Washington, D.C.: Smithsonian Institution Press.

Brewer D., B. K. MacKay. 2001. *Wrens, Dippers, and Thrashers.* New Haven: Yale University Press.

Briggs, S. A., and J. H. Criswell. 1978. Gradual silencing of spring in Washington: selective reduction of species of birds found in three woodland areas over the past 30 years. *Atlantic Naturalist* 32:19–26.

Broadbent J. 1998. *Environmental Politics in Japan: Networks of Power and Protest*. New York: Cambridge University Press.

Brommer J. E., and A. P. Møller. 2010. Range margins, climate change and ecology. Pages 249–274 in A. P. Møller, W. Fiedler, and P. Berthold, ed. *Effects of Climate Change on Birds*. New York: Oxford University Press.

Brown M. L. 2000. *The Wild East: A Biography of the Great Smoky Mountains*. Gainesville: University Press of Florida.

Bruns H. 1960. The economic importance of birds in forests. *Bird Study* 7:193–208.

Burke D. M., and E. Nol. 1998. Influence of food abundance, nest-site habitat, and forest fragmentation on breeding Ovenbirds. *Auk* 115:96–104.

Burney D. A., G. S. Robinson, and L. P. Burney. 2003. *Sporormiella* and the late Holocene extinctions in Madagascar. *Proceedings of the National Academy of Sciences, USA* 100:10800–10805.

Burnham C. R. 1988. The restoration of the American chestnut: Mendelian genetics may solve a problem that has resisted other approaches. *American Scientist* 76:478–487.

Butcher G. S., W. A. Niering, W. J. Barry, and R. H. Goodwin. 1981. Equilibrium biogeography and the size of nature preserves: an avian case study. *Oecologia* 49:29–37.

Butin E., N. P. Havill, J. S. Elkinton, and M. E. Montgomery. 2004. Feeding preference of three lady beetle predators of the hemlock woolly adelgid (Homoptera: Adelgidae). *Journal of Economic Entomology* 97:1635–1641.

Caughley G. 1981. Overpopulation. Pages 7–19 in P. A. Jewell, S. J. Holt, and D. Hart, ed. *Problems in Management of Locally Abundant Wild Mammals*. New York: Academic Press.

Chambers S. M. 2010. A perspective on the genetic composition of eastern coyotes. *Northeastern Naturalist* 17:205–210.

Charmantier A., R. H. McCleery, L. R. Cole, C. Perrins, L. E. B. Kruuk, and B. C. Sheldon. 2008. Adaptive phenotypic plasticity in response to climate change in a wild bird population. *Science* 320:800–803.

Cheah C. A. S.-J. 2010. Connecticut's threatened landscape: natural enemies for biological control of invasive species. *Frontiers of Plant Science* 57:5–16.

Cheah C. A. S.-J. 2011. *Sasajiscymnus* (=Pseudoscymnus) *tsugae*, a ladybeetle from Japan. Pages 43–52 in B. P. Onken and R. C. Reardon, ed. *Implementation and Status of Biological Control of the Hemlock Woolly Adelgid*. U.S. Forest Service Publication FNTET-2011-04, Forest Health Technology Enterprise Team, USDA Forest Service, www.fs.fed.us/foresthealth/technology/.

Chesser R. T., R. C. Banks, F. K. Barker, et al. 2010. Fifty-first supplement to the American Ornithologists' Union check-list of North American birds. *Auk* 127:726–744.

Ching K. K. 1991. Temperate deciduous forests in East Asia. Pages 539–555 in E. Röhrig and B. Ulrich, ed. *Ecosystems of the World 7: Temperate Deciduous Forests*. New York: Elsevier.

Chiver I., L. J. Ogden, and B. J. Stutchbury. 2011. Hooded Warbler (*Wilsonia citrina*). A. Poole, ed. *The Birds of North America Online*, Cornell Laboratory of Ornithology, http://bna.birds.cornell.edu/bna/species/110/articles/introduction.

Cieślak M. 1985. Influence of forest size and other factors on breeding bird species number. *Ekologia Polska* 33:103–121.

Cogbill C. V., J. Burk, and G. Motzkin. 2002. The forests of presettlement New England, USA: spatial and compositional patterns based on town proprietor surveys. *Journal of Biogeography* 29:1279–1304.

Comisky L., A. A. Royo, and W. P. Carson. 2005. Deer browsing creates rock refugia gardens on large boulders in the Allegheny National Forest, Pennsylvania. *American Midland Naturalist* 154:201–206.

Connor E. F., and E. D. McCoy. 1979. The statistics and biology of the species-area relationship. *American Naturalist* 113:791–833.

Costa S. D. 2011. Insect-killing fungi for HWA management: current status. Pages 107–115 in B. P. Onken and R. C. Reardon, ed. *Implementation and Status of Biological Control of the Hemlock Woolly Adelgid*. U.S. Forest Service Publication FNTET-2011-04, Forest Health Technology Enterprise Team, USDA Forest Service, www.fs.fed.us/foresthealth/technology/.

Costello C. A., M. Yamasaki, P. J. Pekins, W. B. Leak, and C. D. Neefus. 2000. Songbird response to group selection harvests and clearcuts in a New Hampshire northern hardwood forest. *Forest Ecology and Management* 127:41–54.

Côté S. D., T. P. Rooney, J. Tremblay, C. Dussault, and D. M. Waller. 2004. Ecological impacts of deer overabundance. *Annual Review of Ecology, Evolution, and Systematics* 35:113–147.

Cox T. R., R. S. Maxwell, P. D. Thomas, and J. J. Malone. 1985. *This Well-Wooded Land: Americans and Their Forests from Colonial Times to the Present*. Lincoln: University of Nebraska Press.

Cramp S., C. M. Perrins, and D. J. Brooks. 1985. *Handbook of the Birds of Europe, the Middle East and North Africa: The Birds of the Western Palearctic*. New York: Oxford University Press.

Cronon W. 1983. *Changes in the Land: Indians, Colonists, and the Ecology of New England*. New York: Hill and Wang.

Dalgeish H. J., and R. K. Swihart. 2012. American chestnut past and future: implications of restoration for resource pulses and consumer populations of eastern U.S. forests. *Restoration Ecology* 20:490–497.

Davis M. B. 1969. Climatic changes in southern Connecticut recorded by pollen deposition at Rogers Lake. *Ecology* 50:409–422.

Davis M. B. 1983. Quaternary history of deciduous forests of eastern North America and Europe. *Annals of the Missouri Botanical Garden* 70:550–563.

deCalesta D. S. 1994. Effect of white-tailed deer on songbirds within managed forests in Pennsylvania. *Journal of Wildlife Management* 58:711–718.

DeGraaf R. M., and M. Yamasaki. 2001. *New England Wildlife: Habitat, Natural History, and Distribution*. Hanover, N.H.: University Press of New England.

Delcourt H. R. 2002. *Forests in Peril: Tracking Deciduous Trees from Ice-Age Refuges into the Greenhouse World*. Blacksburg, Va.: McDonald & Woodward.

Delcourt P. A., and H. R. Delcourt. 1987. *Long-term Forest Dynamics of the Temperate Zone: A Case Study of Late-Quaternary Forests in Eastern North America*. New York: Springer-Verlag.

Delcourt P. A., and H. R. Delcourt. 2004. *Prehistoric Native Americans and Ecological Change: Human Ecosystems in Eastern North America Since the Pleistocene*. New York: Cambridge University Press.

Denevan W. M. 1992. The pristine myth: the landscape of the Americas in 1492. *Annals of the Association of American Geographers* 82:369.

DeNitto G. A. 2003. Assessing the pest risks of wood imports into the United States of America. *New Zealand Journal of Forestry Science* 33:399–410.

Dettmers R. 2003. Status and conservation of shrubland birds in the northeastern U.S. *Forest Ecology and Management* 185:81–93.

Ditchkoff S. S., and F. A. Servello. 1998. Litterfall: an overlooked food source for wintering white-tailed deer. *Journal of Wildlife Management* 62:250–255.

Dodds K. J., and D. A. Orwig. 2011. An invasive urban forest pest invades natural environments: Asian longhorned beetle in northeastern U.S. hardwood forests. *Canadian Journal of Forest Research* 41:1729–1742.

Donahue B. 2004. *The Great Meadow: Farmers and the Land in Colonial Concord*. New Haven: Yale University Press.

Dong H., Y. Li, Q. Wang, and G. Yao. 2011. Impacts of invasive plants on ecosystems in natural reserves in Jiangsu of China. *Russian Journal of Ecology* 42:133–137.

Dubos R. J. 1972. *A God Within*. New York: Scribner.

Dubos R. J. 1976. Symbiosis between the earth and humankind. *Science* 193:459–462.

Eaton S. W. 1995. Northern Waterthrush (*Parkesia noveboracensis*). A. Poole, ed. *The Birds of North America Online*, Cornell Laboratory of Ornithology, http://bna.birds .cornell.edu/bna/species/182/articles/introduction.

Ehrenfeld J. G., P. Kourtev, and W. Huang. 2001. Changes in soil functions following invasions of exotic understory plants in deciduous forests. *Ecological Applications* 11:1287–1300.

Ehrlich P. R., D. S. Dobkin, and D. Wheye. 1994. *The Birdwatcher's Handbook: A Guide to the Natural History of the Birds of Britain and Europe*. New York: Oxford University Press.

Ellis B., K. R. Johnson, and R. E. Dunn. 2003. Evidence for an in situ early Paleocene rainforest from Castle Rock, Colorado. *Rocky Mountain Geology* 38:73–100.

Elvin M. 2004. *The Retreat of the Elephants: An Environmental History of China*. New Haven: Yale University Press.

Eschtruth A. K., and J. J. Battles. 2008. Acceleration of exotic plant invasion in a forested ecosystem by a generalist herbivore. *Conservation Biology* 23:388–399.

Faaborg J., M. Brittingham, T. Donovan, and J. Blake. 1995. Habitat fragmentation in the temperate zone. Pages 357–380 in T. E. Martin and D. M. Finch, ed. *Ecology and Management of Neotropical Migratory Birds*. New York: Oxford University Press.

Faaborg J., R. T. Holmes, A. D. Anders, et al. 2010. Conserving migratory land birds in the New World: do we know enough? *Ecological Applications* 20:398–418.

Falcon-Lang, H. J., R. A. MacRae, and A. Z. Csank. 2004. Palaeoecology of Late Creta-ceous polar vegetation preserved in the Hansen Point Volcanics, NW Ellesmere Island, Canada. *Palaeogeography, Palaeoclimatology, Palaeoecology* 212:45–64.

Faliński J. B. 1988. Regeneration and fluctuation in the Białowieża Forest (NE Po-land). *Vegetatio* 77:115–128.

Faliński J. B., and K. Falińska. 1986. *Vegetation Dynamics in Temperate Lowland Prime-val Forests: Ecological Studies in Białowieża Forest*. Boston: Dr. W. Junk; Distribu-tors for the U.S. and Canada, Kluwer Academic Publishers.

Fazey I., and J. Fischer. 2009. Assisted colonization is a techno-fix. *Trends in Ecology and Evolution* 24:475.

Fenton M. B. 2012. Bats and white-nose syndrome. *Proceedings of the National Acad-emy of Sciences, USA* 109:6794–6795.

Fike J., and W. A. Niering. 1999. Four decades of old field vegetation development and the role of *Celastrus orbiculatus* in the northeastern United States. *Journal of Vegetation Science* 10:483–492.

Fiorillo A. R., and R. A. Gangloff. 2000. Theropod teeth from the Prince Creek Formation (Cretaceous) of northern Alaska, with speculations on arctic dino-saur paleoecology. *Journal of Vertebrate Paleontology* 20:675–682.

Ford H. A. 1987. Bird communities on habitat islands in England. *Bird Study* 34:205–218.

Ford T. B., D. E. Winslow, D. R. Whitehead, and M. A. Koulol. 2001. Reproductive success of forest-dependent songbirds near an agricultural corridor in south-central Indiana. *Auk* 118:864–873.

Forrest R., M. Schreurs, and R. Penrod. 2008. A comparative history of U.S. and Japanese environmental movements. Pages 13–37 in P. P. Karan and U. Sug-anuma, ed. *Local Environmental Movements: A Comparative Study of the United States and Japan*. Lexington: University Press of Kentucky.

Foster D. R. 1992. Land-use history (1730–1990) and vegetation dynamics in cen-tral New England, USA. *Journal of Ecology* 80:753–772.

Foster D. R. 2002. Thoreau's country: a historical-ecological perspective on con-servation in the New England landscape. *Journal of Biogeography* 29:1537–1555.

Foster, D. R., and J. D. Aber. 2004. *Forests in Time: The Environmental Consequences of 1000 Years of Change in New England*. New Haven: Yale University Press.

Foster D. R., and E. R. Boose. 1992. Patterns of forest damage resulting from cata-strophic wind in central New England, USA. *Journal of Ecology* 80:79–98.

Foster D. R., S. Clayden, D. A. Orwig, et al. 2002. Oak, chestnut and fire: climatic and cultural controls of long-term forest dynamics in New England, USA. *Jour-nal of Biogeography* 29:1359–1379.

Foster D. R., and G. Motzkin. 2003. Interpreting and conserving the openland habitats of coastal New England: insights from landscape history. *Forest Ecology and Management* 185:127–150.

Frank B., and C. Battisti. 2005. Area effect on bird communities, guilds and spe-cies in a highly fragmented forest landscape of central Italy. *Italian Journal of Zoology* 72:297–304.

Frelich L. E. 2002. *Forest Dynamics and Disturbance Regimes: Studies from Temperate Evergreen-Deciduous Forests.* New York: Cambridge University Press.

Friesen L. E., M. D. Cadman, and R. J. MacKay. 1999. Nesting success of neotropical migrant songbirds in a highly fragmented landscape. *Conservation Biology* 13:338–346.

Friesen L. E., P. F. J. Eagles, and R. J. MacKay. 1995. Effects of residential development on forest-dwelling neotropical migrant songbirds. *Conservation Biology* 9:1408–1414.

Fukui D., T. Hirao, M. Murakami, and H. Hirakawa. 2011. Effects of treefall gaps created by windthrow on bat assemblages in a temperate forest. *Forest Ecology and Management* 261:1546–1552.

Fuller D. Q., and L. Qin. 2009. Water management and labour in the origins and dispersal of Asian rice. *World Archaeology* 41:88–111.

Fuller J. L. 1998. Ecological impact of the mid-Holocene hemlock decline in southern Ontario, Canada. *Ecology* 79:2337–2351.

Fuller J. L., D. R. Foster, Jason S. McLachlan, and N. Drake. 1998. Impact of human activity on regional forest composition and dynamics in central New England. *Ecosystems* 1:76–95.

Fuller R. J. 2000. Influence of treefall gaps on distributions of breeding birds within interior old-growth stands in Białowieża Forest, Poland. *Condor* 102:267–274.

Fuller R. J., K. J. Gaston, and C. P. Quine. 2007b. Living on the edge: British and Irish woodland birds in a European context. *Ibis* 149:53–63.

Fuller R. J., and R. M. A. Gill. 2001. Ecological impacts of increasing numbers of deer in British woodland. *Forestry* 74:193–199.

Fuller R. J., and C. Rose. 1995. *Bird Life of Woodland and Forest.* New York: Cambridge University Press.

Fuller R. J., K. W. Smith, P. V. Grice, et al. 2007a. Habitat change and woodland birds in Britain: implications for management and future research. *Ibis* 149:261–268.

Fuller T. K., and S. DeStefano. 2003. Relative importance of early-successional forests and shrubland habitats to mammals in the northeastern United States. *Forest Ecology and Management* 185:75–79.

Gale G. A., L. A. Hanners, and S. R. Patton. 1997. Reproductive success of Worm-eating Warblers in a forested landscape. *Conservation Biology* 11:246–250.

George, T. L., and L. A. Brand. 2002. The effects of habitat fragmentation on birds in coast redwood forests. *Studies in Avian Biology* 25:92–102.

George T. S. 2001. *Minamata: Pollution and the Struggle for Democracy in Postwar Japan.* Cambridge: Harvard University Asia Center.

Gibbs J. N. 2003. Protecting Europe's forests: how to keep out both known and unknown pathogens. *New Zealand Journal of Forestry Science* 33:411–419.

Gienapp P., R. Leimu, and J. Merilä. 2007. Responses to climate change in avian migration time: microevolution versus phenotypic plasticity. *Climate Research* 35:25–35.

Gil L., P. Fuentes-Utrilla, A. Soto, M. T. Cervera, and C. Collada. 2004. English elm is a 2,000-year-old Roman clone. *Nature* 431:1053.

Gill J. L., J. W. Williams, S. T. Jackson, et al. 2009. Pleistocene megafaunal col- lapse, novel plant communities, and enhanced fire regimes in North America. *Science* 326:1100–1103.

Gill J. L., J. W. Williams, S. T. Jackson, et al. 2012. Climatic and megaherbivory controls on late-glacial vegetation dynamics: a new, high-resolution, multi- proxy record from Silver Lake, Ohio. *Quaternary Science Reviews* 34:66–80.

Gill R. M. A., and V. Beardall. 2001. The impact of deer on woodlands: the effects of browsing and seed dispersal on vegetation structure and composition. *For- estry* 74:209–218.

Gill R. M. A., and R. J. Fuller. 2007. The effects of deer browsing on woodland structure and songbirds in lowland Britain. *Ibis* 149:119–127.

Głowaciński Z., and O. Järvinen. 1975. Rate of secondary succession in forest bird communities. *Ornis Scandinavica* 6:33–40.

Godin A. J. 1977. *Wild Mammals of New England.* Baltimore: Johns Hopkins Univer- sity Press.

Goetsch C., J. Wigg, A. Royo, et al. 2011. Chronic over browsing and biodiversity collapse in a forest understory in Pennsylvania: results from a 60-year-old deer exclusion plot. *Journal of the Torrey Botanical Society* 138:220–224.

Graham A. 1972a. *Floristics and Paleofloristics of Asia and Eastern North America.* New York: Elsevier.

Graham A. 1972b. Outline of the origin and historical recognition of floristic af- finities between Asia and eastern North America. Pages 1–16 in A. Graham, ed. *Floristics and Paleofloristics of Asia and Eastern North America.* New York: Elsevier.

Graham A. 1999. *Late Cretaceous and Cenozoic History of North American Vegetation.* New York: Oxford University Press.

Graham R. 1990. Evolution of new ecosystems at the end of the Pleistocene. Pages 54–60 in L. D. Agenbroad, J. I. Mead, and L. W. Nelson, ed. *Megafauna and Man: Discovery of America's Heartland.* Hot Springs: Mammoth Site of Hot Springs, South Dakota, Inc., and Flagstaff: Northern Arizona University.

Grayson D. K. 2008. Holocene underkill. *Proceedings of the National Academy of Sci- ences, USA* 105:4077–4078.

Grayson D. K., and D. J. Meltzer. 2003. A requiem for North American overkill. *Journal of Archaeological Science* 30:585–593.

Greenberg C. H., and J. D. Lanham. 2001. Breeding bird assemblages of hurricane- created gaps and adjacent closed canopy forest in the southern Appalachians. *Forest Ecology and Management* 154:251–260.

Gregory R. D., P. Vorisek, A. Van Strien, et al. 2007. Population trends of wide- spread woodland birds in Europe. *Ibis* 149:78–97.

Grimm E. C., and G. L. Jacobson Jr. 2004. Late-Quaternary vegetation history of the eastern United States. Pages 381–402 in A. R. Gillespie, S. C. Porter, and B. F. Atwater, ed. *The Quaternary Period in the United States.* New York: Elsevier.

Grixti J. C., L. T. Wong, S. A. Cameron, and C. Favret. 2009. Decline of bumble bees (*Bombus*) in the North American Midwest. *Biological Conservation* 142:75–84.

Grove A. T., and O. Rackham. 2001. *The Nature of Mediterranean Europe: An Ecological History.* New Haven: Yale University Press.

Guthrie D. 1990. Late Pleistocene faunal revolution: a new perspective on the extinction debate. Pages 42–53 in L. D. Agenbroad, J. I. Mead, and L. W. Nelson, ed. *Megafauna and Man: Discovery of America's Heartland.* Hot Springs: Mammoth Site of Hot Springs, South Dakota, Inc., and Flagstaff: Northern Arizona University.

Gwinner E. 1986. Internal rhythms in bird migration. *Scientific American* 254:84–92.

Gwinner E. 1996. Circadian and circannual programmes in avian migration. *Journal of Experimental Biology* 199:39–48.

Haack R. A., F. Hérard, J. Sun, and J. J. Turgeon. 2010. Managing invasive populations of Asian longhorned beetle and citrus longhorned beetle: a worldwide perspective. *Annual Review of Entomology* 55:521–546.

Hagan, J.M. 1992. Conservation biology when there is no crisis—yet. *Conservation Biology* 6:475–476.

Hajek A. E. 1999. Pathology and epizootiology of *Entomophaga maimaiga* infections in forest Lepidoptera. *Microbiology and Molecular Biology Reviews* 63:814–835.

Hajek A. E. 2004. *Natural Enemies: An Introduction to Biological Control.* New York: Cambridge University Press.

Hall B., G. Motzkin, D. R. Foster, et al. 2002. Three hundred years of forest and land-use change in Massachusetts, USA. *Journal of Biogeography* 29:1319–1335.

Hamel P. B. 2000. Cerulean Warbler (*Dendroica cerulea*). A. Poole, ed. *The Birds of North America Online,* Cornell Laboratory of Ornithology http://bna.birds.cornell.edu/bna/species/511/articles/introduction.

Haney J. C., and C. P. Schaadt. 1996. Functional roles of eastern old growth in promoting forest bird diversity. Pages 76–88 in M. D. David, ed. *Eastern Old-Growth Forests: Prospects for Rediscovery and Recovery.* Washington, D.C.: Island Press.

Harris L. D. 1984. *The Fragmented Forest: Island Biogeography Theory and the Preservation of Biotic Diversity.* Chicago: University of Chicago Press.

Harrison S. P., G. Yu, H. Takahara, and I. C. Prentice. 2001. Palaeovegetation: diversity of temperate plants in East Asia. *Nature* 413:129.

Havill N., and M. E. Montgomery. 2008. The role of arboreta in studying the evolution of host resistance to the hemlock woolly adelgid. *Arnoldia* 65:2–9.

Havill N., M. Montgomery, and M. Keena. 2011. Hemlock woolly adelgid and its hemlock hosts: a global perspective. Pages 3–14 in B. P. Onken and R. C. Reardon, ed. *Implementation and Status of Biological Control of the Hemlock Woolly Adelgid.* U.S. Forest Service Publication FNTET-2011-04, Forest Health Technology Enterprise Team, USDA Forest Service, www.fs.fed.us/foresthealth/technology/.

Healy W. 1997. Influence of deer on the structure and composition of oak forests in central Massachusetts. Pages 249–266 in W. J. McShea, H. B. Underwood, and J. H. Rappole, ed. *The Science of Overabundance: Deer Ecology and Population Management.* Washington, D.C.: Smithsonian Institution Press.

Hebard F. V. 2001. Backcross breeding program produces blight-resistant American chestnuts. *Ecological Restoration* 19:252–254.

Hejl S. J., J. A. Holmes, and D. E. Kroodsma. 2002. Winter Wren (*Troglodytes hiemalis*). A. Poole, ed. *The Birds of North America Online,* Cornell Laboratory of Ornithology, http://bna.birds.cornell.edu/bna/species/623/articles/intro duction.

Heldbjerg H., and T. Fox. 2008. Long-term population declines in Danish trans-Saharan migrant birds. *Bird Study* 55:267–279.

Heltzel J. M., and P. J. Leberg. 2006. Effects of selective logging on breeding bird communities in bottomland hardwood forests in Louisiana. *Journal of Wildlife Management* 70:1416–1424.

Hemond H. F., W. A. Niering, and R. H. Goodwin. 1983. Two decades of vegetation change in the Connecticut Arboretum Natural Area. *Bulletin Torrey Botanical Club* 110:184–194.

Henry J. D., and J. M. A. Swan. 1974. Reconstructing forest history from live and dead plant material: an approach to the study of forest succession in southwest New Hampshire. *Ecology* 55:772–783.

Herbeck L. A., and D. R. Larsen. 1999. Plethodontid salamander response to silvicultural practices in Missouri Ozark forests. *Conservation Biology* 13:623–632.

Hewson C. M., A. Amar, J. A. Lindsell, et al. 2007. Recent changes in bird populations in British broadleaved woodland. *Ibis* 149:14–28.

Hewson C. M., and D. G. Noble. 2009. Population trends of breeding birds in British woodlands over a 32-year period: relationships with food, habitat use and migratory behaviour. *Ibis* 151:464–486.

Hickey J. J. 1937. Bird-Lore's first breeding-bird census. *Bird Lore* 39:373–374.

Hickling R., D. B. Roy, J. K. Hill, et al. 2006. The distributions of a wide range of taxonomic groups are expanding polewards. *Global Change Biology* 12:450–455.

Higuchi H. 1998. Host use and egg color of Japanese cuckoos. Pages 80–93 in S. I. Rothstein and S. K. Robinson, ed. *Parasitic Birds and Their Hosts: Studies in Coevolution.* New York: Oxford University Press.

Higuchi H., and E. Morishita. 1999. Population declines of tropical migratory birds in Japan. *Actinia* 12:51–59.

Higuchi H., Y. Tsukamoto, S. Hanawa, and M. Takeda. 1982. Relationship between forest areas and the number of bird species. *Strix* 1:70–78.

Hino T. 1990. Palaearctic deciduous forests and their bird communities: comparisons between East Asia and West-Central Europe. Pages 87–94 in A. Keast, ed. *Biogeography and Ecology of Forest Bird Communities.* The Hague, Netherlands: SPB Academic Publishing.

Hinsley S. A., J. E. Carpenter, R. K. Broughton, et al. 2007. Habitat selection by Marsh Tits *Poecile palustris* in the U.K. *Ibis* 149:224–233.

Hinsley S. A., R. Pakeman, P. E. Bellamy, and I. Newton. 1996. Influences of habitat fragmentation on bird species distributions and regional population sizes. *Proceedings of the Royal Society B: Biological Sciences* 263:307–313.

Holmes R. T. 2011. Avian population and community processes in forest ecosystems: long-term research in the Hubbard Brook Experimental Forest. *Forest Ecology and Management* 262:20–32.

Holt C. A., R. J. Fuller, and P. M. Dolman. 2010. Experimental evidence that deer browsing reduces habitat suitability for breeding Common Nightingales *Luscinia megarhynchos*. *Ibis* 152:335–346.

Holt C. A., R. J. Fuller, and P. M. Dolman. 2011. Breeding and post-breeding responses of woodland birds to modification of habitat structure by deer. *Biological Conservation* 144:2151–2162.

Hoodless A. N., and G. J. M. Hirons. 2007. Habitat selection and foraging behaviour of breeding Eurasian Woodcock *Scolopax rusticola*: a comparison between contrasting landscapes. *Ibis* 149:234–249.

Hopkins J. J., and K. J. Kirby. 2007. Ecological change in British broadleaved woodland since 1947. *Ibis* 149:29–40.

Horsley S. B., S. L. Stout, and D. S. DeCalesta. 2003. White-tailed deer impact on the vegetation dynamics of a northern hardwood forest. *Ecological Applications* 13:98–118.

Horton T. 2010. The continuing saga of the American chestnut. *American Forests* 115:32–37.

House of Japan. 2011. Oita City to reintroduce wolves. www.houseofjapan.com/local /oita-city-to-reintroduce-wolves.

Huddle N., M. Reich, and N. Stiskin. 1975. *Island of Dreams: Environmental Crisis in Japan*. New York: Autumn Press.

Hudson C. M. 1997. *Knights of Spain, Warriors of the Sun: Hernando de Soto and the South's Ancient Chiefdoms*. Athens: University of Georgia Press.

Huntley B., and I. C. Prentice. 1993. Holocene vegetation and climates of Europe. Pages 136–168 in H. E. Wright Jr., J. E. Kutzbach, T. Webb III, et al., ed. *Global Climates Since the Last Glacial Maximum*. Minneapolis: University of Minnesota Press.

Hurrell J. W., and K. E. Trenberth. 2010. Climate change. Pages 9–29 in A. P. Møller, W. Fiedler, and P. Berthold, ed. *Effects of Climate Change on Birds*. New York: Oxford University Press.

Hutchinson G. E. 1965. *The Ecological Theater and the Evolutionary Play*. New Haven: Yale University Press.

Ikeda T. 2002. Araiguma. Page 70 in M. Okimasa and W. Izumi, ed. *Handbook of Alien Species in Japan*. Tokyo: Chijin Shokan.

Imanishi S. 2002. The drastic decline of breeding population of Brown Shrike *Lanius cristatus superciliosus* at Nobeyama Plateau in central Japan. *Journal of the Yamashina Institute of Ornithology* 34:228–231.

Ito T. Y., M. Shimoda, and S. Takatsuki. 2009. Productivity and foraging efficiency of the short-grass (*Zoysia japonica*) community for sika deer. Pages 145–157 in D. R. McCullough, S. Takatsuki, and K. Kaji, ed. *Sika Deer: Biology and Management of Native and Introduced Populations*. New York: Springer.

Ito T. Y., and S. Takatsuki. 2009. Home range, habitat selection, and food habits of the sika deer using the short-grass community on Kinkazan Island, northern Japan. Pages 159–170 in D. R. McCullough, S. Takatsuki, and K. Kaji, ed. *Sika Deer: Biology and Management of Native and Introduced Populations*. New York: Springer.

Jacobs D. E. 2007. Toward development of silvical strategies for forest restoration of American chestnut (*Castanea dentata*) using blight-resistant hybrids. *Biological Conservation* 137:497–506.

Johnson C. 2009. Megafaunal declines and fall. *Science* 326:1072–1073.

Johnson K. R., and B. Ellis. 2002. A tropical rainforest in Colorado 1.4 million years after the Cretaceous-Tertiary boundary. *Science* 296:2379–2383.

Jones, M. K., and X. Liu. 2009. Origins of agriculture in East Asia. *Science* 324:730–731.

Juzwik J., T. C. Harrington, W. L. MacDonald, and D. N. Appel. 2008. The origin of *Ceratocystis fagacearum*, the oak wilt fungus. *Annual Review of Phytopathology* 46:13–26.

Kadoya T., S. Suda, and I. Washitani. 2009. Dragonfly crisis in Japan: a likely consequence of recent agricultural habitat degradation. *Biological Conservation* 142:1899–1905.

Kalland A., and P. J. Asquith. 1997. Japanese perceptions of nature: ideas and illusions. Pages 1–35 in P. J. Asquith and A. Kalland, ed. *Japanese Images of Nature: Cultural Perspectives.* Richmond, Surrey: Curzon Press.

Karasawa K., S. Yamane, S. Koshikawa, and S. Takinoiri. 1991. *Toshin nio keru karasuno shudan negura no kotaisuuchosa* [Count of roosting crows in inner Tokyo]. *Urban Birds* 8:17–25.

Karell P., K. Ahola, T. Karstinen, et al. 2011. Climate change drives microevolution in a wild bird. *Nature Communications* 2:doi: 10.1038/ncomms1213.

Kawaji N. 1994. Lower predation rates on artificial ground nests than arboreal nests in western Hokkaido. *Japanese Journal of Ornithology* 43:1–9.

Kays R. W., A. Curtis, and J. J. Kirchman. 2010. Rapid adaptive evolution of northeastern coyotes via hybridization with wolves. *Biology Letters* 6:89–93.

Kays R. W., M. E. Gompper, and J. C. Ray. 2008. Ecology of eastern coyotes based on large-scale estimates of abundance. *Ecological Applications* 18:1014–1027.

Kellert S. R. 1991. Japanese perceptions of wildlife. *Conservation Biology* 5:297–308.

Kerr A. 2001. *Dogs and Demons: Tales from the Dark Side of Japan.* New York: Hill and Wang.

Kershner B., and R. T. Leverett. 2004. *The Sierra Club Guide to the Ancient Forests of the Northeast.* San Francisco: Sierra Club Books.

King D. I., R. M. DeGraaf, and C. R. Griffin. 2001. Productivity of early successional shrubland birds in clearcuts and groupcuts in an eastern deciduous forest. *Journal of Wildlife Management* 65:345–350.

Kirby K. J. 2001. The impact of deer on the ground flora of British broadleaved woodland. *Forestry* 74:219–229.

Kitchener A. C., and N. Yamaguchi. 2010. Biogeography, morphology and taxonomy. Pages 53–86 in R. L. Tilson and P. J. Nyhus, ed. *Tigers of the World: The Science, Politics, and Conservation of Panthera tigris.* Boston: Elsevier/Academic Press.

Kizlinski M. L., D. A. Orwig, R. C. Cobb, and D. R. Foster. 2002. Direct and indirect ecosystem consequences of an invasive pest on forests dominated by eastern hemlock. *Journal of Biogeography* 29:1489–1503.

Kleijn D., R. A. Baquero, Y. Clough, et al. 2006. Mixed biodiversity benefits of agri-environment schemes in five European countries. *Ecology Letters* 9:243–254.

Knight J. 2003. *Waiting for Wolves in Japan: An Anthropological Study of People-Wildlife Relations.* New York: Oxford University Press.

Kobori H., and R. B. Primack. 2003. Participatory conservation approaches for *satoyama*, the traditional forest and agricultural landscape of Japan. *Ambio* 32:307–311.

Kovach A. I., M. K. Litvaitis, and J. A. Litvaitis. 2003. Evaluation of fecal mtDNA analysis as a method to determine the geographic distribution of a rare lagomorph. *Wildlife Society Bulletin* 31:1061–1065.

Kress S. W. 1983. The use of decoys, sound recordings and gull control for re-establishing a tern colony in Maine. *Colonial Waterbirds* 6:185–196.

Kricher J. C. 2011. *Tropical Biology.* Princeton: Princeton University Press.

Kricher J. C., and G. Morrison. 1988. *A Field Guide to Eastern Forests: North America.* Boston: Houghton Mifflin.

Krusic R. A., M. Yamasaki, C. D. Neefus, and P. J. Pekins. 1996. Bat habitat use in White Mountain National Forest. *Journal of Wildlife Management* 60:625–631.

Kuijper D. P. J., J. P. G. M. Cromsigt, B. Jędrzejewska, et al. 2010b. Bottom-up versus top-down control of tree regeneration in the Białowieża Primeval Forest, Poland. *Journal of Ecology* 98:888–899.

Kuijper D. P. J., B. Jędrzejewska, B. Brzeziecki, et al. 2010a. Fluctuating ungulate density shapes tree recruitment in natural stands of the Białowieża Primeval Forest, Poland. *Journal of Vegetation Science* 21:1082–1098.

Kulikoff A. 1986. *Tobacco and Slaves: The Development of Southern Cultures in the Chesapeake, 1680–1800.* Williamsburg, Va.: Institute of Early American History and Culture, and Chapel Hill: University of North Carolina Press.

Kurosawa R. 2009. Disturbance-induced bird diversity in early successional habitats in the humid temperate region of northern Japan. *Ecological Research* 24:687–696.

Kurosawa R., and R. A. Askins. 1999. Differences in bird communities on the forest edge and in the forest interior: are there forest-interior specialists in Japan? *Journal of the Yamashina Institute for Ornithology* 31:63–79.

Kurosawa R., and R. A. Askins. 2003. Effects of habitat fragmentation on birds in deciduous forests in Japan. *Conservation Biology* 17:695–707.

Kyle C. J., A. R. Johnson, B. R. Patterson, et al. 2006. Genetic nature of eastern wolves: Past, present and future. *Conservation Genetics* 7:273–287.

Labandeira, C. C., K. R. Johnson, and P. Wilf. 2002. Impact of the terminal Cretaceous event on plant-insect associations. *Proceedings of the National Academy of Sciences, USA* 99:2061–2066.

Latham E. L., and R. E. Ricklefs. 1993. Continental comparisons of temperate zone tree species diversity. Pages 294–314 in R. Ricklefs E. and D. Schluter, ed. *Species Diversity in Ecological Communities.* Chicago: University of Chicago Press.

Laurent E. L. 2001. Mushi. *Natural History* 110:70–75.

Lawler J. J., A. S. Ruesch, J. D. Olden, and B. H. McRae. 2013. Projected climate-driven faunal movement routes. *Ecology Letters* 2013 doi:10.1111/ele.12132.

Leck C. F., B. G. Murray Jr., and J. Swineboard. 1988. Long-term changes in the breeding bird populations of a New Jersey forest. *Biological Conservation* 46:145–157.

Lehmkuhl, J. F., L. F. Ruggiero, and P. A. Hall. 1991. Landscape-scale configurations of forest fragmentation and wildlife richness and abundance in the southern Washington Cascade Range. Pages 425–442 in L. F. Ruggiero, K. B. Aubry, A. B. Carey, and M. H. Huff, technical coordinators, *Wildlife and Vegetation of Unmanaged Douglas-Fir Forests*. U.S. Forest Service General Technical Report PNW-285. Portland, Ore.: Pacific Northwest Research Station.

Leopold A. 1949. *A Sand County Almanac, and Sketches Here and There*. New York: Oxford University Press.

Leopold A. (L. B. Leopold, ed.) 1953. *Round River: From the Journals of Aldo Leopold*. New York: Oxford University Press.

Leverett R. 1996. Definitions and history. Pages 3–17 in M. D. Davis, ed. *Eastern Old-Growth Forests: Prospects for Rediscovery and Recovery*. Washington, D.C.: Island Press.

Lewis J. G. 2005. The *Forest Service and the Greatest Good: A Centennial History*. Durham, N.C.: Forest History Society.

Li H. 1952. Floristic relationships between eastern Asia and eastern North America. *Transactions of the American Philosophical Society, New Series* 42:371–429.

Liebhold A. M., E. G. Brockerhoff, L. J. Garrett, et al. 2012. Live plant imports: the major pathway for forest insect and pathogen invasions of the U.S. *Frontiers in Ecology and the Environment* 10:135–143.

Liebhold A. M., and D. G. McCullough. 2011. Forest insects. *In* D. Simberloff and M. Rejmánek, ed. *Encyclopedia of Biological Invasions*. Berkeley: University of California Press.

Liebhold A. M., W. L. MacDonald, D. Bergdahl, and V. C. Mastro. 1995. Invasion by exotic forest pests: a threat to forest ecosystems. *Forest Science Monographs* 30:1–58.

Litvaitis J. A. 1993. Response of early successional vertebrates to historic changes in land use. *Conservation Biology* 7:866–873.

Litvaitis J. A. 2003. Are pre-Columbian conditions relevant baselines for managed forests in the northeastern United States? *Forest Ecology and Management* 185:113–126.

Litvaitis J. A., J. P. Tash, M. K. Litvaitis, et al. 2006. A range-wide survey to determine the current distribution of New England cottontails. *Wildlife Society Bulletin* 34:1190–1197.

Litvaitis M. K., and J. A. Litvaitis. 1996. Using mitochondrial DNA to inventory the distribution of remnant populations of New England cottontails. *Wildlife Society Bulletin* 24:725–730.

Litwin T. S., and C. R. Smith. 1992. Factors influencing the decline of Neotropical migrants in a northeastern forest fragment: isolation, fragmentation or mosaic effects? Pages 483–496 in J. M. Hagan III and D. W. Johnston, ed. *Ecology and Conservation of Neotropical Migrant Landbirds*. Washington, D.C.: Smithsonian Institution Press.

Liu X., H. V. Hunt, and M. K. Jones. 2009. River valleys and foothills: changing archaeological perceptions of north China's earliest farms. *Antiquity* 83:82–95.

Loeb S. C., and J. M. O'Keefe. 2006. Habitat use by forest bats in South Carolina in relation to local, stand, and landscape characteristics. *Journal of Wildlife Management* 70:1210–1218.

Lorimer C. G., and A. S. White. 2003. Scale and frequency of natural disturbances in the northeastern U.S.: implications for early successional forest habitats and regional age distributions. *Forest Ecology and Management* 185:41–64.

Lu H., J. Zhang, K. Liu, et al. 2009. Earliest domestication of common millet (*Panicum miliaceum*) in East Asia extended to 10,000 years ago. *Proceedings of the National Academy of Sciences, USA* 106:7367–7372.

Lynch J. F. 1989. Distribution of overwintering nearctic migrants in the Yucatan Peninsula. 1. General patterns of occurrence. *Condor* 91:515–544.

Lynch J. F., and R. F. Whitcomb. 1978. Effects of insularization of the eastern deciduous forest on avifaunal diversity and turnover. Pages 461–489 *in* A. Marmelstein, ed. *Classification, Inventory and Analysis of Fish and Wildlife Habitat: Proceedings of a Symposium*. Washington, D.C.: U.S. Fish and Wildlife Service OBS-78/76.

Maekawa F. 1974. Origin and characteristics of Japan's flora. Pages 33–85 *in* M. Numata, editor. *The Flora and Vegetation of Japan*. New York: Kodansha; Elsevier Scientific.

Mann C. 2005. *1491: new revelations of the Americas before Columbus*. New York: Vintage Books.

Marquis D. A. 1981. Effect of deer browsing on timber production in Allegheny hardwood forests of northwestern Pennsylvania, USDA Forest Service Research Paper NE-475:1–10. Newtown Square, Pa.: Northeastern Research Station.

Martin P. S. 1967. Prehistoric overkill: the global model. Page 453 *in* P. S. Martin and H. E. Wright, ed. *Pleistocene Extinctions; The Search for a Cause*. New Haven: Yale University Press.

Martin P. S. 2005. *Twilight of the Mammoths: Ice Age Extinctions and the Rewilding of America*. Berkeley: University of California Press.

Martin P. S., and D. W. Steadman. 1999. Prehistoric extinctions on islands and continents. Page 394 *in* R. D. E. MacPhee, ed. *Extinctions in Near Time: Causes, Contexts, and Consequences*. New York: Kluwer Academic/Plenum Publishers.

Martin T. E., and J. Clobert. 1996. Nest predation and avian life-history evolution in Europe versus North America: a possible role of humans? *American Naturalist* 147:1028–1046.

Mason R. J. 1999. Whither Japan's environmental movement? An assessment of problems and prospects at the national level. *Pacific Affairs* 72:187–207.

Mason W. L. 2007. Changes in the management of British forests between 1945 and 2000 and possible future trends. *Ibis* 149:41–52.

Matthysen E. 1999. Nuthatches (*Sitta europaea*: Aves) in forest fragments: demography of a patchy population. *Oecologia* 119:501–509.

Matthysen E., and F. Adriaensen. 1998. Forest size and isolation have no effect on reproductive success of Eurasian Nuthatches (*Sitta europaea*). *Auk* 115:955–963.

McCabe T. R., and R. E. McCabe. 1997. Recounting whitetails past. Pages 11–26 *in* W. J. McShea, H. B. Underwood, and J. H. Rappole, ed. *The Science of Overabundance: Deer Ecology and Population Management.* Washington, D.C.: Smithsonian Institution Press.

McCartan L., B.H. Tiffany, J. A. Wolfe, et al. 1990. Late Tertiary floral assemblage from upland gravel deposits of the southern Maryland coastal plain. *Geology* 18:311–314.

McClure M. S. 1995. *Diapterobates humeralis* (Oribatida: Ceratozetidae): an effective control agent of hemlock woolly adelgid (Homoptera: Adelgidae) in Japan. *Environmental Entomology* 24:1207–1215.

McClure M. S., and C. A. S.-J. Cheah. 1999. Reshaping the ecology of invading populations of hemlock woolly adelgid, *Adelges tsugae* (Homoptera: Adelgidae), in eastern North America. *Biological Invasions* 1:247–254.

McDonald R., D. L. Mausel, S. M. Salom, and L. T. Kok. 2011. A case study of a release of the predator *Laricobius nigrinus* Fender against hemlock woolly adelgid, *Adelges tsugae*, Annand, at the urban community forest interface: Hemlock Hill, Lees-McRae College, Banner Elk, North Carolina. Pages 170–177 *in* B. P. Onken and R. C. Reardon, ed. *Implementation and Status of Biological Control of the Hemlock Woolly Adelgid.* U.S. Forest Service Publication FNTET-2011-04, Forest Health Technology Enterprise Team, USDA Forest Service, www.fs.fed.us/foresthealth /technology/.

McGarigal K., and W. C. McComb. 1995. Relationships between landscape structure and breeding birds in the Oregon Coast Range. *Ecological Monographs* 65:235–260.

McKenney D. W., J. H. Pedlar, K. Lawrence, et al. 2007. Potential impacts of climate change on the distribution of North American trees. *Bioscience* 57:939–948.

McNamee T. 1996. The Return of Il Lupo. *Natural History* 105 (12):50–59.

McNaughton S. J. 1984. Grazing lawns: animals in herds, plant form, and coevolution. *American Naturalist* 124:863–886.

McShea W. J., W. Healy, P. Devers, et al. 2007. Forestry matters: decline of oaks will impact wildlife in hardwood forests. *Journal of Wildlife Management* 71:1717–1728.

McShea W. J., M. V. McDonald, E. S. Morton, et al. 1995. Long-term trends in habitat selection by Kentucky Warblers. *Auk* 112:375–381.

Mech L. D., and L. Boitani. 2003. Wolf social behavior. Pages 1–34 *in* L. D. Mech and L. Boitani, ed. *Wolves: Behavior, Ecology, and Conservation.* Chicago: University of Chicago Press.

Meier A. J., S. P. Bratton, and D. C. Duffy. 1996. Biodiversity in the herbaceous layer and salamanders in Appalachian primary forests. Pages 49–64 *in* M. D. Davis, ed. *Eastern Old-Growth Forests: Prospects for Rediscovery and Recovery.* Washington, D.C.: Island Press.

Mellars P., and S. C. Reinhardt. 1979. The early postglacial settlement of northern Europe: an ecological perspective. Pages 234–293 *in* P. Mellars, ed. *The Early Postglacial Settlement of Northern Europe: An Ecological Perspective.* Pittsburgh: University of Pittsburgh Press.

Menzel A., T. H. Sparks, N. Estrella, et al. 2006. European phenological response to climate change matches the warming pattern. *Global Change Biology* 12:1969–1976.

Menzel M. A., T. C. Carter, J. M. Menzel, et al. 2002. Effects of group selection silviculture in bottomland hardwoods on the spatial activity patterns of bats. *Forest Ecology and Management* 162:209–218.

Milecka K., A. M. Noryśkiewicz, and G. Kowalewski. 2009. History of the Białowieża primeval forest, NE Poland. *Studia Quaternaria* 26:25–39.

Mitchell F. J. G., and E. Cole. 1998. Reconstruction of long-term successional dynamics of temperate woodland in Białowieża Forest, Poland. *Journal of Ecology* 86:1042–1059.

Miyaki M., and K. Kaji. 2004. Summer forage biomass and the importance of litterfall for a high-density sika deer population. *Ecological Research* 19:405–409.

Miyaki M., and K. Kaji. 2009. Shift to litterfall as year-round forage for sika deer after a population crash. Pages 171–180 in D. R. McCullough, S. Takatsuki, and K. Kaji, ed. *Sika Deer: Biology and Management of Native and Introduced Populations.* New York: Springer.

Miyawaki S., and I. Washitani. 2004. Invasive alien plant species in riparian areas of Japan: the contribution of agricultural weeds, revegetation species and aquacultural species. *Global Environmental Research* 8:89–101.

Mönkkönen M., and D. A. Welsh. 1994. A biogeographical hypothesis on the effects of human caused landscape changes on the forest bird communities of Europe and North America. *Annales Zoologici Fennici* 31:61–70.

Montgomery M. E., and M. A. Keena. 2011. *Scymnus (Neopullus)* lady beetles from China. Pages 53–76 in B. P. Onken and R. C. Reardon, ed. *Implementation and Status of Biological Control of the Hemlock Woolly Adelgid.* U.S. Forest Service Publication FNTET-2011-04, Forest Health Technology Enterprise Team, USDA Forest Service, www.fs.fed.us/foresthealth/technology/.

Mörtberg U. 2001. Resident bird species in urban forest remnants: landscape and habitat perspectives. *Landscape Ecology* 16:193–203.

Namba T., Y. Yabuhara, K. Yukinari, and R. Kurosawa. 2010. Changes in the avifauna of the Hokkaido University campus, Sapporo, detected by a long-term census. *Ornithological Science* 9:37–48.

Nash R. 2001. *Wilderness and the American Mind.* New Haven: Yale University Press.

Neilson R. P., L. F. Pitelka, A. M. Solomon, et al. 2005. Forecasting regional to global plant migration in response to climate change. *BioScience* 55:749–759.

Newman J. A., M. Anand, H. A. L. Henry, et al. 2011. *Climate Change Biology.* Cambridge, Mass.: CAB International.

Nichols D. J., and K. R. Johnson. 2008. *Plants and the K-T Boundary.* Cambridge, U.K.: Cambridge University Press.

Niemelä P., and W. J. Mattson. 1996. Invasion of North American forests by European phytophagous insects. *Bioscience* 46:741–753.

Niering W. A., and R. H. Goodwin. 1962. Ecological studies in the Connecticut Arboretum Natural Area. I. Introduction and a survey of vegetation types. *Ecology* 43:41–54.

Niklasson M., E. Zin, T. Zielonka, et al. 2010. A 350-year tree-ring fire record from Białowieża Primeval Forest, Poland: implications for Central European lowland fire history. *Journal of Ecology* 98:1319–1329.

Norton C. J., Y. Kondo, A. Ono, et al. 2010. The nature of megafaunal extinctions during the MIS 3–2 transition in Japan. *Quaternary International* 211:113–122.

Oliver C. D. 1981. Forest development in North America following major disturbances. *Forest Ecology and Management* 3:153–168.

Onken B. P., and R. C. Reardon. 2011. An overview and outlook for biological control of hemlock woolly adelgid. Pages 222–228 in B. P. Onken and R. C. Reardon, ed. *Implementation and Status of Biological Control of the Hemlock Woolly Adelgid.* U.S. Forest Service Publication FNTET-2011-04, Forest Health Technology Enterprise Team, USDA Forest Service, www.fs.fed.us/foresthealth/technology/.

Opdam P., G. Rijsdijk, and F. Hustings. 1985. Bird communities in small woods in an agricultural landscape: effects of area and isolation. *Biological Conservation* 34:333–352.

Opler P. A. 1978. Insects of American chestnut: possible importance and conservation concern. Pages 83–85 in J. McDonald, ed. *Proceedings of the American Chestnut Symposium*, Morgantown: West Virginia University Press.

Orwig D. A., D. R. Foster, and D. L. Mausel. 2002. Landscape patterns of hemlock decline in New England due to the introduced hemlock woolly adelgid. *Journal of Biogeography* 29:1475–1487.

Owen-Smith N. 1987. Pleistocene extinctions: the pivotal role of megaherbivores. *Paleobiology* 13:351–362.

Paillet F. L. 2002. Chestnut: history and ecology of a transformed species. *Journal of Biogeography* 29:1517–1530.

Paquet J.-Y., X. Vandevyvre, L. Delahaye, and J. Rondeux. 2006. Bird assemblages in a mixed woodland-farmland landscape: the conservation value of silviculture-dependant open areas in plantation forest. *Forest Ecology and Management* 227:59–70.

Parker A. G., A. S. Goudie, E. E. Anderson, et al. 2002. A review of the mid-Holocene elm decline in the British Isles. *Progress in Physical Geography* 26:1–45.

Parker, G. 1995. *Eastern Coyote: The Story of Its Success.* Halifax, Nova Scotia: Nimbus Publishing.

Parmesan C., and G. Yohe. 2003. A globally coherent fingerprint of climate change impacts across natural systems. *Nature* 421:37–42.

Parrish J. M., J. T. Parrish, J. H. Hutchison, and R. A. Spicer. 1987. Late Cretaceous vertebrate fossils from the North Slope of Alaska and implications for dinosaur ecology. *Palaios* 2:377–389.

Parshall T., and D. R. Foster. 2002. Fire on the New England landscape: regional and temporal variation, cultural and environmental controls. *Journal of Biogeography* 29:1305–1317.

Paton P. W. 1994. The effect of edge on avian nest success: how strong is the evidence? *Conservation Biology* 8:17–26.

Perrins C. M., and R. Overall. 2001. Effect of increasing numbers of deer on bird populations in Wytham Woods, central England. *Forestry* 74:299–309.

Peterken G. F. 1996. *Natural Woodland: Ecology and Conservation in Northern Temperate Regions*. New York: Cambridge University Press.

Peterken G. F., and M. Game. 1984. Historical factors affecting the number and distribution of vascular plant species in the woodlands of central Lincolnshire. *Journal of Ecology* 72:155–182.

Peterson C. J., and S. A. Pickett. 1995. Forest reorganization: a case study in an old-growth forest catastrophic blowdown. *Ecology* 76:763–774.

Petranka J. W., M. E. Eldridge, and K. E. Haley. 1993. Effects of timber harvesting on southern Appalachian salamanders. *Conservation Biology* 7:363–370.

Phillips M. K., V. G. Henry, and B. T. Kelly. 2003. Restoration of the red wolf. Pages 272–288 in L. D. Mech and L. Boitani, ed. *Wolves: Behavior, Ecology, and Conservation*. Chicago: University of Chicago Press.

Pierce D. S. 2000. *The Great Smokies: From Natural Habitat to National Park*. Knoxville: University of Tennessee Press.

Pigott C. D. 1975. Natural regeneration of Tilia cordata in relation to forest-structure in the forest of Białowieża, Poland. *Philosophical Transactions of the Royal Society of London: Series B, Biological* 270:151–179.

Pimm S., and R. A. Askins. 1995. Forest losses predict bird extinctions in eastern North America. *Proceedings of the National Academy of Sciences, USA* 92:9343–9347.

Pinto, B. 2010. Brief historical ecology of northern Portugal during the Holocene. *Environment and History* 16:3–42.

Porneluzi P., J. C. Bednarz, L. J. Goodrich, et al. 1993. Reproductive performance of territorial Ovenbirds occupying forest fragments and a contiguous forest in Pennsylvania. *Conservation Biology* 7:618–622.

Prather J. W., and K. G. Smith. 2003. Effects of tornado damage on forest bird populations in the Arkansas Ozarks. *Southwestern Naturalist* 48:292–297.

Primack R. B., H. Higuchi, and A. J. Miller-Rushing. 2009. The impact of climate change on cherry trees and other species in Japan. *Biological Conservation* 142:1943–1949.

Primack R. B., H. Kobori, and S. Mori. 2000. Dragonfly pond restoration promotes conservation awareness in Japan. *Conservation Biology* 14:1553–1554.

Prothero D. R. 2009. *Greenhouse of the Dinosaurs: Evolution, Extinction, and the Future of Our Planet*. New York: Columbia University Press.

Pulido F., P. Berthold, G. Mohr, and U. Querner. 2001. Heritability of the timing of autumn migration in a natural bird population. *Proceedings of the Royal Society B: Biological Sciences* 268:953–959.

Pyne S. J. 1982. *Fire in America: A Cultural History of Wildland and Rural Fire*. Princeton: Princeton University Press.

Qian H., and R. E. Ricklefs. 1999. A comparison of the taxonomic richness of vascular plants in China and the United States. *American Naturalist* 154:160–181.

Qian H., and R. E. Ricklefs. 2000. Large-scale processes and the Asian bias in species diversity of temperate plants. *Nature* 407:180–182.

Rackham O. 1980. *Ancient Woodland: Its History, Vegetation and Uses in England*. London: E. Arnold.

Rackham O. 1986. *The History of the Countryside*. London: J.M. Dent.

Rackham O. 2006. *Woodlands.* London: HarperCollins Publishers.

Rappole J. H. 1995. *The Ecology of Migrant Birds. A Neotropical Perspective.* Washington, D.C.: Smithsonian Institution Press.

Ricciardi A., and D. Simberloff. 2009. Assisted colonization is not a viable conservation strategy. *Trends in Ecology and Evolution* 24:248–253.

Ripple W. J., and R. L. Beschta. 2003. Wolf reintroduction, predation risk, and cottonwood recovery in Yellowstone National Park. *Forest Ecology and Management* 184:299–313.

Ripple W. J., and R. L. Beschta. 2007. Restoring Yellowstone's aspen with wolves. *Biological Conservation* 138:514–519.

Robbins C. S. 1979. Effect of forest fragmentation on bird populations. Pages 198–212 in R. M. DeGraaf and K. E. Evans, ed. *Management of North-Central and Northeastern Forests for Nongame Birds: Workshop Proceedings.* U. S. Forest Service General Technical Report NC-51, St. Paul, Minn.: North Central Forest Experiment Station.

Robbins C. S., D. K. Dawson, and B. A. Dowell. 1989. Habitat area requirements of breeding forest birds of the middle Atlantic states. *Wildlife Monographs* 103:1–34.

Robbins C. S., B. A. Dowell, D. K. Dawson, et al. 1992. Comparison of Neotropical migrant landbird populations wintering in tropical forest, isolated forest fragments, and agricultural habitats. Pages 207–220 in J. M. Hagan III and D. W. Johnston, ed. *Ecology and Conservation of Neotropical Migrant Landbirds,* Washington, D.C.: Smithsonian Institution Press.

Robinson D. W., and S. K. Robinson. 1999. Effects of selective logging on forest bird populations in a fragmented landscape. *Conservation Biology* 13:58–66.

Robinson G. S., L. P. Burney, and D. A. Burney. 2005. Landscape paleoecology and megafaunal extinction in southeastern New York state. *Ecological Monographs* 75:295–315.

Robinson S. K., S. I. Rothstein, M. C. Brittingham, et al. 1995a. Ecology and behavior of cowbirds and their impact on host populations. Pages 428–460 in T. E. Martin and D. M. Finch, ed. *Ecology and Management of Neotropical Migratory Birds.* New York: Oxford University Press.

Robinson S. K., F. R. Thompson III, T. M. Donovan, et al. 1995b. Regional forest fragmentation and the nesting success of migratory birds. *Science* 267:1987–1990.

Röhrig E. 1991. Temperate deciduous forests in Mexico and Central America. Pages 371–375 in E. Röhrig and B. Ulrich, ed. *Ecosystems of the World 7: Temperate Deciduous Forests.* New York: Elsevier.

Rooney T. P., and W. J. Dress. 1997. Species loss over sixty-six years in the ground layer vegetation of Heart's Content, an old-growth forest in Pennsylvania USA. *Natural Areas Journal* 17:297–305.

Rooney T. P., and D. M. Waller. 2003. Direct and indirect effects of white-tailed deer in forest ecosystems. *Forest Ecology and Management* 181:165–176.

Root T. L., J. T. Price, K. R. Hall, et al. 2003. Fingerprints of global warming on wild animals and plants. *Nature* 421:57–60.

Rose F. 1992. Temperate forest management: its effects on bryophyte and lichen floras and habitats. Pages 211–233 in J. Bates and A. Farmer, ed. *Bryophytes and Lichens in a Changing Environment.* New York: Clarendon Press, Oxford University Press.

Rosenberg K. V., and M. G. Raphael. 1986. Effects of forest fragmentation on vertebrates in Douglas-fir forests. Pages 263–272 in J. Verner, M. L. Morrison, and C. J. Ralph, ed. *Wildlife 2000: Modeling Habitat Relationships of Terrestrial Vertebrates.* Madison: University of Wisconsin Press.

Roth R. R., and R. K. Johnson. 1993. Long-term dynamics of a Wood Thrush population breeding in a forest fragment. *Auk* 110:37–48.

Royer D. L., C. P. Osborne, and D. J. Beerling. 2005. Contrasting seasonal patterns of carbon gain in evergreen and deciduous trees of ancient polar forests. *Paleobiology* 31:141.

Royo A. A., R. Collins, M. B. Adams, et al. 2010. Pervasive interactions between ungulate browsers and disturbance regimes promote temperate forest herbaceous diversity. *Ecology* 91:93–105.

Ruddiman W. F. 2008. *Earth's Climate: Past and Future.* New York: W.H. Freeman.

Ruddiman W. F., and J. E. Kutzbach. 1991. Plateau uplift and climatic change. *Scientific American* 266 (March):66–75.

Runkle J. R. 1982. Patterns of disturbance in some old-growth mesic forests in eastern North America. *Ecology* 63:1533–1546.

Runkle J. R. 1996. Central mesophytic forests. Pages 161–177 in M. D. Davis, ed. *Eastern Old-Growth Forests: Prospects for Rediscovery and Recovery.* Washington, D.C.: Island Press.

Runkle J. R. 2000. Canopy tree turnover in old-growth mesic forests of eastern North America. *Ecology* 81:554–567.

Runte A. 1987. *National Parks: The American Experience,* Lincoln: University of Nebraska Press.

Russell D. A. 2009. *Islands in the Cosmos: The Evolution of Life on Land.* Bloomington: Indiana University Press.

Russell E. W. B. 1983. Indian-set fires in the forests of the northeastern United States. *Ecology* 64:78–88.

Russell F. L., D. B. Zippin, and N. L. Fowler. 2001. Effects of white-tailed deer (*Odocoileus virginianus*) on plants, plant populations and communities: a review. *American Midland Naturalist* 146:1–26.

Rutberg A. T. 1997. The science of deer management: an animal welfare perspective. Pages 37–54 in W. J. McShea, H. B. Underwood, and J. H. Rappole, ed. *The Science of Overabundance: Deer Ecology and Population Management.* Washington, D.C.: Smithsonian Institution Press.

Saito T., T. Yamanoi, and K. Kaiho. 1986. End-Cretaceous devastation of terrestrial flora in the boreal Far East. *Nature* 323:253–255.

Sanderson F. J., P. F. Donald, D. J. Pain, et al. 2006. Long-term population declines in Afro-Palearctic migrant birds. *Biological Conservation* 131:93–105.

Sands, J. P., S. J. DeMaso, L. A. Brennan, and M. J. Schnupp, ed. 2012. *Wildlife Science: Connecting Research with Management.* Boca Raton, Fla.: Taylor & Francis.

Sarjeant W. A. S., and P. J. Currie. 2001. The "great extinction" that never happened: the demise of the dinosaurs considered. *Canadian Journal of Earth Sciences* 38:239–247.

Sauer J. R., J. E. Hines, and J. Fallon. 2008. *The North American Breeding Bird Survey, Results and Analysis, 1966–2007*. Version 5.15.2008. Retrieved from www.mbr -pwrc.usgs.gov/bbs/bbs.html. USGS Patuxent Wildlife Research Center, Laurel, Md.

Schmaltz J. 1991. Deciduous forests of southern South America. Pages 557–578 in E. Röhrig and B. Ulrich, ed. *Ecosystems of the World 7: Temperate Deciduous Forests*. New York: Elsevier.

Schreurs M. A. 2002. *Environmental Politics in Japan, Germany, and the United States*. New York: Cambridge University Press.

Schulte P., L. Alegret, I. Arenillas, et al. 2010. The Chicxulub asteroid impact and mass extinction at the Cretaceous-Paleogene boundary. *Science* 327:1214–1218.

Seaton K. 1996. The Nature Conservancy's preservation of old growth. Pages 274–283 in M. D. Davis, ed. *Eastern Old-Growth Forests: Prospects for Rediscovery and Recovery*. Washington, D.C.: Island Press.

Selva S. B. 1996. Using lichens to assess ecological continuity in northeastern forests. Pages 35–48 in M. D. Davis, ed. *Eastern Old-Growth Forests: Prospects for Rediscovery and Recovery*. Washington, D.C.: Island Press.

Shaffer L. 1992. *Native Americans Before 1492: The Moundbuilding Centers of the Eastern Woodlands*. Armonk, N.Y.: M.E. Sharpe.

Shapiro J. 2001. *Mao's War Against Nature: Politics and the Environment in Revolutionary China*. New York: Cambridge University Press.

Sheldon B. C. 2010. Genetic perspectives on the evolutionary consequences of climate change in birds. Pages 149–168 in A. P. Møller, W. Fiedler, and P. Berthold, ed. *Effects of Climate Change on Birds*. New York: Oxford University Press.

Shidei T. 1974. Forest vegetation zones. Pages 87–124 in M. Numata, ed. *The Flora and Vegetation of Japan*. New York: Kodansha, Elsevier Scientific.

Shimatani I. K., and Y. Kubota. 2011. The spatio-temporal forest patch dynamics inferred from the fine-scale synchronicity in growth chronology. *Journal of Vegetation Science* 22:334–345.

Short K. 2000. *Nature in Tokyo*. New York: Kodansha International.

Shu J., W. Wang, L. Jiang, and H. Takahara. 2010. Early Neolithic vegetation history, fire regime and human activity at Kuahuqiao, Lower Yangtze River, East China: new and improved insight. *Quaternary International* 227:10–21.

Signor P. W., III, and J. H. Lipps. 1982. Sampling bias, gradual extinction patterns and catastrophes in the fossil record. Pages 291–296 in L. T. Silver and P. H. Schultz, ed. *Geological Implications of Impacts of Large Asteroids and Comets on the Earth*. Boulder, Colo.: Geological Society of America.

Silander J. A., Jr., and D. M. Klepeis. 1999. The invasion ecology of Japanese barberry (*Berberis thunbergii*) in the New England landscape. *Biological Invasions* 1:189–201.

Silverberg R. 1968. *Mound Builders of Ancient America: The Archaeology of a Myth*. Athens: Ohio University Press.

Simberloff D., and P. Stiling. 1996. How risky is biological control? *Ecology* 77:1965–1974.

Skelton P. W., R. A. Spicer, S. P. Kelley, and I. Gilmour. 2003. *The Cretaceous World.* New York: Open University; Cambridge University Press.

Stahle D. W. 1996. Tree rings and ancient forest history. Pages 321–343 in M. D. Davis, ed. *Eastern Old-Growth Forests: Prospects for Rediscovery and Recovery.* Washington, D.C.: Island Press.

Stuart A. J. 1999. Late Pleistocene megafaunal extinctions: a European perspective. Pages 257–269 in R. D. E. MacPhee, ed. *Extinctions in Near Time: Causes, Contexts, and Consequences.* New York: Kluwer Academic/Plenum Publishers.

Sweet A. R., and D. R. Braman. 2001. Cretaceous-Tertiary palynofloral perturbations and extinctions within the *Aquilapollenites* phytogeographic province. *Canadian Journal of Earth Sciences* 38:249–269.

Sykes M. T., I. C. Prentice, and W. Cramer. 1996. A bioclimatic model for the potential distributions of north European tree species under present and future climates. *Journal of Biogeography* 23:203–233.

Takatsuki S. 2009. Effects of sika deer on vegetation in Japan: a review. *Biological Conservation* 142:1922–1929.

Takatsuki S., and T. Gorai. 1994. Effects of sika deer on the regeneration of a *Fagus crenata* forest on Kinkazan Island, northern Japan. *Ecological Research* 9:115–120.

Takatsuki S., and T. Y. Ito. 2009. Plants and plant communities on Kinkazan Island, northern Japan, in relation to sika deer herbivory. Pages 125–143 in D. R. McCullough, S. Takatsuki, and K. Kaji, ed. *Sika Deer: Biology and Management of Native and Introduced Populations.* New York: Springer.

Takeuchi K., R. D. Brown, I. Washitani, et al. 2003. *Satoyama: The Traditional Rural Landscape of Japan.* New York: Springer.

Tanai T. 1972. Tertiary history of vegetation in Japan. Pages 235–255 in A. Graham, ed. *Floristics and Paleofloristics of Asia and Eastern North America.* New York: Elsevier.

Temple S. A., and J. R. Cary. 1988. Modeling dynamics of habitat-interior bird populations in fragmented landscapes. *Conservation Biology* 2:340–347.

Terborgh J., L. Lopez, P. Nuñez, M. Rao, et al. 2001. Ecological meltdown in predator-free forest fragments. *Science* 294:1923–1926.

Thirgood J. V. 1981. *Man and the Mediterranean Forest: A History of Resource Depletion.* New York: Academic Press.

Thomas P., and J. R. Packham. 2007. *Ecology of Woodlands and Forests: Description, Dynamics and Diversity.* New York: Cambridge University Press.

Thompson F. R., III. 2007. Factors affecting nest predation on forest songbirds in North America. *Ibis* 149:98–109.

Thompson F. R., III, T. M. Donovan, R. M. DeGraaf, et al. 2002. A multi-scale perspective of the effects of forest fragmentation on birds in eastern forests. *Studies in Avian Biology* 24:8–19.

Thornton I. W. B. 1996. *Krakatau: The Destruction and Reassembly of an Island Ecosystem.* Cambridge: Harvard University Press.

Thorson R. M. 2002. *Stone by Stone: The Magnificent History in New England's Stone Walls*. New York: Walker & Company.

Tiffney B. H. 1985a. Perspectives on the origin of the floristic similarity between eastern Asia and eastern North America. *Journal of the Arnold Arboretum* 66:73–94.

Tiffney B. H. 1985b. The Eocene North Atlantic land bridge: its importance in Tertiary and modern phytogeography of the Northern Hemisphere. *Journal of the Arnold Arboretum* 66:243–273.

Tilghman N. G. 1989. Impacts of white-tailed deer on forest regeneration in northwestern Pennsylvania. *Journal of Wildlife Management* 53:524–532.

Tingley M. W., D. A. Orwig, R. Field, and G. Motzkin. 2002. Avian response to removal of a forest dominant: consequences of hemlock woolly adelgid infestations. *Journal of Biogeography* 29:1505–1516.

Tomiatojc L. 2000. Did White-backed Woodpeckers ever live in Britain? *British Birds* 93:452–456.

Totman C. D. 1989. *The Green Archipelago: Forestry in Preindustrial Japan*. Berkeley: University of California Press.

Totman C. D. 2000. *A History of Japan*. Malden, Mass.: Blackwell.

Tremblay J., I. Thibault, C. Dussault, et al. 2005. Long-term decline in white-tailed deer browse supply: can lichens and litterfall act as alternative food sources that preclude density-dependent feedbacks. *Canadian Journal of Zoology* 83:1087–1096.

Tsuji Y., and S. Takatsuki. 2004. Food habits and home range use by Japanese macaques on an island inhabited by deer. *Ecological Research* 19:381–388.

Tsutsui W. M. 2003. Landscapes in the dark valley: toward an environmental history of wartime Japan. *Environmental History* 8:294–311.

Tyrrell L. E., and T. R. Crow. 1994. Structural characteristics of old-growth hemlock-hardwood forests in relation to age. *Ecology* 75:370–386.

Ueta M. 1998. Crow-related low nesting success of small birds in Tokyo area. *Strix* 16:67–71.

van Dorp D., and P. F. M. Opdam. 1987. Effects of patch size, isolation and regional abundance on forest bird communities. *Landscape Ecology* 1:59–73.

Vera F. W. M. 2000. *Grazing Ecology and Forest History*. New York: CABI.

Vines G. 2002. Gladerunners. *New Scientist* 175:35.

Visser M. E., L. J. T. Holleman, and P. Gienapp. 2006. Shifts in caterpillar biomass phenology due to climate change and its impact on the breeding biology of an insectivorous bird. *Oecologia* 147:164–172.

Vitousek P. M., C. M. D'Antonio, L. L. Loope, and R. Westbrooks. 1996. Biological invasions as global environmental change. *American Scientist* 84:468–478.

Vitt P., K. Havens, and O. Hoegh-Guldberg. 2009. Assisted migration: part of an integrated conservation strategy. *Trends in Ecology and Evolution* 24:473–474.

vonHoldt, B. M., J. P. Pollinger, D. A. Earl, et al. 2011. A genome-wide perspective on the evolutionary history of enigmatic wolf-like canids. *Genome Research* 21:1294–1305.

Wada T. 1994. Effects of height of neighboring nests on nest predation in the Rufous Turtle-Dove (*Streptopelia orientalis*). *Condor* 96:812–816.

Waldram M. S., W. J. Bond, and W. D. Stock. 2008. Ecological engineering by a mega-grazer: white rhino impacts on a South African savanna. *Ecosystems* 11:101–112.

Walker B. L. 2005. *The Lost Wolves of Japan*. Seattle: University of Washington Press.

Wappler T., E. D. Currano, P. Wilf, et al. 2009. No post-Cretaceous ecosystem depression in European forests? Rich insect-feeding damage on diverse middle Palaeocene plants, Menat, France. *Proceedings of the Royal Society B: Biological Sciences* 276:4271–4277.

Ward A. I. 2005. Expanding ranges of wild and feral deer in Great Britain. *Mammal Review* 35:165–173.

Warnecke L., J. M. Turner, T. K. Bollinger, et al. 2012. Inoculation of bats with European *Geomyces destructans* supports the novel pathogen hypothesis for the origin of white-nose syndrome. *Proceedings of the National Academy of Sciences, USA* 109:6999–7003.

Watanabe T. 2008. The management of mountain natural parks by local communities in Japan. Pages 259–268 in P. P. Karan and U. Suganuma, ed. *Local Environmental Movements: A Comparative Study of the United States and Japan*. Lexington: University Press of Kentucky.

Way J. G., L. Rutledge, T. Wheeldon, and B. N. White. 2010. Genetic characterization of eastern "coyotes" in eastern Massachusetts. *Northeastern Naturalist* 17:189–204.

Webb T., III, P. J. Bartlein, S. P. Harrison, and K. H. Anderson. 1993. Vegetation, lake levels and climate in eastern North America for the past 18,000 years. Pages 415–467 in H. E. Wright Jr., J. E. Kutzbach, T. Webb III, et al., ed. *Global Climates Since the Last Glacial Maximum*. Minneapolis: University of Minnesota Press.

Webster, J. R., K. Morkeski, C. A. Wojculewski, et al. 2012. Effects of hemlock mortality on streams in the southern Appalachian Mountains. *American Midland Naturalist* 168:112–131.

Wen, J. 1999. Evolution of eastern Asian and eastern North American disjunct distributions in flowering plants. *Annual Review of Ecology and Systematics* 30:421–455.

Wesołowski T. 1983. The breeding ecology and behaviour of Wrens *Troglodytes troglodytes* under primaeval and secondary conditions. *Ibis* 125:499–515.

Wesołowski T. 2007. Primeval conditions—what can we learn from them? *Ibis* 149:64–77.

Wesołowski T., and L. Tomialojć. 1997. Breeding bird dynamics in a primaeval temperate forest: long-term trends in Białowieża National Park (Poland). *Ecography* 20:432–453.

Whitaker D. M., and I. G. Warkentin. 2010. Spatial ecology of migratory passerines on temperate and boreal forest breeding grounds. *Auk* 127:471–484.

Whitcomb R. F., C. S. Robbins, J. F. Lynch, et al. 1981. Effects of forest fragmentation on avifauna of the eastern deciduous forest. Pages 125–205 in R. L. Burgess and D. M. Sharpe, ed. *Forest Island Dynamics in Man-Dominated Landscapes*. New York: Springer-Verlag.

White R. H. 1998. Changes in avian communities along successional gradients caused by tornadoes in hardwood forests. M.S. thesis, Empire State College, Saratoga Springs, New York.

Wilcove D. S. 1988. Changes in the avifauna of the Great Smoky Mountains, 1947–1983. *Wilson Bulletin* 100:256–271.

Wilf P., C. C. Labandaira, K. R. Johnson, and B. Ellis. 2006. Decoupled plant and insect diversity after the end-Cretaceous extinction. *Science* 313:1112–1115.

Williams M. 1989. *Americans and Their Forests: A Historical Geography.* New York: Cambridge University Press.

Williams M. 2003. *Deforesting the Earth: From Prehistory to Global Crisis.* Chicago: University of Chicago Press.

Wilson J. D., A. D. Evans, and P. V. Grice. 2009. *Bird Conservation and Agriculture.* New York: Cambridge University Press.

Wilson P. J., S. Grewal, I. D. Lawford, et al. 2000. DNA profiles of the eastern Canadian wolf and the red wolf provide evidence for a common evolutionary history independent of the gray wolf. *Canadian Journal of Zoology* 78:2156–2166.

Wilson P. J., S. Grewal, F. F. Mallory, and B. N. White. 2009. Genetic characterization of hybrid wolves across Ontario. *Journal of Heredity* 100:S80–S89.

Wing S. L. 2004. Mass extinctions in plant evolution. Pages 61–97 in P. D. Taylor, ed. *Extinctions in the History of Life.* New York: Cambridge University Press.

Wing S. L., G. J. Harrington, F. A. Smith, et al. 2005. Transient floral change and rapid global warming at the Paleocene-Eocene boundary. *Science* 310:993–996.

Wolfe J. A. 1972. An interpretation of Alaskan Tertiary floras. Pages 201–233 in A. Graham, ed. *Floristics and Paleofloristics of Asia and Eastern North America.* New York: Elsevier.

Wolfe J. A. 1979. Temperature parameters of humid to mesic forests of eastern Asia and relation to forests of other regions of the northern hemisphere and Australasia. *U.S. Geological Survey Professional Paper* 1106:1–37.

Wolfe J. A. 1987. Late Cretaceous-Cenozoic history of deciduousness and the terminal Cretaceous event. *Paleobiology* 13:215–226.

Wolfe J. A., and G. R. Upchurch Jr. 1986. Vegetation, climatic and floral changes at the Cretaceous-Tertiary boundary. *Nature* 324:148–152.

Wolfe J. A., and G. R. Upchurch Jr. 1987. North American nonmarine climates and vegetation during the Late Cretaceous. *Palaeogeography, Palaeoclimatology, Palaeoecology* 61:33–77.

Woolhouse M. E. J. 1983. The theory and practice of the species-area effect, applied to the breeding birds of British woods. *Biological Conservation* 27:315–332.

Woolhouse M. E. J. 1987. On species richness and nature reserve design: an empirical study of U.K. woodland avifauna. *Biological Conservation* 40:167–178.

Wormworth J., and Ç. Sekercioğlu. 2011. *Winged Sentinels: Birds and Climate Change.* New York: Cambridge University Press.

Wunderle J. M., and R. B. Waide. 1993. Distribution of overwintering nearctic migrants in the Bahamas and Greater Antilles. *Condor* 95:904–933.

Xu, J. 2011. China's new forests aren't as green as they seem. *Nature* 477:371.

Yalden D. W., and U. Albarella. 2009. *The History of British Birds.* New York: Oxford University Press.

Yamashita A., J. Sano, and S. Yamamoto. 2002. Impact of a strong typhoon on the structure and dynamics of an old-growth beech (*Fagus crenata*) forest, Southwestern Japan. *Folia Geobotanica* 37:5–16.

Yamauchi K., S. Yamazaki, and Y. Fujimaki. 1997. Breeding habitats of *Dendrocopos major* and *D. minor* in urban and rural areas. [In Japanese with English summary]. *Japanese Journal of Ornithology* 46:121–131.

Yamaura Y., T. Amano, T. Koizumi, et al. 2009. Does land-use change affect biodiversity dynamics at a macroecological scale? A case study of birds over the past 20 years in Japan. *Animal Conservation* 12:110–119.

Young D., and M. Young. 2005. *The Art of the Japanese Garden,* New Clarendon, Vt.: Tuttle Publishing.

Zheng, Y. F., G. P. Sun, L. Qin, et al. 2009. Rice fields and modes of rice cultivation between 5000 and 2500 BC in east China. *Journal of Archaeological Science* 36:2609–2616.

Zuckerberg B., A. M. Woods, and W. F. Porter. 2009. Poleward shifts in breeding bird distributions in New York state. *Global Change Biology* 15:1866–1883.

Index

Page numbers in italics refer to illustrations

Acadia National Park, Maine, 210
"*Acer*" *arcticum* complex, 10
Acorns, 77, 151, 155, 201
Adelgid, hemlock woolly, 186–187, 188,
    192, 193–195, 196, 197, 198–199
Adena culture, Ohio, 46
Adirondack Forest Preserve, 68
Adirondack Park, New York, 160, 210, 211
After Armageddon (Vriesen), 17
Agriculture: abandoned farmland,
    63–66, 65, 80, 95, 100; in China,
    55–56; in Japan, 51; Mississippian
    culture, 46, 47–48, 52; and nature
    conservation, 220–221, 235–236;
    origin and spread of, 40–43; sustain-
    able methods of, 64–65
Alder, 25
Allegany State Park, New York, 97, 107
Allegheny National Forest, Pennsylvania,
    82–83, 148
Alligator River National Wildlife Refuge,
    157
American Chestnut Foundation, 199, 200
American Ornithologists' Union, 131
Ancient forests: defined, 93. *See also*
    Old-growth forests
Anemone, 1; wood, 27
Animal populations: in canopy gaps,
    80–82; and empty forests, 53–54;
    North American/Japanese compari-
    son, 6–7; in old-growth forests,
    91–92, 92; in regenerated forests,
    61–62, 64; in scrub-shrub habitat,
94–103; in wilderness areas, 215.
    See also Birds; Grazing animals;
    Predators
Ant, leafcutter, 145
Antelope, 138
Aphid, woolly alder, 194
Appalachian Mountains, 24, 25, 26
Appleman Lake, Indiana, 36
Apricot, Japanese, 227
Arbutus, trailing, 4
Arctic forest. *See* Cretaceous Arctic forest
Ash, 22, 25, 36, 74, 145; emerald ash
    borer, 187; European mountain, 76;
    white, 64
Aspen, 31, 83, 157
Asteroid impact, 12–18, 31
Audubon Naturalist Society, 106
Aurochs, 77, 88, 89

Backcrossing, 199, 200
Bamboo, dwarf, 60
Barberry: common, 204; Japanese,
    137–138, 203–204
Barley, 40, 43
Basswood, 48
Bats: in canopy gaps, 81–82; white-nose
    syndrome in, 203
Battles, John, 204
Bear: Asian black, 62; black (American
    black), 116, 141; brown, 235; short-
    faced, 34
Beaver, 157; European, 74; giant, 37
Beaver dams, 157

Beech, 22, 44, 48, 50–51, 60, 137;
    American, 28, 83–84, 88, 145, 147,
    148, 149–150; European, 72, 177;
    Japanese, 87; southern, 9
Bee colonies, honey harvesting from, 73
Beetles: Asian long-horned, 187, 191–192;
    dung, 89; elm bark, 202; lady,
    193–195, 196, 197; Laricobius, 195
Bering Strait, 5
Berthold, Peter, 173
Białowieża Forest, Poland, 71–78, 79,
    81, 88, 101, 103, 132, 134, 161, 223
Bighorn Basin, Wyoming, 20
Biological control programs, 193–198
Birch, 10, 31, 72, 74, 75, 146, 187; black,
    155; Japanese cherry, 87; yellow,
    83, 88
Birds: breeding territory, 110–112, 113;
    brood parasitism, 109–110, 113,
    120–121, 123–124, 130; in canopy
    gaps, 80–81; census of, 106, 107, 114;
    deer grazing impact on, 150–151,
    163–164; and ecosystem structure,
    27–28; and forest regeneration, 61,
    105–106; migratory, 6, 108, 116–117,
    122, 134, 170–173; nest predation,
    109, 112, 121, 124, 128, 130, 132, 133,
    135–136; North American/East Asian
    differences, 6–7; in old-growth
    forests, 71, 91–92; in open habitat, 88;
    predators of, 109, 121–122; primitive,
    11; protection of breeding habitat,
    116–117; range shifts of, 174–175, 176;
    scientific names, 244–246; in scrub-
    shrub habitat, 97–103, 99; seasonal
    timing of, 169–171, 170; and winter
    habitat loss, 107, 122, 134
Birds, distribution in forest fragments:
    in Europe, 125–130, 132, 133–134;
    forest area/bird density rate, 114–116,
    115, 118, 123, 128–129; in Japan,
    117–123; in North America, 108–114,
    116; patterns of, 123–125; winter
    wrens, 131–133

Bison, 73, 75, 76, 77, 138, 159, 207;
    American, 156, 209; European, 74,
    88, 161
Blackberry, 138, 146, 147, 163; Allegheny, 83
Blackcap, 164, 172, 173
Black stem grain rust, 204
Blowdowns/windstorms, 79, 82–84,
    85–86, 100
Blueberry, 204
Boar, wild, 62, 144, 162
Bobcat, 96
Bobiec, Andrzej, 79
Braun, E. Lucy, 24, 26
Breeding programs, disease-resistance,
    198–201
British forests: bird population in,
    101–102, 126–127, 129, 131–132, 134,
    163–165; clearing of, 45–46, 52;
    conservation approach to, 220–221;
    deer grazing in, 163–165; manage-
    ment of, 57, 218–220; old-growth, 90;
    pathogenic fungi in, 185; protection
    of, 57; shrub-scrub habitat in,
    101–102; wolf eradication in, 161–162
Browsing animals. See Grazing animals
Bubonic plague, 45, 46
Buffalo: Murr water, 89; short-horned
    water, 42
Building regulations, 58
Bumblebee, 27, 237
Bunting: indigo, 99; Japanese reed
    (ochre-rumped bunting), 230;
    meadow, 102
Bur cucumber, 205
Burnham, Charles, 199
Burning, 36, 38–39
Burnweed, 83
Butterfly, 237; harvester, 194
Butternut canker, 187

Cahokia mound-building center, Illinois,
    46–47
Canopy gaps: animal population in,
    80–82; and natural disturbances, 72,

79–80, 86–87, 105; and plant diversity, 152; shrub/grass invasion of, 138; size of, 80, 94

Capercaillie, 126, 129

Carbon dioxide, atmospheric, 176, 177, 179

Castle Rock, Colorado, 19, 20

Cat: as bird predator, 109, 121; saber-toothed, 34

Catalpa, 4, 5

Catbird, gray, 98, 100

Caterpillars, 107, 169–170, 171, 186, 194, 197

Cathedral Pines, Connecticut, 82

Caughley, Graeme, 151, 154

Cenozoic Period, 12

Ceratopsians, 11

Charcoal deposits, 11, 35, 36, 39, 42, 48, 88

Charmantier, A., 170

Chat, yellow-breasted, 99, 100, 101

Cheah, Carole, 194, 196, 197

Cherry, 22, 83; black, 146, 147; Japanese mountain, 167–168; pin, 146; tree flowering, 167, 168, 228

Chestnut, 50–51; American, 28, 184, 199, 200; blight, 182–183, 198; Chinese, 199, 200; disease-resistant breeding, 198–201, 200; golden, 22; horse, 191; sweet, 183

Chickadee, 114

Chicxulub Crater, as asteroid impact site, 13

Chiffchaff, 164

Chinese forests: clearing of, 49, 52, 56; land use in, 55–56; mammal extinction in, 37; plant diversity in, 5–6; protection of, 49, 56

Chipmunk, eastern, 201

"Clapper tree" hybrid, 199

Clearcut openings, bird species in, 95, 98, 99, 102

Climate change: adaptation of species, 168, 169–178, 180; evidence for, 166–168; and glaciation, 2–3, 5, 23–26; and mammal extinctions, 34–38; and Mississippian decline, 47; and rainforest trees, 20–21; resilient response to, 31, 178, 180; vulnerability to extinctions, 5, 178–180

Climate envelope models, 176–177

Closed-canopy forests, 81, 89, 104, 106, 164

Clovis culture, 34, 35

Coelurosaurs, 11

Cofitachequi mound-building center, South Carolina, 47

Conifers: deciduous, in Cretaceous Period, 8, 9–10; plantations, 58, 59, 59–60

Connecticut College Arboretum, 105, 196

Conservation strategies: combined approach, 7, 234–238; humanized regional landscapes, 216–223; miniaturized natural landscapes, 223–234; priorities of, 237; wilder-ness preservation, 207–216

Coppicing, 57, 60, 61, 101, 102, 122, 164, 219

Corn (maize), 46, 47, 48

Cottontail: eastern, 96–97; New England, 95–97

Cottonwood, 157

Cougar, 140

Cowbird, brown-headed, 99; as brood parasites, 109–110, 123–124, 130, 135; habitat, 109

Coyote, 157–158; colonization routes, 158; and wolf hybridization, 158–161

Crane: red-crowned, 224, 230; white-naped, 224; hooded, 224

Crawford Lake, Ontario, 48

Creeper, brown, 91, 92

Creeper, Virginia, 4, 5

Cretaceous Arctic forest: asteroid impact theory of end of, 12–18; deciduous conifers in, 8–10; dinosaurs in, 10–12; temperatures in, 9

Cretaceous-Tertiary (K-T) boundary, 14–16, 18, 31, 250n37

Crocodile, 17

Crossbill, red, 92

Crow, 109, 124; carrion, 88; hooded, 88;
    jungle, 121

Cryptomeria, 50, 58–59, 59

Cuckoo: little, 120; northern hawk-, 120;
    oriental, 120–121

Cucumber, bur, 205

Cumberland Plateau, 25

Currant shrubs, 189

Cycads, 8, 10

Cypress: bald, 8, 30; hinoki, 50, 58, 59

Daisen Forest Reserve, Japan, 87

Davis, Margaret, 28

Dead wood, decomposing, 71, 78

DeCalesta, David, 150

Deccan Traps, 13

Deciduous forest origins: and asteroid
    impact, 12–18, 31; and climate
    cooling, 21–26; and climate warming,
    20–21; Cretaceous arctic forests,
    8–10; at Cretaceous middle latitudes,
    18–19; and North American/East
    Asian similarities, 2–6; post-glacial
    spread, 26–31; resilience as explana-
    tion for survival, 31–32

Deer: Chinese water, 163; density of,
    137, 140, 146–148, 151–153; fallow,
    162–164; hunting of, 143, 154–156;
    overpopulation defined, 153–154;
    predators of, 139, 140–141, 156–161;
    red, 73, 74, 161, 163; Reeves's muntjac,
    163–164; roe, 74, 163; sika, 137, 139,
    139, 151, 163; white-tailed, 145–156,
    204; winter diet of, 151–152, 155;
    Yabe's giant, 37

Deer grazing: and bird loss, 150–151,
    163–164; in European forests, 75,
    75–76, 161, 163–165; forest ecosystem
    impact of, 137–140, 139; and invasive
    species spread, 204; and plant
    diversity, 148–150, 152; and seedling/
    sapling survival, 75–76, 137, 138,
    145–148, 149, 155

Deer parks, 161

Deforestation. *See* Forest clearing

Delcourt, Hazel, 25

Delcourt, Paul, 25

Denver Basin, Colorado, end of Creta-
    ceous, 16, 17

Diapterobates humeralis, 193

Dinosaurs: in Cretaceous Arctic forest,
    10–12; end-of-Cretaceous extinction,
    12, 13, 18; wetland habitat of, 16;
    Williston Basin fossils, 14

Disease: epidemics, Old World, 62;
    resistance to, breeding programs and,
    198–201. *See also* Pathogens, fungal

Distribution of species, and climate
    change, 173–175, 175

Dog, raccoon, 121

Dogwood, 22

Domesday Book, 45–46

Donahue, Brian, 64

Dooley, Sallie, 187

Dragonfly, 231, 236, 237

Dubos, René, 216–217

Dun, Edwin, 143

Dunnock, 102, 164

Dutch elm disease, 183–185, 202

Eagle-owl, Eurasian, 129

Ecological meltdown, 144

Ecosystem structure, theory of, 27–28

Edmontosaurus, 11

Egg laying, seasonal timing of, 169, 170,
    170, 171

Elder, marsh, 41

Elephant, 33; African, 38; Asian, 37, 49;
    Naumann's, 37; straight-tusked, 37

Elk, 153, 156–157, 159

Ellesmere Island, Canada, 8, 9

Elm, 10, 22; American, 145, 185, 186;
    decline and recovery of, 202; Dutch
    elm disease, 183–185, 202; elm bark
    beetle, 202; English, 185, 186;
    European, 185; wych, 185

Elm bark beetle, 202

Elvin, Mark, 49, 56
Emerald ash borer, 187
Emerson, Ralph Waldo, 208, 214
Empty forests, 53–54
Endangered Species Act (U.S.), 95
English forests. *See* British forests
Entomophaga maimaiga, 197
Eocene Period, climate change in, 20–21, 31
Epping Forest, England, 219
Eschtruth, Anne, 204
Esthetic approach to forest management, 57
European Economic Community (EEC), 189, 190
European forests: bird population in, 125–130, 132, 133–134, 135; clearing of, 44–45, 56–57, 129, 135; conservation approach to, 215–223; deer grazing in, 75, 75–76, 161, 163–165; management of, 57–58, 61, 217–218; old-growth, 71–78, 88–89; pathogenic fungi in, 189–190; predators in, 161–162; timber harvesting in, 44, 57. *See also* British forests
Evelyn, John, 57
Evolutionary adjustment to climate change, 171–173
Extinctions: asteroid impact and, 12–18, 31; climate change and, 178–180; glaciation and, 2, 5, 26, 30; of large mammals, 33–38; limits to resilience, 31–32; North American/East Asian differences in rate of, 3; on oceanic islands, 108; and plant distribution, 3, 5

Farming. *See* Agriculture
Farmland, abandoned, 63–66, 80, 95, 100
Fern parks, 150
Ferns, 146, 149–150, 152; common wood, 148; at end of Cretaceous, 14, 17, 18; hay-scented, 148
Fig, 20

Finches, 92
Fir, 2, 26, 176
Fire frequency: and forest succession, 84; in hunting-gathering societies, 38–40; and large mammal decline, 35–36; in old growth forests, 73, 87–88
Fisher, 64
Flowering times, and warming trend, 167–168, 168
Flycatcher: Acadian, 99, 187; blue-and-white, 118; Narcissus, 61; pied, 126, 171; tyrant, 6; willow, 157
Flying squirrel, southern, 64
Foraminifera, 13
Ford, Hugh, 125, 127
Forest animals. *See* Animal populations; Birds; Predators
Forest clearing (deforestation): in Britain, 45–46, 52; burning, 38–40; in China, 49, 56; deer grazing and, 137–140; in Europe, 44–45, 56–57, 129, 135; halted by bubonic plague, 45, 46; in Japan, 49–52, 53, 60; in North America, 62–63, 66, 209; North America, Mississippian culture and, 46–48, 52, 62; in Southeast Asia, 122. *See also* Tree harvesting
Forest fragments. *See* Birds, distribution in forest fragments
Forest management: for biological diversity, 93, 98, 99, 135–136; central core model in, 104; coppicing, 57, 60, 61, 101, 102, 164, 219; esthetic approach to, 57; forest rides, 219–220; hunting reserves, 57, 73; origins of, 55–58; pollarding, 60, 61, 218–219; rotational harvesting, 58, 102, 212; selective harvesting/thinning, 60, 70, 81, 98–100, 104, 217–218; sustainable, 57–62, 61, 68, 69, 208–209, 212; tree plantations, 57, 59–60, 68, 69, 101, 218, 223
Forest openings, 89, 102, 111, 137. *See also* Canopy gaps

Forest protection. *See* Forest management; Wilderness preservation
Forest regeneration: and bird populations, 61, 105–106; in China, 56; and deer browsing, 145–146; and empty forests, 53–54; in Japan, 50–51, 55, 60–62; after natural disturbances, 82–87, 92; in North America, 52, 63–66; and resilience, 69
Forest reserves, national, 67, 68, 73–74, 210, 212–213, 214–215, 233
Forest rides, 219–220
Forest Service, U.S., 68, 145, 212–213
Foster, David, 64
Fox, red, 121
France, forest management in, 57
Fuller, Robert, 81
Fungi, scientific names, 243

Gardens, Japanese, 226, 227, 229
Garlic mustard, 204
German forests: clearing of, 57; management of, 57–58, 69
Gifford, Sanford Robinson, 211
Ginkgo, 5, 8, 9–10
Glaciation: and deciduous tree decline, 23–26; and extinctions, 2, 5, 26, 30
Global warming, 166–167, 176, 179, 181
Gnatcatcher, blue-gray, 99
Goldenrod, 205
Gooseberry, 189
Goosefoot, 41
Goshawk, northern, 126
Grass, silver, 138
Grasses/grassland: and burning, 38–39; and deer grazing, 137, 138, 147, 149–150; forest rides, 219–220; grazing lawn, 138; Miocene, 21–22; national grasslands, 67; *Zoysia,* 138, 139
Grazing animals: and bird loss, 150–151; forest ecosystem impacts of, 137–140, 139; forest openings maintained by, 89, 138; and oak regeneration, 77–78,

88; and plant diversity loss, 148–150; predators of, 139, 140–141, 156–162; and seedling/sapling survival, 75, 75–76, 137, 138, 145–148, 149
Grazing lawn, 138, 139, 140
Great Smoky Mountains National Park, 68, 103, 107, 157, 184, 195, 210, 212
Greek civilization, and forest clearing, 43–44
Green fertilizer, 59, 60
Greenhouse gases, 177, 179, 181
Greenland, 5
Grosbeak, masked, 118
Gum, black, 29
Gymnosperms, Cretaceous, 8
Gypsy moth, 186, 188, 191, 196–198

Hackberry, 25
Hadrosaurs (duck-billed dinosaurs), 10–11
Hangzhou Bay, China, 42
Harris, Larry, 104
Hawk, 112; red-shouldered, 64
Hawk-cuckoo, northern, 120
Hazel, 44
Healy, William, 154–155
Heart's Content Scenic Area, 148
Hemlock, 22, 28, 30, 91, 92, 105; Carolina, 192; decline and recovery of, 201–202; disease-resistant breeding, 198; eastern, 88, 148, 192, 193; northern Japanese, 193; southern Japanese, 193; western, 195
Hemlock woolly adelgid, 186–187, 188, 192, 193–195, 196, 197, 198–199
Hickory, 22, 25, 26, 28, 30; shagbark, 64
Hideyoshi, Toyotomi, 51
Hiei, Mount, 119
High-yield approach to forest management, 57–58
Higuchi, Hiroyoshi, 118
Hippopotamus, 36–37, 89
Hiram Fox Wildlife Management Area, Massachusetts, 107

Hoge Veluwe woodland, Netherlands, 169–170
Holly, 61
Hominids, 34
Hopewell culture, Ohio, 46
Hop-hornbeam, 36
Hornbeam, 36, 44, 163; European, 74; recruitment rate, 75–76
Horses, 89
Hotokenuma Marsh, Japan, 230–231
Huckleberry, 204
Hudson River School, 208
Hui-Lin Li, 5
Hunting: deer, 143, 154–156; forest reserves, 57, 73; overkill hypothesis, 33–36
Hunting-gathering societies: fire use in, 38–40; and mammal extinctions, 33–38; Mississippian, 46, 47; transition to agriculture, 41–42
Hutchinson, G. Evelyn, 7
Hybridization, for disease-resistance, 199–200, 200

Industrialization, 49
Insects: in canopy gaps, 82; chestnut specialized, 201; conservation of, 231, 236, 237; end-of-Cretaceous extinctions, 14, 18, 20; introduced, economic cost of, 188; in Japanese culture, 229, 234; range shifts, 175; scientific names, 243; and seasonal timing, 169
Insects, herbivorous, 182; Asian long-horned beetle, 187, 191–192; control of, 188–198; gypsy moth, 186, 191, 196–198; hemlock woolly adelgid, 186–187, 192, 193–195, 196, 198–199
Invasive plants, 202–205, 215
Invasive species. See Insects, herbivorous; Pathogens, fungal
Irrigation, 41, 42, 43, 52, 56, 235
Island hypothesis, 108
Ito, Kensuke, 193

Jackdaws, 88
Japan: conservation movement in, 230–234; nature appreciation in, 223–229, 227, 228, 233, 236–237
Japanese forests: bird population in, 6–7, 102–103, 117–123, 130, 133; clearing of, 50, 51–52, 53, 60, 137–140; deer grazing in, 137–140; fragmentation of, 117–123; invasive plants in, 204–205; and mammal extinction, 37; management of, 58–62; natural disturbances in, 86–87; pathogenic fungi in, 186; predator loss in, 143–145; regeneration of, 50–51, 55, 60–62; scrub-shrub, 102–103; timber harvesting in, 50–51, 52, 53, 58, 60, 122, 233; wolf population in, 141–143
Japan Wolf Association, 144
Jay, Eurasian, 118, 127

Kano Sanraku, 225
Kappa Bridge, Japan, 228–229
Katano Kamoike Nature Reserve, Japan, 229
Kellert, Stephen, 233
Kinai Basin, Japan, 49–51
Kingfisher, common, 224
Kinkazan Island, Japan, 137–140, 140, 147
Kiritappu Wetland, Hokkaido, Japan, 230
Knight, John, 259n26
Krakatau island volcanic eruption, 14
K-T boundary, 14–16, 18, 31, 250n37
Kurosawa, Reiko, 102, 120, 230
Kyle, C.J., 161
Kyoto, Japan, 119

Labandeira, Conrad, 18
Land bridges, 5, 33
Landscape art: in Japan, 224, 225; in North America, 208, 211
Land trusts, 213–214
Larch, 59

Laricobius nigrinus, 195
Lawngrass, Japanese (*Zoysia japonica*), 138, 139
Leaf miner, 20
Leaf-mining moths, 201
Leaf shape, relation to climate, 9
Lemurs, 37
Leopold, Aldo, 209, 212, 214, 216, 237
Leucothoe grayana, 137
Lichens, 90–91
Lime, 71, 75, 76, 77; small-leaved, 74
Limestone outcrops, 215
Linnets, 101
Litvaitis, John, 96
Live plant imports, regulation of, 189, 206
Locust, black, 205; honey, 145
Logging. *See* Tree harvesting
Lynx, 74; Eurasian, 140

Macaque, Japanese, 62, 137
Magnolia, 5, 30
Maize, 48
Mammals: large, extinction of, 33–38; scientific names, 247–248
Mammoth, 33, 34; woolly, 37
Man and the Mediterranean Forest (Thirgood), 43–44
Mann, Charles, 43
Maple, 10, 22, 25, 28, 31, 75, 146, 187, 192; Norway, 74; red, 64, 147; striped, 147; sugar, 48, 64, 88, 145
Marshall, Robert, 212–213
Marten, Japanese, 121
Martin, Paul, 33, 36
Mason, Robert, 231
Mastodon, 36, 37
Matthysen, Erik, 128
Mayflower, Canada, 149
Maymont Park, Richmond, Virginia, 187
McClure, Mark, 193
Meadowview Research Farms (Virginia), 200
Megafossils, 18

Mesolithic birds, in Britain, 129
Mesophytic forest, 24–25, 26, 85
Mice, white-footed, 201
Migratory behavior, genetic, 171–173
Millet cultivation, 42
Miocene Period: climate change in, 2–3, 21, 31; grasslands of, 21–22; plant diversity in, 2, 22–23
Mississippian culture, and forest clearing, 46–48, 52, 62
Molecular genetics techniques, 200
Mongoose, 193
Monkey, howler, 145
Montgomery, Michael, 194
Moon Crossing Bridge, Kyoto, 227, 228
Moose, 74, 159
Moth: American chestnut, 201; caterpillars, 107, 169–170, 171, 186, 197; gypsy, 186, 188, 191, 196–198; leaf-mining, 201; scallop-shell, 146; winter, 169
Mound-building cultures, 46–48, 62
Mountain lion, 157
Mouse, white-footed, 201
Mud Buttes, North Dakota, 15
Muir, John, 208, 214
Muntjac, Reeves's, 163–164
Murai, Hidenori, 118
Myotis, little brown, 203

Nanoplankton, calcareous, 13
Nash, Roderick, 207
National Audubon Society, 106, 213
National parks and forests, 67, 68, 69, 73–74, 207–208, 210, 212–213, 214–215, 233
Natural disturbances, 78; and biological diversity, 152; and bird habitats, 100; and canopy gaps, 72, 79–80, 86–87, 105; fires, 73, 87–88; forest regeneration after, 82–87, 92; frequency of, 84; and old-growth forests, 103–104; openings created by, 94–95; in wilderness areas, 215
Natural experiments, 6

Nature Conservancy, 82, 116–117, 213, 235–236

Nematode, pine wood, 186

Nest boxes, 218

Nest predation, 109–110, 112, 121, 124, 128, 130, 132, 133, 135–136

Netherlands forests, bird population in, 128

Nettle, stinging, 76

Nichols, D. J., 15

Nightingale, 164–165

Ninna-ji Temple (Kyoto), 227

Nishi Honganji Temple (Kyoto), 224

Nobunaga, Oda, 51

North American forests: bird population in, 106–114, 126, 130; burning in, 38–40; canopy gaps in, 80; clearing of, 46–48, 52, 62–63, 66, 68; closed-canopy, 81; deer population in, 145–156; East Asia, comparison with, 1–7, 4; herbivorous insects in, 186–187, 191–192, 195; invasive plants in, 205; and mammal extinction, 33–38, 209; national parks/forests, 67, 68; natural disturbances in, 82–83, 85–86; old-growth, 78–79, 85, 87–88, 91; pathogenic fungi in, 183, 185; predators in, 157–161; regeneration of, 52, 63–66; scrub-shrub habitat, 96–101. See also Cretaceous Arctic forest; Wilderness preservation

Nuthatch, Eurasian, 127, 128

Oak, 22, 26, 28, 31, 36, 44, 50–51, 60, 71, 72, 75, 105; acorn production, 77, 151, 155; black, 64; evergreen, 50; and fire disturbance, 87–88; and grazing animals, 77–78, 88, 163; gypsy moth caterpillars in, 186; pedunculate (English), 74, 77, 177; white, 13

Oak-chestnut forests, 183

Oak wilt, 189–190

Old-growth forests: biological diversity in, 72–73, 89–93; browsing/grazing animals in, 75, 75–76, 77–78; canopy gaps in, 79–80; definitions of, 93–94; fire disturbance in, 73, 87–88; as information source, 70–71, 103; and natural disturbances, 103–104; small patches of, 103; structural characteristics of, 78–79; young/old trees in, 72–73

Ophiostoma novo-ulmi, 185

Ophiostoma ulmi, 185

Opler, Paul, 201

Oriole, Eurasian golden, 126

Ovenbird, 99, 105, 106, 114

Overpopulation, defined, 153–154

Owl: barred, 92; Eurasian eagle-, 129

Ox, musk, 12

Ozark Mountains, Missouri, 91

Paleocene Period, 31; rainforest, 19–20; transition from Cretaceous, 14, 15, 16

Paleogene Period, 3

Paleolithic Period, 37

Parasitic fungus. See Pathogens, fungal

Parasitism, brood, 109–110, 120–121, 123–124, 130

Partners in Flight Program, 116

Partridgeberry, 4, 5

Passive sampling hypothesis, 258n65

Pathogens, fungal: as biological pest control agent, 195–196, 197; chestnut blight, 182–183, 198; control of, 188–198; Dutch elm disease, 183–185, 202; economic cost of, 188; oak wilt, 189–190; pine wilt disease, 186; white-nose syndrome, 203; white pine blister rust, 187, 189, 198

Periwinkle, common, 64

Petersham, Massachusetts, 64, 65

Photosynthesis, effect of winter darkness, 9

Pigeon: Japanese green, 61; passenger, 209

Pinchot, Gifford, 208

Pine, 22, 36, 72, 75; jack, 26, 28; Japanese
    red, 50, 60, 186; longleaf, 63; pine wilt
    disease, 186, 189; pitch, 87; red, 24,
    26, 28; Scots, 73; sugar, 189; western
    white, 189, 198; white, 28, 48, 63, 82,
    92, 187, 189; white pine blister rust,
    187, 189, 198
Pine wilt disease, 186, 189
Pipe, Indian, 1
Pipit, meadow, 102
Pisgah Forest, New Hampshire, 86
Pisgah National Forest, North Carolina,
    91, 92
Plant diversity: and asteroid impact,
    18–19; and climate cooling, 20–21,
    23–26; and deer grazing, 148–150,
    149, 152; island hypothesis of, 108;
    and natural disturbances, 84; North
    American/East Asian parallels, 1–7, 4;
    in old-growth forests, 72–73, 89–91;
    Paleocene rain forest, 19–20
Plant genera: common ancestor for, 5;
    distribution of, 3, 4, 5
Plants, scientific names, 239–243
Plato, 44
Pleistocene Period: glaciation in, 2–3, 21,
    23–26, 29, 31, 174; large mammal
    extinctions in, 34–35
Pliocene Period, 21
Plymouth colony, Massachusetts, 62
Poison ivy, 22
Pollarding, 60, 61, 218–219
Pollen records, 14, 15, 16, 18, 20, 22, 24,
    25–26, 28, 29, 30, 35, 36, 39, 40, 42,
    44, 45, 47, 48, 70, 72, 77, 87, 88, 89,
    174, 176, 178, 180, 201, 202
Poplar, 28, 74
Poverty Point, Louisiana, 41
Powerline corridors, scrub-shrub habitat
    on, 95, 98, 100
Prairie-dog, 138
Predators: as biological control agents,
    193, 194; control of, 153; of deer,
    140–141, 156–162; dinosaurs, 12;

introduced, 121–122; loss of, 139, 141,
    143–145, 159, 161–162; nest, 109–110,
    112, 121, 124, 128, 132, 133, 135–136;
    reintroduction of, 156–161, 165, 215
"Prehistoric overkill" hypothesis,
    33–38
Prothero, Donald, 13
Pyne, Stephen, 39

Quabbin Reservoir, Massachusetts,
    154–155
Quail, common, 71

Rabbits, in scrub-shrub habitat, 95–97
Rabies, in wolves, 143–144
Raccoons, 109, 121–122, 124, 205
Rackham, Oliver, 220
Ragweed, 48; giant, 205
Rainforests: and climate change, 20–21;
    empty forests, 53–54; in Paleocene
    Period, 19
Range shifts, and climate change,
    173–175, 175, 176
Redstart, American, 27, 92
Redwood, 130; dawn, 8, 9–10, 19, 22;
    coast, 10, 71, 208
Retreat of the Elephants (Elvin), 49
Rhinoceros: white, 38; woolly, 37
Rhododendron, pontic, 204–205
Rice cultivation, 41–43, 49, 52
Robbins, Chandler, 114
Robin, Siberian blue, 120
Rogers Lake, Connecticut, 28–29
Roman Empire, forest clearing in,
    44–45
Rondeau Provincial Park, Canada, 155
Rooks, 88
Rooney, Thomas, 150
Rotational timber harvesting, 58, 102
Royo, Alejandro, 151
Runkle, James, 85
Rust: black stem grain, 204; white pine
    blister, 187, 189, 198
Rutberg, Alan, 154

Salamanders, 92, 93, 108; *Plethodon*, 91
Sandy soils, 87, 216
Saplings. *See* Seedlings and saplings
Sasajiscymnus tsugae, 193–194, 196
Sassafras, 30
Satoyama landscape, 235
Scientific names, 239–248
Scrub-shrub habitat, 94–103
Seal, Solomon's, 149
Seasonal timing, of birds, 169–171, 170
Seedlings and saplings: in canopy gaps,
    79–80, 86–87; and deer grazing,
    75–76, 137, 138, 145–146, 147, 150, 155;
    in forest regeneration, 83–84; in
    scrub-shrub habitat, 98; survival
    factors for, 76–77; in young forests,
    100
Selective harvesting, 60, 70, 81, 98–100,
    104, 217–218
Selva, Steven, 90–91
Selway-Bitterroot Wilderness Area,
    Montana, 213, 215–216
Sequoia, 8, 22
Serow, Japanese, 144
Serpentine bedrock, 216
Shenandoah National Park, 210, 212
Ship timber, shortages of, 44
Shiretoko National Park, Hokkaido,
    86–87, 103, 233
Shrike: brown, 102; bull-headed, 102;
    great gray, 102; red-backed, 101;
    thick-billed, 102
Shrubs: deer grazing, 137, 147–151,
    163–164; scrub-shrub habitat, 94–103
Sierra Club, 208
Silver grass, 138
Skylark (Eurasian sky lark), 102
Sloth, ground, 36, 37
Smallpox epidemic, 62
Solomon's seal, 149
Soto, Hernando de, 47
Sparrow: Eurasian tree, 224; field, 100
Speedwell; ivyleaf, 205; Persian, 205
Sporomiella fungus, 35, 36

Spruce, 28, 36, 75, 177; Norway, 71, 74,
    76–77, 102
Squash, 48
Squirrel: Eastern gray, 201, 205; Eurasian
    red, 205; Japanese giant flying-, 61;
    southern flying, 64
Steep Rock Association, 193
Stiltgrass, Japanese, 204
Stonechat, European, 102
Stone tools, 39, 62
Stone walls, of abandoned farms, 63–64,
    65
Sunflower, 41
Svenning, Jens-Christian, 89
Sweden, bird population in, 127, 132
Sweetgum, 22, 30
Sycamore, 10, 22

Tamarack, 10
Tanager, scarlet, 114
Tarpan, 77, 88
Teal, Baikal, 224, 229
Tern, Arctic, 110
Thirgood, J. V., 43–44
Thompson, Frank, 123
Thoreau, Henry David, 208, 214
Thrasher, brown, 100
Thrush: song, 164; Swainson's, 91; wood,
    106, 107, 117
Tianluoshan, China, 41
Tidal marshes, 179
Tiger, 140–141
Tikal National Park, Guatemala, 93, 94
Timber harvesting. *See* Tree harvesting
Tionesta Natural Area, Pennsylvania,
    82–83, 85, 103
Tit: coal, 118, 127; crested, 127; great,
    169–171, 170; long-tailed, 164; marsh,
    214; willow, 127
Titmouse, 114; tufted, 99
Tokyo Port Wild Bird Park, 229–230
Tomakomai Experimental Forest, Japan,
    82
Totman, Conrad, 50

Towhee: eastern, 98, 100; rufous-sided, 99

Transcendentalist view of wilderness, 208

Treecreeper, short-toed, 127

Tree harvesting: clearcut logging, 71, 89–90, 91, 95, 97–100, 102, 146, 147; in Europe, 44, 57; in Japan, 50–51, 52, 53, 58, 60, 122, 233; in North America, 63, 66, 68; rotational, 58, 102; selective, 60, 70, 81, 98–100, 104, 217–218; wood shortages, 44, 49, 50, 57, 58, 68. *See also* Forest clearing (deforestation)

Tree plantations, 57, 59–60, 68, 69, 101, 218, 223

Tree range shift, and climate change, 176–178, 180

Triceratops, 11, 14, 16

Triceratops Swamp (Vriesen), 16

Trillium, red, 149

Tucker (Walter C.) Preserve, Ohio, 145–146

Tuliptree, 22, 30

Tupelo, 30; black, 29

Twilight in the Adirondacks (Gifford), 211

Two Wagtails (Kano Sanraku), 225

Tyrannosaurus, 14

Understory plants, deer grazing of, 145–151, 152

Urano, Tadahisa, 193

Urbanization, 49

Velociraptor, 11

Vera, Frans, 77–78, 88–89, 220

Veronica, 205

Violet, 1, 27, 108

Vireo, 6, 27; red-eyed, 99, 105, 106, 114; white-eyed, 100

Vriesen, Jan, 16, 17

Walker, Brett, 142

Walnut, 10

Wapiti, 156

Warbler, 6; Bachman's, 81; black-and-white, 99, 106; Blackburnian, 91; black-throated blue, 111–112, 113; black-throated green, 187; blue-winged, 98, 100; cerulean, 114, 117; chestnut-sided, 98; Eastern crowned leaf, 120; European, 172; garden, 101, 102, 164; golden-winged, 101; Gray's grasshopper-, 102; hooded, 80–81, 92, 99; Japanese bush-, 102, 120; Japanese marsh (marsh grassbird), 230; Kentucky, 151; magnolia, 91; melodious, 102; pine, 99; prairie, 99, 100; willow, 164; worm-eating, 99

Warming, global, 166–167, 176, 179, 181

Water buffalo: Murr, 89; short-horned, 42

Waterthrush, northern, 117

Wesołowski, Tomas, 132, 133

Wheat, 40, 204

Whinchat, 71

White Mountain National Forest, 107

White-nose syndrome, 203

White pine blister rust, 187, 189, 198

Whitethroat, 101; common, 102; lesser, 102

Wildebeest, 138

Wilderness Act (U.S.), 213

Wilderness and the American Mind (Nash), 207

Wilderness areas, 212–213, 214–216, 235

Wilderness preservation: cultural importance of, 207–209, 214; Dubos on, 217; in Europe and Asia, 235; national parks and forests, 67, 68, 73–74, 207–208, 210, 211, 212–213, 214–216; scientific justification for, 209; and second-growth forests, 209–210; small reserves, 213–214, 215, 216

Wildflowers, 90, 108, 147–148, 149, 150, 152

Williams, Michael, 57

Williston Basin, U.S. and Canada, 14, 17, 18

Willow, 74, 157, 163

Windstorms/blowdowns, 79, 82–84, 85–86, 100

Wingnut tree, 22

Winter dormancy, 9, 10

Winter habitat loss, bird, 107, 122, 134

Witch hazel, 4, 5

Wolf: and coyote hybridization, 158–161; as deer predator, 140, 141, 156–157; dire, 34; eastern, 158–159; eradication, 7, 143–145, 161–162; gray, 74, 139, 144, 156, 159; in Japanese culture, 141–143; red, 157, 159, 160; reintroduction of, 156, 157, 160

Wolf pack, 141

Wood: dead and decomposing, 71, 78; product imports, regulation of, 189; production, local, 206; shortages, 44, 49, 50, 57, 58, 68. See also Tree harvesting

Wood-boring insects, 188

Woodcock, Eurasian, 219

Wood pasture, 218, 220

Woodpecker, 71, 218; black, 129, 130; downy, 105; great spotted, 127; ivory-billed, 81; Japanese green, 118; pileated, 64; pygmy, 118; white-backed, 129

Wren: Eurasian, 131, 132, 135; Pacific, 131, 133; winter, 92, 130–133

Wytham Woods, England, 163–164, 169, 170

Xerxes Society, 237

Yamaura, Yuichi, 122

Yangtze Delta, China, 41–42

Yellowhammer, 101, 102

Yellowstone National Park, 68, 156, 210

Yellowthroat, common, 98

Yosemite Valley, 68

Younger Dryas period, 34, 36, 72, 179

Zanthoxylum piperitum, 138

Zelkova, 22

Zoysia grassland, 138, 139